STRUCTURED ANALYSIS METHODS

FOR COMPUTER INFORMATION SYSTEMS

URS CONSULTANTS
3605 WARRENSVILLE CENTER ROAD
CLEVELAND, OHIO 44122-5203
(216) 283-4000

AN INTERNATIONAL
PROFESSIONAL SERVICES
ORGANIZATION

JEFFREY M. DAVIS
COMPUTER SYSTEMS ANALYST

STRUCTURED ANALYSIS METHODS

FOR COMPUTER INFORMATION SYSTEMS

Lavette C. Teague, Jr.   Christopher W. Pidgeon
California State Polytechnic University, Pomona

SCIENCE RESEARCH ASSOCIATES, INC.
Chicago, Henley-on-Thames, Sydney, Toronto

A Subsidiary of IBM

IN MEMORIAM

John Leeds Pidgeon (1924–1983)
Lavette Cox Teague (1899–1975)

To our fathers, for whose examples of modest integrity, devoted pursuit of useful knowledge, and uncommon common sense we are deeply grateful, this small addition to the "inventory of time" is affectionately dedicated.

Project Editor	Richard Myers
Acquisition Editors	Michael Carrigg & Terry Baransy
Project Coordinator	Cathy Rundell
Copy Editors	Tom Szalkiewicz & Cordell Cooper
Designer	Mark Jaensch
Illustrators	Jim McDermott & Randall Miyake
Cover Design	© Janet Bollow Associates
Cover Photo	© Don Carroll, The Image Bank West
Composition	Graphic Typesetting Service, Los Angeles

Library of Congress Cataloging in Publication Data

Teague, Lavette C., 1934–
    Structured analysis methods for computer information systems.

    Includes index.
    1. System design.   2. System analysis.
    I. Pidgeon, Christopher W., 1950–   .   II. Title.
    QA76.9.S88T4 1984   003   83-20121
    ISBN 0-574-21495-X

Printed in the United States of America.

10 9 8 7 6 5 4 3 2

PREFACE

Within the past decade there have been significant advances in the theory and practice of developing computer information systems. They support information system development not only with new tools and techniques but also with a new conceptual framework; thus these new methods are called "structured."

This book presents current tools and techniques of systems analysis—data flow diagrams, the data dictionary, transform descriptions, and data base description—based upon the data flow modeling approach of Edward Yourdon and his associates. The book provides an understanding of the information system life cycle in relation to systems analysis. The discussion is grounded in general systems theory and the essentials of systems analysis in order to stress principles that will outlast the anticipated rapid changes in specific tools and techniques. The book also stresses an organized, top-down approach to problem solving as it demonstrates the use of structured analysis techniques to model business information systems. The data flow models are presented as an integral part of understanding and communicating system requirements rather than as *ex post facto* and therefore irrelevant documentation. Other important areas of emphasis include non-technical dimensions of business decision-making that affect information systems and the crucial role of communication skills—the oral skills of interviewing and presentation as well as skills in written and graphic communication.

The book received its impetus from the Data Processing Management Association's Model Curriculum in Computer Information Systems. It is intended for use in CIS-4, the introductory systems analysis course, for which one of the authors served on the panel of experts. However, there is somewhat less attention here to detailed system design than that suggested in the model curriculum as published, for we prefer to defer those topics to a system design course.

The book consists of four parts. Part I (Chapters 1–5) begins with a broad discussion of systems analysis, gradually narrowing the focus to the activities and methods of structured analysis. It introduces some important concepts from general systems theory and applies them to describing business systems and information systems, presents an information system life cycle, identifies the participants in the development of computer information systems, and provides an overview of structured methods for analyzing computer information systems.

Part II (Chapters 6–11) focuses on the tools and techniques for system description in terms of data flows, data stores, and transformations. It presents the conventions for data flow diagrams, the system dictionary, Structured English, decision trees, and decision tables. It also explains an object-relationship approach to logical data base description. Part II concludes by describing how to prepare a set of data flow diagrams and how to evaluate and refine a system model. These chapters also introduce the FastFood Store system, from which examples appear throughout Parts II and III.

Part III (Chapters 12–16) shows how these structured analysis methods are applied in the process of stating users' requirements for a computer information system. Using the FastFood Store as an example, it illustrates the development of physical and logical models of the current system and then the derivation of logical and physical models of a new system. It explains what must be done to complete the structured specification, discusses the process of changing the specification, and describes alternatives to complete, custom software development.

Part IV (Chapters 17–21) places the products and process of structured analysis within the context of computer information system development, giving a fuller description of the portions of the life cycle that precede and follow analysis. It also includes a discussion of methods of information gathering and reporting. The book concludes with an overview of alternative approaches to modeling system requirements, a survey of current trends in system development, and some of the impacts of these trends on the future of system development.

ACKNOWLEDGMENTS

We owe special thanks to Terry Baransy for his initial faith in us and for his continuing encouragement and support while we were writing this book. Thanks are due to Michael Carrigg for assuming editorial responsibilities after SRA moved its College Division to Chicago. The patience and perseverance of the staff of Graphic Typesetting Service, Los Angeles, are sincerely appreciated. We also appreciate the perceptive comments of those who reviewed the manuscript for the publisher: Dr. Alan L. Eliason, University of Oregon; Prof. Myron H. Goldberg, Pace University; Dr. Donald G. Golden, Cleveland State University; Dr. Earl E. Halvas, Western Michigan University; Carolyn M. Johnson, General Health, Inc.; and Prof. Alford J. Nanni, Jr., Boston University. In addition, we are grateful to our colleagues and students at California State Polytechnic University, Pomona, for their helpful suggestions based on classroom use of the manuscript. Our thanks to them all for helping us make this a better book.

Pomona, California   Lavette C. Teague, Jr.
July 12, 1984   Christopher W. Pidgeon

CONTENTS

PART I: INTRODUCTION TO ANALYSIS OF COMPUTER INFORMATION SYSTEMS 1

Chapter 1: Basic Systems Concepts 5

Coping with Ill-Structured Complexity 5
What is a System? 7
   System Structures 8
   Hierarchies of Systems 10
   System Boundaries 13
Modeling and Representation 15
   The Need for System Models 15
   Characteristics of a Good Model 16
   Representation 17
   Notation 18
   Generating a System Model 18
   The Complementarity of Analysis and Synthesis 19
   Limitations of System Models 19
Summary 20

Chapter 2: Computer Information Systems for Business 24

Computer Information Systems for Business 24
A Systems View of Business 25
Why Businesses Need and Use Information 26

Problem-solving and Decision-making   28
Functions of an Information System   30
    Communication   30
    Storage   30
    Transformation   31
Information Processing Systems   31
    Manual and Automated Systems   32
Describing a Computer Information System   32
    A Matter of Emphasis   33
    Physical and Logical System Descriptions   33
What Is Information?   34
Summary   35

Chapter 3:   The Information System Life Cycle   39

Overview of the Information System Life Cycle   40
Major Phases in the Life Cycle   42
    1. Identify User Needs   42
    2. Establish User Requirements   43
    3. Determine System Hardware and System Environment   43
    4. Design the System   43
    5. Develop the System Acceptance Tests   44
    6. Construct the System   44
    7. Integrate with the Using Organization   44
    8. Operate, Modify and Enhance the System   44
The Life Cycle: Limitations and Fallacies   45
Summary   45

Chapter 4:   Participants in Systems Analysis   48

Roles and Functions of the User   49
    Types of Users   49
    Functions and Responsibilities of the Users   51
Roles and Functions of the Analyst   52
    A Scenario   52
    Functions and Responsibilities of the Analyst   52
    Required Analysis Skills   53
Users and Analysts--A Synergism   54
Summary   54

Chapter 5:   Specifying User Requirements   56

Goals of Structured Analysis   56
The Process of Structured Analysis   58
    Characteristics of the Process   58
    The Activities of Structured Analysis   59
Products of Structured Analysis   61

Data Flow Diagrams   61
System Dictionary   61
Data Base Description   61
The Complete Specification   61
Characteristics of the Structured Specification   62
Tools of Structured Analysis   62
Techniques of Structured Analysis   62
Summary   63

PART II: TOOLS AND TECHNIQUES OF STRUCTURED ANALYSIS   65

Chapter 6:   Data Flow Diagrams   67

Reading Data Flow Diagrams: A Dialog   67
Components of the Data Flow Diagram   76
Basic Definitions   77
Symbols   77
Data Flow   77
Data Store   78
Transform   79
Naming System Components   80
Sets of Data Flow Diagrams   80
Hierarchical Structure   80
Balancing   80
Numbering Conventions   80
Primitive Transformations   84
Conventions for Data Stores   84
How to View a Data Flow Diagram   86
Data Flow Diagrams vs. Program Flowcharts   86
What a Data Flow Diagram Does Not Show   89
Relationship Among the Parts of a System Description   90
Summary   91

Chapter 7:   The System Dictionary   94

Goals of the Analysis Phase System Dictionary   94
The Organization of the System Dictionary   95
Entries in the System Dictionary   95
Describing Data Structure   96
The Constituents of Data Structure   96
The Data Structuring Operations   96
The Notation for Describing Data Structure   98
Describing Nested Data   99
Entries in the Data Dictionary   100
The Data Flow Entry   101
The Data Store Entry   102

The Data Element Entry   103
Performing Data Decomposition   104
   A Comprehensive Example   104
   Decomposition and Data Elements   109
   Some Data Decomposition Guidelines   110
The Origin/Destination Dictionary   112
The Transform Dictionary   112
Summary   113

Chapter 8:   Transform Description   118

Goals and Objectives for Transform Descriptions   118
Some Typical Questions   120
Alternatives for Policy Description   121
Structured English   121
   What Is Structured English?   121
   What Is Structured About Structured English?   121
   Expressing the Control Structures   126
   Combining the Control Structures   130
   Indicating the Scope of the Control Structures   130
   The English Parts of Structured English   133
   Some Structured English Examples   134
   Guidelines for Structured English   136
Decision Tables   138
   Relation of Decision Variables to the Data Dictionary   140
   Constructing a Decision Table   141
Decision Trees   144
   Constructing a Decision Tree   144
   When to Use a Table and When to Use a Tree   146
Hybrid Transform Descriptions   147
Summary   148

Chapter 9:   Logical Data Base Description   153

The Data Base Environment   154
   Users and User Views   154
   Data Bases and Data Base Management Systems   155
   Roles and Responsibilities in the Data Base Environment   157
Components of a Logical Data Base Description   159
   Objects, Attributes, and Values   159
   Objects, Attributes, and Values in the Data Dictionary   161
   Relationships   161
   Relationships in the Data Dictionary   163
   Logical Operations on Objects and Relationships   163
   Operations and Conditions in the Structured Specification   165
Logical Data Base Access   166

Testing the Logical Data Base Description   168
Some Typical Questions   172
Summary   173

Chapter 10:   Drawing Data Flow Diagrams   176

Getting Started—The Initial Data Flow Diagram   177
Continuing the Data Flow Diagram   179
Completing the Data Flow Diagrams   180
  Completing the System Description   181
Principles for Drawing Data Flow Diagrams   182
  Connections Among Components   182
  Balancing from Level to Level   182
  Hierarchy and Decomposition   183
  Data Store Conventions   185
  Conservation of Data Flow   188
  Emphasis on Logical Rather than Physical Description   191
  Naming Components   191
  Minimum Redundancy   192
Some Helpful Hints   193
  Where to Show Origins and Destinations   193
  Show the Big Picture First   193
  The Number of Transforms in Diagram 0   194
  Avoid Aggregation Which Conceals or Confuses   195
  Minimize the Number of Interfaces   195
Evaluating Data Flow Diagrams   195
Summary   195

Chapter 11:   Evaluation and Refinement of a System Description   197

Criteria for Evaluating a System Description   197
  Completeness   197
  Consistency   199
  Correctness   199
  Communication   200
Refinement of Data Flow Diagrams   201
  Independence of the Transformations   201
  Aggregation of Related Transforms   204
Walkthroughs   207
  What Is a Walkthrough?   208
  Why Have Walkthroughs?   208
  Who Should Participate?   209
Conduct of the Walkthrough   212
User Walkthroughs   213
Quality Control Checklists for System Descriptions   213
Summary   216

PART III: APPLYING THE TOOLS: THE ANALYSIS PROCESS 223

Chapter 12: Modeling the Current System 229

What Is Needed? 229
Developing the Physical Current System Description 231
  Working from the Outputs Backwards to the Inputs 232
  Stimulus-Response: Working from the Inputs Forward to the Outputs 233
  Refining the Initial Physical Current System Description 234
Paring the Current System Description to Essentials 234
  Essential Transformations 234
  Essential Data Stores 235
  The Essential Data Dictionary 237
  Essential Transform Descriptions 237
  Constructing the Logical Current System Description 237
The FastFood Store—Current System Description 237
  Developing a Refined Physical Current System Description 237
  Deriving a Logical Data Base Description for the Accounts Payable Subsystem 247
  Parts of the Logical Current System Description 263
Summary 269

Chapter 13: Modeling the New System 271

Establishing the New System Requirements 272
Defining Alternative Scopes of Automation 274
Selecting the Best Alternative 276
The FastFood Store—New System Descriptions 277
  Identifying the Domain of Change 278
  Modifying the System Description Inside the Domain of Change 283
  Defining Automation Alternatives 288
  Selecting the Best Alternative 291
Summary 296

Chapter 14: Completing the Structured Specification 298

Critical Detailed Requirements 298
  1. Editing and Error Checking 299
  2. Error Messages 299
  3. Transforms for System Start-Up and Shutdown 299
  4. Building and Maintaining Stores for Reference Data 299
  5. More Data Dictionary Detail 299
  6. Critical Control Flows 300
  7. Critical Timing and Response Requirements 302
  8. Inflexible Input/Output Formats 302
  9. Conversion Requirements 302

10. Audit Requirements and System Controls   304
11. Security and Backup   305
12. Quantitative and Testable Performance Targets   305
Packaging the System Description   305
An Outline for the Complete Structured Specification   306
Summary   310

Chapter 15: Coping with Changing Requirements   312

The Fact of Change   312
Sources of Change in Business Systems   313
The Impact of Business Change on Information System Requirements   315
Changing the Structured Specification Incrementally   317
Management Issues in Changing Requirements   320
Summary   321

Chapter 16: Alternative Processes for Systems Analysis   323

Factors Influencing the Analysis Process   323
The Users' Profile   324
The Analyst's Profile   324
The Project Profile   324
When Not to Look at the Current System   325
Short Cuts in Producing the Current System Description   326
Strategies for Producing the New System Description   326
System Requirements for Packaged Software   327
Prototyping   327
Summary   329

PART IV: THE CONTEXT FOR STRUCTURED ANALYSIS   331

Chapter 17: Identifying User Needs   333

Questions Addressed   333
The Activities of Identifying User Needs   335
Identify Any Current Deficiencies   335
Establish New System Objectives   336
Identify Acceptable Systems   336
Prepare the Feasibility Report   337
The Feasibility Report   337
Summary   339

Chapter 18: Information Gathering, Management, and Reporting   341

Gathering Information   342
Information Sources   342
Information Gathering Methods and Techniques   342

Evaluating Information   346
Information Management   347
Manual Methods of Information Management   347
Automated Aids to Information Management   348
Reporting Information   349
Modes of Communication   349
Principles of Communication   350
Summary   352

Chapter 19: Computer Information System Design   354

What Is Design?   354
The Goals of Computer Information System Design   356
The Design Phase of the System Life Cycle   356
Tools and Techniques of Structured Design   358
Design Methods   358
Design Tools   359
Techniques for Generating an Initial Design   362
Iteration and Refinement   365
Roles of Users and Analysts During System Design   365
The Structured Specification as the Basis for Design   365
Summary   366

Chapter 20: System Construction, Testing, Installation, and Operation   369

Develop the System Acceptance Tests   369
Construct the System   372
Integrate with the Using Organization   373
Conversion   373
Preparation for System Change   374
Operate, Modify, and Enhance the System   374
Post-Implementation Review   374
Defining Requirements for System Enhancement   375
Summary   375

Chapter 21: The Future of Structured System Development   377

Alternative System Modeling Techniques   377
Data-Flow Techniques   378
Data-Structure Techniques   380
Control-Flow Techniques   382
Trends in System Development   384
User-driven Computing   385
Fourth-Generation Languages   386
The Information Center   386
Automating System Development   387
The Impact on Users   390

The System Owner   390
The Responsible User   391
The Hands-on User   391
The Beneficial User   391
The Impact on Analysts   392
The Challenge of the Future   392
Summary   392

Glossary   395

Index   406

PART I

INTRODUCTION TO THE ANALYSIS OF COMPUTER INFORMATION SYSTEMS

The analysis of computer information systems is a process of learning about and describing users' requirements for an information processing system. This process involves the need to cope systematically with the complexities of the world while taking human limitations into account. Systems analysis requires limiting the area of study, abstracting essential features of reality, modeling a real system to show the relationships among its components, and subdividing a complex whole into parts of intelligible and manageable size.

An important conceptual foundation for the selective study of complex relationships is known as general systems theory. Chapter 1 presents aspects of general systems theory that are especially relevant to systems analysis, using a variety of examples from many fields of human endeavor.

In Chapter 2 these general system concepts are applied to the description of computer information systems. Any computer information processing system can be modeled with three types of components—information flow, information storage, and transformations of information. In a data flow model, the relationships among these components are depicted in data flow diagrams, supplemented by detailed definitions of the structure of the data and by a specification of the procedures for transforming the data.

Although data flow methods can be used to describe all types of information processing systems, this book emphasizes their application to information processing in business organizations. Manufacturing businesses are heavily dependent on information to monitor and manage their performance as well as to support management decisions; service businesses are almost exclusively information-processing organizations. Chapter 2 also examines the organizational context for business information systems, surveying the information needs of businesses as well as major applications of computers to business information systems.

As a process of determining users' requirements, systems analysis is part of a larger cycle of system development and use. Requirements analysis is preceded by a request to change an information processing system and an evaluation of the desirability and feasibility of the proposed change. The result of analysis—a statement of users' requirements—poses a problem to be solved by system designers. This solution is then programmed, tested, and delivered to the system users for integration into their organization and its way of operating. Chapter 3 presents the management and developmental context for systems analysis—the information system life cycle.

The statement of requirements for the proposed system is produced through the joint efforts of systems analysts and the various types of users to be supported by it. Computer information systems promise increasing productivity, job satisfaction, and enhanced leisure when used wisely. Their successful integration into business organizations requires careful planning, management support, technical expertise, and continuing interaction and cooperation by all the participants in the process. As an agent for effective change, a systems analyst with both technical and interpersonal skills has an important role to play. Chapter 4 identifies the various participants in systems analysis and system development and describes their respective roles in the process. It also summarizes the skills required by an analyst throughout the process.

Chapter 5 completes this introduction to the analysis of computer information systems. It characterizes structured analysis in terms of its goals, its process, its products, its tools and techniques. It presents an overview of the activities necessary to establish users' requirements for an information processing system. It also introduces the major parts of the document that specifies the users' requirements—the structured specification.

1 Basic Systems Concepts

Systems analysis can be one of the most complex of human activities. It can be complex intellectually (in order to understand a system and organize it conceptually), but it can also be complex practically (in order to coordinate and organize the efforts and interactions of the participants in the activities of analysis). When analyzing systems in general and computer information systems in particular, we need to cope with the complexity inherent in our world in spite of our human limitations. This chapter presents basic concepts from general systems theory and shows how they are useful in dealing with complexity.

COPING WITH ILL-STRUCTURED COMPLEXITY

Many complex situations that people encounter are not only complicated; they are also unstructured, or so poorly organized that their structure is not apparent. Complexity and lack of structure also characterize the environment for most systems analysis. Systems analysts are often called upon to investigate an unfamiliar situation and recommend what should be done. The problems requiring investigation may be unknown, and the analysts may have little background in the specific problem area to be studied. They must learn as their work proceeds. Methods of systems analysis must provide ways to overcome lack of structure and understanding as well as ways to cope with complexity.

To deal with complexity rationally and wisely, we must first organize and understand it. We must discover—or impose—some kind of order. Finding or creating order and structure permits us to understand a situation, communicate that understanding, and then take appropriate action.

For centuries thinkers have debated about the extent to which the world we experience incorporates some inherent structure and the extent to which we impose patterns of order on our experience in order to perceive and understand it. Human language, so intimately connected to human thought, is one of the great organizers of experience and a principal means of communicating and sharing experience as well as understanding it. Human civilization is characterized and supported by social, political, and economic structures for organizing common efforts toward societal goals. Science describes the order of the natural world. Technology manages and changes the natural world to obtain food, shelter, clothing, to convert energy, and to construct networks of transportation and communication. Art reveals the beauty of nature and rearranges the natural world in new and imaginative patterns to heighten our awareness of our existence.

Tension between chaos and order, rapid change and stability, randomness and purpose, confusion and meaning, is the background for human understanding, decision, and action.

In the past, change occurred slowly. Traditional solutions to traditional problems were handed down from one generation to another. The structures of society and technology had time to evolve in close adaptation to their environment.

Life is more complicated today. Many situations whose past structure was determined by tradition through slow evolution are now the result of conscious planning and decision. We are confronted by new problems for which there are no traditional solutions. The pace of change is too rapid for solutions to evolve; they must be created, or designed, as each new problem arises.

Christopher Alexander has contrasted unselfconscious cultures, in which people learn "informally, through imitation and correction" with cultures like ours, in which design is "taught academically according to explicit rules." When confronted with rapid change and new complexity, the unselfconscious cultures are unable to make good design decisions:

> The Slovakian peasants used to be famous for the shawls they made. These shawls were wonderfully colored and patterned, woven of yarns which had been dipped in homemade dyes. Early in the twentieth century aniline dyes were made available to them. And at once the glory of the shawls was spoiled; they were no longer delicate and subtle, but crude. This change can not have come about because the new dyes were inferior. They were as brilliant and the variety of colors was much greater than before. Yet somehow the new shawls turned out vulgar and uninteresting.
>
> . . . .These craftsmen . . . made beautiful shawls by standing in a long tradition, and by making minor changes whenever something seemed to need improvement. But once presented with more complicated choices, their apparent mastery and judgment disappeared. Faced with the complex unfamiliar task of actually inventing forms from scratch, they were unsuccessful.*

*Christopher Alexander, **Notes on the Synthesis of Form,** Harvard University Press, Cambridge, Mass., 1964, p. 53, 54. Reprinted by permission.

In creating new computer information systems, as well as in many other areas, we need methods for making good decisions in spite of change and complexity. To cope with complexity we must also take into account not only the fluid and disorganized nature of the world around us, but the limitations and strengths of the human mind as well. Psychologists of perception tell us that we can maintain in our short-term memory only a very small number of "chunks" of information (about 5 to 9) simultaneously. They also point out that we can absorb far more information visually in a fixed period of time than we can verbally.

How do we cope with complexity? We structure, and we simplify. We simplify in a variety of ways—by limiting the extent of our interest; by selecting the important or essential features of a situation and ignoring the rest; by breaking up what is large until it becomes of a manageable size; by examining things iteratively, a few at a time, until we have considered everything of interest; by reviewing and refining our ideas to improve them; and by using visual modes of thought whenever possible. Before further describing and illustrating these means of simplification, it is important to introduce and define some terms that will enable us to think and talk more effectively about systems of all kinds.

WHAT IS A SYSTEM?

One of the most general and powerful notions of order and structure is that of a system. The Greek roots of the word system mean "standing together."

A *system* may be defined as an interrelated set of components that are viewed as a whole. The components work together to perform a function or to achieve an objective. The function and objective of the system are usually closely related to the point of view and interests of the person looking at the system.

Thus a system has **components,** the basic parts or elements that make up the system. It also has a **structure,** which denotes the way in which the components are organized. There are several useful ways of thinking about the structure of a system. Sometimes it is described as the arrangement of the parts, sometimes as the relationships that link the components, and sometimes as the rules that describe the interactions among the components.

The *function* of a system is the task, activity, or work that the system performs.

The *objective* of a system is the human goal or purpose that the system serves in carrying out its function.

Many systems perform multiple functions and serve several objectives.

Consider the following examples of systems:

The solar system comprises our sun and nine major planets, their moons, and a belt of minor planets or asteroids between Mars and Jupiter. The behavior of this system follows the physical laws of gravity. The position and motion of each planet relative to the sun is described by Kepler's laws, which state that

1. The orbit of each planet is an ellipse with the sun at one focus.

2. A planet moves along its elliptical path so that a line joining the center of the planet to the center of the sun sweeps out equal areas in equal times.

3. The square of the time it takes for a planet to complete one orbit is proportional to the cube of its mean distance from the sun.

A baseball game has as its major components the two teams; the umpires; the playing field; the ball, gloves, masks, and other equipment; the actions of the players; and the score. The rules of baseball define the valid ways in which these components may be arranged and may interact. Because the rules define the permissible relationships among the components (the system structure), they are an essential part of the system. The function of the game is playing baseball; in this case, what the system does is defined only in terms of the system itself. A baseball game may accomplish many objectives: to provide exercise and camaraderie for amateur players, to provide employment for professional players and managers (as well as hot-dog vendors), to entertain the fans, to provide an occasion for betting, and so on.

A bicycle consists of two wheels, a frame with handlebars, a seat, pedals, and a chain to connect the pedals to the wheels. The function of a bicycle is to transform the circular motion of the cyclist's legs into linear motion. The principal objective is to transport the rider from place to place. However, a bicycle may also provide exercise, recreation, sport, or competition.

System Structures

A few familiar structural patterns—hierarchical, matrix, and network—occur in many frequently encountered systems. However, as the examples above should make clear, there are a great many systems whose structures cannot be adequately described in terms of these three relatively simple forms of organization.

A *hierarchical system structure* is one in which every system component except one is subordinate to exactly one immediate superordinate. A familiar example is the management structure for a corporation such as that shown by the organization chart in Figure 1-1. Each person in the company reports directly to only one person, but a supervisor may be responsible for several subordinates. Figure 1-2 shows a state map. In this state each city is contained in only one county, and the counties in turn are contained in the state. This figure illustrates why some hierarchical systems are called *nested.*

Figure 1-3* shows the interior and exterior of a typical bay of the nave of a Gothic cathedral and a section through the nave. The bay comprises piers, arches, colonettes, vaults, ribs and buttresses.

The ribs span the space enclosed by the bay. The vaults fill the spaces between the ribs. The arches carry the load of the vaults and walls to the piers. The piers (clusters of colonettes) transmit the load to the foundations, which rest upon the ground. The buttresses provide additional support to prevent the vaulting from spreading horizontally. The geometry of each part helps make visible its role in the building structure; each element has its place in the hierarchy.

A hierarchical structure is sometimes called a *tree structure,* because it can always be represented in the form of a tree. A hierarchy tree is usually upside-down,

---

*Sir Banister Fletcher, *A History of Architecture*, 18th edition, © 1975 The Royal Institute of British Architects and The University of London, p. 600. Reprinted by permission of The Athlone Press.

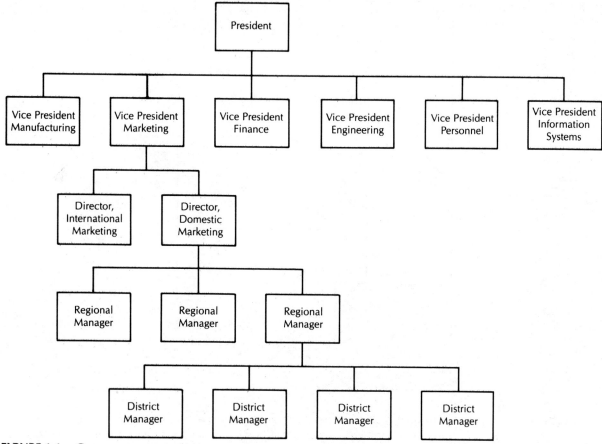

FIGURE 1-1. Organization chart.

as in Figure 1-4, with the **root** at the top and the **branches** extending down; the elements at the ends of the branches are sometimes called **leaves**.

A **matrix** structure is one in which the location of the components is determined by a combination of two or more factors. The matrix form of organization is found in many businesses. In architectural and engineering firms this structure is based on both project assignment and professional disciplines (see Figure 1-5). Each project has a manager, while each discipline has a department head responsible for allocating personnel and providing them with additional technical expertise and support.

A **network** structure consists of a set of **nodes** (or points) connected by **arcs** (or links or lines). The arcs permit **flows** between pairs of nodes in specified directions. An electrical circuit diagram (Figure 1-6)* and an interstate highway map (Figure 1-7) are examples of networks.

*Wai-Kai Chen, **Linear Networks and Systems**, Brooks Cole Engineering Division, Wadsworth, Inc., Monterey, California, © 1983, p.33. Reprinted by permission of PWS Publishers, Boston, Mass.

FIGURE 1-2. Nested hierarchy: a state map.

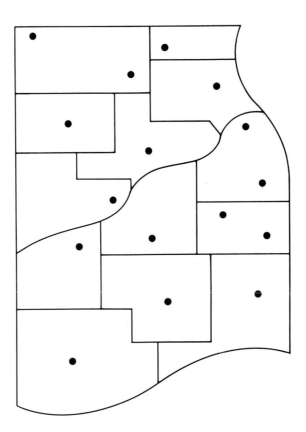

Hierarchies of Systems

How do we select and isolate the system we wish to investigate from the complexity of the environment in which it is embedded and with which it interacts? Systems themselves come in hierarchies. Every system can be viewed as part of a more inclusive system. There is always a temptation to expand our area of study to obtain a more comprehensive perspective. On the other hand, it is always possible to shrink the focus of our study to look at finer and finer detail. Thus the problem becomes unmanageable, or we fail to solve it because we never finish studying it. These dilemmas which occur in physical design as well as in systems analysis, are captured in John Eberhard's story of the doorknob.

One anxiety inherent in design methods is the hierarchical nature of complexity. This anxiety moves in two directions, escalation and infinite regression. I will use a story, "The Warning of the Doorknob," to illustrate the principles of escalation.

This has been my experience in Washington when I had money to give away. If I gave a contract to a designer and said, "The doorknob to my office doesn't have much imagination, much design content. Will you design me a new doorknob?" He would say "Yes," and after we establish a price he goes away. A week later he comes back and says, "Mr. Eberhard, I've been thinking about that doorknob. First, we ought to ask ourselves whether a doorknob is the best way of opening and closing a door." I say, "Fine, I believe in imagination, go to it." He comes back later and says, "You know, I've been thinking about your problem, and the only reason you want a doorknob is

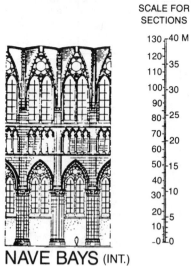

NAVE BAYS (INT.)

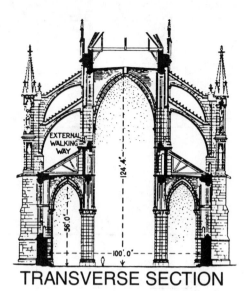

TRANSVERSE SECTION

NAVE BAYS (EXT.)

FIGURE 1-3. Structure of a bay of Rheims cathedral (interior elevation and section).

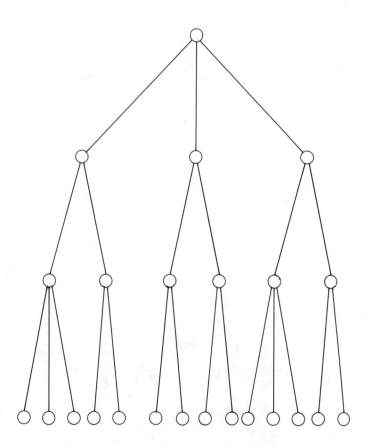

FIGURE 1-4. A tree structure.

FIGURE 1-5. A matrix structure.

		Project 1	Project 2	Project 3	Project 4	Project 5
DISCIPLINE	Architecture					
	Structural					
	Mechanical					
	Electrical					
	Interiors					

that you presume you want a door to your office. Are you *sure* that a door is the best way of controlling egress, exit, and privacy?" "No, I'm not sure at all." "Well I want to worry about that problem." He comes back a week later and says, "The only reason we have to worry about the aperture problem is that you insist on having four walls around your office. Are you sure that is the best way of organizing this space for the kind of work you do as a bureaucrat?" I say, "No, I'm not sure at all." Well, this escalates until (and this has literally happened in two contracts, although not through this exact process) our physical designer comes back and he says with a very serious face, "Mr. Eberhard, we have to decide whether capitalistic democracy is the best way to organize our country before I can possibly attack your problem."

On the other hand is the problem of infinite regression: If this man faced with the design of the doorknob had said, "Wait. Before I worry about the doorknob, I want

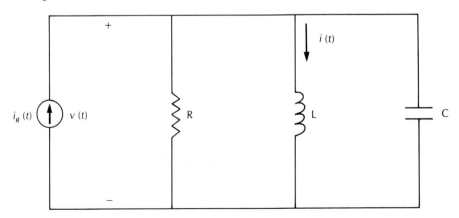

FIGURE 1-6. Network structure: an electrical circuit diagram.

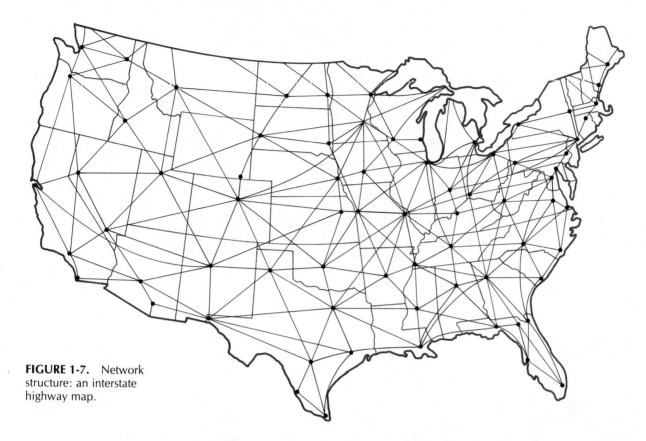

FIGURE 1-7. Network structure: an interstate highway map.

to study the shape of man's hand and what man is capable of doing with it," I would say, "Fine." He would come back and would say, "The more I thought about it, there's a *fit* problem. What I want to study first is how metal is formed, what the technologies are for making things with metal in order that I can know what the real parameters are for fitting the hand." "Fine." But then he says, "You know I've been looking at metal-forming and it all depends on metallurgical properties. I really want to spend three or four months looking at metallurgy so that I can understand the problem better." "Fine." After three months he'll come back and say, "Mr. Eberhard, the more I look at metallurgy, the more I realize that it is atomic structure that's really at the heart of this problem." And so, our physical designer is in atomic physics from the doorknob. That is one of our anxieties, the hierarchical nature of complexity.*

System Boundaries

You have to draw the line somewhere. It is undesirable, if not impossible, to be concerned with the entire world and to try to solve all its problems at once. In any specific situation the scope of our system should be sufficiently inclusive and our system description should be sufficiently detailed to serve our purpose; but the more we extend the scope, and the more detail we include, the more complex the

*John P. Eberhard, "We Ought to Know the Difference," in **Emerging Methods in Environmental Design and Planning,** Gary T. Moore, editor, The MIT Press, Cambridge, Mass., 1970, pp. 364–365. Reprinted by permission.

system becomes. Yet there is always the temptation to enlarge the scope and increase the amount of detail, as Eberhard's story has illustrated.

In practice, where to locate the system boundary is often one of the most difficult decisions an analyst has to make. The **boundary** (Figure 1-8) defines the limits of the system. Inside the boundary is the system, the part of the world to be studied, described, structured, and, usually, changed. Outside the boundary is the **environment** or **context**, the part of the world to be ignored except for a few important interactions at the boundary between the system and the environment. A diagram, such as Figure 1-8, which shows the system, its environment, and, perhaps, selected interactions between them is called a **system context diagram** or **context diagram.**

The term **interface** is often used to describe an interaction—between system and environment, between two systems, or between two components of a system, as shown in Figures 1-9 and 1-10. The basic geometric analogy for an interface is that of Figure 1-9, where the area of interaction appears as a zone of contact. In Figure 1-10, however, the interfaces or connections between components are shown as lines or arcs; in this case, the representation has lost the underlying image of touching.

FIGURE 1-8.   A system context diagram.

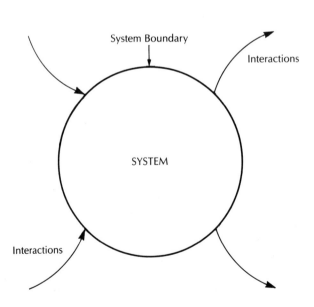

ENVIRONMENT

System Boundary

Interactions

SYSTEM

Interactions

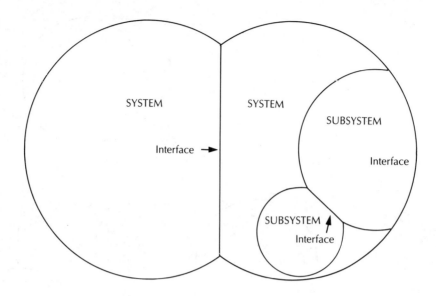

FIGURE 1-9. System and subsystem interfaces.

MODELING AND REPRESENTATION

Once we have identified a system as being of interest to us, limited the area of interest by defining the system boundary, and established which interactions between the system and its environment are important, we must decide how to describe the system. We must reduce the complexity of this system description by further simplification.

Producing a simple system description involves **abstraction** and **selectivity**. The two major approaches to constructing or presenting system descriptions are **aggregation** and **decomposition**. The resulting description is usually a **system model**, communicated and manipulated through one or more **representations**.

The Need for System Models

System descriptions, or models, serve several purposes in system studies. The most basic of these is understanding the system—grasping conceptually what the components are and how they interact to carry out the system functions and objectives. Closely related is the use of a model to communicate this understanding of a system to others. The more clearly we understand, the better we can communicate; and clearer communication can result in improved understanding. We use models to study the behavior of a system because it is usually impossible or impractical to deal with the system directly. This is especially useful in understanding the behavior of new systems or predicting the future performance of existing systems. Engineers

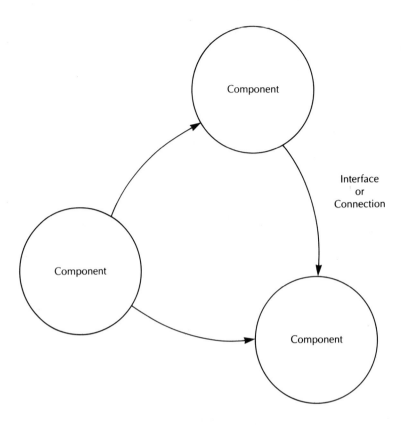

FIGURE 1-10. Interfaces as connections.

model the structural system of a multistory office building before it is constructed to study the effect of wind and earthquake forces. Economists investigate the effects of changes in levels of inflation, production, and employment. Models may be used to study alternative proposals for system change and select the best.

Characteristics of a Good Model

Every model is an abstraction from reality—a selection of system characteristics that are relevant to the purpose for which the model was constructed. In modeling a building, an interior designer will be interested in the color and surface texture of a wall, a lighting engineer in its ability to transmit or reflect light, an air conditioning engineer in its thermal conductivity, and a structural engineer in its weight and the load it can carry.

A good model is one appropriate to its purpose. Some models, such as the engineer's model of a building structure, are primarily oriented to simulating or predicting system behavior: these are called *performance models*. They are judged by how well they approximate system performance. Other models, such as the plans and specifications for a building, are oriented toward system construction. These are often called *specifications* and are evaluated by their adequacy for realizing or constructing the system, primarily in terms of completeness and consistency. For example, construction documents for a building are complete if all the components of the building are described and are consistent—for example, if beams, ducts, and lights are not shown occupying the same space.

Good models also support the process of systems analysis and design. This is usually a highly interactive process, so the models must be easy to manipulate, change, and throw away. In this respect a mathematical model of a building structure, such as a set of simultaneous equations, may be preferable to a graphical model. A quick sketch of the exterior of a building may be better for an architect than a three-dimensional scale model, or a diagram of the locations of emergency exits may be preferable to a photograph of the interior of an airplane. Graphic models strongly aid the imaging or visualization of a system. They permit a simultaneous perception of the system in contrast to verbal models or descriptions, which must be perceived sequentially in time, and thus take advantage of the greater rate of information transfer through the eye.

Good models are also felt to be **natural.** They incorporate properties in such a way as to emphasize the inherent similarities between the model and the essentials of the system being modeled.

The best models also become **transparent** to their users. That is, the users are able to "see through" the model, so that when they manipulate the model they appear to be interacting directly with the system; the model does not get in their way. To be sure, this transparency is partly the result of skill and much practice on the part of the user, but a natural system description that supports the visual imagination greatly facilitates achieving transparency.

Representation

A **representation** of a model is a graphical or physical way of displaying or demonstrating the components and relationships in the model. Sometimes a model may be represented in more than one way. For example, a set of rules for making decisions in given situations may be depicted in the form of a table, as in Figure 1-11, or that of a tree, as in Figure 1-12. The two representations are equivalent in that both show the actions to be taken for each combination of conditions. Because of this, either representation may be converted into the other, yet, as we might expect,

FIGURE 1-11. Structure of a decision table.

	1	2	3	4	5	6	7	8	9	10	11	12	13	14	15	16	17	18
Recent Sales Activity	F	N	S	F	N	S	F	N	S	F	N	S	F	N	S	F	N	S
Inventory Level	H	H	H	O	O	O	L	L	L	H	H	H	O	O	O	L	L	L
Perishable	Y	Y	Y	Y	Y	Y	Y	Y	Y	N	N	N	N	N	N	N	N	N
% Change	0	−10	−15	0	0	−5	+5	+5	0	+5	−5	−10	+10	0	−10	+15	+5	0

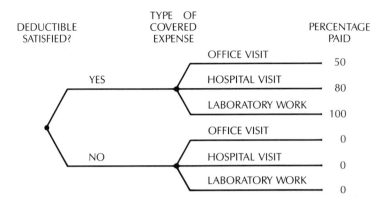

FIGURE 1-12. Structure of a decision tree.

sometimes one representation will be preferable because of how we intend to use it. (Chapter 8 presents a detailed discussion of decision trees and tables for structured analysis.) Or, as shown in Figure 1-4 and 1-2, a hierarchical structure may be represented either as a tree structure or as a set of nested contours.

Notation

Sometimes there are also alternative notations. A *notation* is a symbolic way of describing a system. For example, in describing the moves in a chess game, different sets of symbols may be used to denote the pieces and the squares of the chess board. There is the traditional English notation and the algebraic notation that is supplanting it. (See Exercise 6 for details.)

Generating a System Model

There are two principal approaches to generating a system model: *aggregation* and *decomposition.*

Aggregation. *Aggregation* is a process of assembling a system using a set of elementary components, often called *primitives*. This "kit-of-parts" approach is like fabricating a building out of bricks, mortar, lumber, and nails, or stringing a necklace of beads, or organizing complex patterns of information out of 0s and 1s. It proceeds from the parts to the whole and is thus frequently characterized as *bottom-up*. (Think of constructing the foundation of a building first or looking at an organization chart from the most detailed, or bottom, level. Classically, this process is known as *synthesis,* from Greek roots meaning "put together," whose Latin equivalent is *composition*. Thus synthesizing a chemical compound and composing a symphony are both described as bottom-up activities in which something new is created.

Decomposition. *Decomposition,* on the other hand, proceeds from the whole to the parts. This approach *partitions* a system into its constituent elements, much as an artist rents an empty loft and organizes the space into distinct areas for painting, exhibiting paintings, cooking, eating, sleeping, and bathing. A developer takes a tract of land and *subdivides* it into lots. Decomposition is characterized as a *top-down* approach. An organization chart is read beginning with the chairman of the board. The classical term for decomposition is *analysis;* the best English approximation of the Greek is probably "break up," but the preposition "ana" also connotes "from above" or "top-down." A television picture is decomposed into red, green, and blue images for transmission; a literary critic analyzes the structure of a poem. Analysis is most often conceptual rather than physical. We describe system structure by distinguishing, isolating, or identifying the components.

The Complementarity of Analysis and Synthesis

Clearly, analysis and synthesis proceed in opposite directions and take opposite points of view, yet in practice they are complementary. In many situations system models are generated through a combination of analysis and synthesis. In this sense, what is frequently called *systems analysis* requires synthesis, and *system design,* often considered a synthetic activity, requires analysis. When we have a job to do, we work bottom-up, top-down, and from the middle in both directions. The result is what matters. If we master both analytic and synthetic techniques, we can apply each of them to generating system models as we choose.

Aggregation and decomposition have been discussed in relation to generation of models. The same approaches are useful in presenting a system model. However, the model, once developed, exists independently of the developer. From the completed model it is impossible to tell which approach was used to generate it. Moreover, the most effective way to present a system description does not necessarily match the approach used in generation.

Limitations of System Models

Expertise in the systems field is built on a broad knowledge of system models as well as facility in applying them to specific situations. Because every model is an abstraction, a description of reality selected to carry out an objective, it is limited. It is limited by the aspects of reality omitted from the model, the relationships among system components ignored by the model, and the degree to which the model approximates reality. The systems expert is aware of these limitations. It can be just as important to know what a model does *not* show as what it shows. It is important not to apply a model outside its intended domain without being aware of the situation. Thus throughout this book we will emphasize the limitations of the system models presented. Expertise is also based on the ability to see a system from multiple perspectives, to see the whole as well as the parts, and to exercise sound judgment.

SUMMARY

Complexity and lack of structure characterize the environment for most systems analysis. Finding or creating order and structure permits people to understand a situation, communicate that understanding, and then take appropriate action.

Thinking about the world in terms of systems is one of the most general and powerful ways in which human beings provide order and structure. A system is a whole comprising interrelated components organized according to a structure. Most systems perform one or more functions and fulfill human purposes or objectives. Common types of structures include hierarchies, matrices, and networks.

Our world can be viewed as a hierarchy of nested systems—capable of infinite expansion on the one hand and divisible into subsystems of smaller and smaller scope on the other. Establishing system boundaries to limit the area of study is one of the most difficult tasks for systems analysts. The system boundary separates a system from its context, or environment, and locates the interfaces between the system and the environment.

A system model is a selective, abstract description of a system. Models are used to understand existing systems, to simulate or predict future system behavior, and to realize or construct new systems. System models are generated through two complementary processes—analysis and synthesis. Analysis works from the top down, decomposing or partitioning the whole into its constituent parts. Synthesis works from the bottom up, constructing the whole by combining or aggregating primitive elements.

The simplifications introduced by system models allow human beings to think and act effectively in complex situations in spite of human limitations. Yet models have their own limitations. The wise user respects them when exploiting the conceptual and practical power of systems thinking.

REVIEW

1-1. State some human limitations that make it difficult to deal with ill-structured complexity.

1-2. List seven ways of coping with ill-structured complexity.

1-3. Define the following terms:
   a. system.
   b. component.
   c. system structure.
   d. system objective.
   e. system function.
   f. system boundary.
   g. environment of a system.
   h. interface.
   i. subsystem.

1-4. Name three types of system structure and give an example of each.

1-5. What is the difference between a system model and its representation?

1-6. What is the difference between a representation and a notation?

1-7. List four characteristics of a good model.

1-8. What is the difference between aggregation and decomposition?

1-9. What type of model is used to
   a. Describe or predict system behavior?
   b. Present a consistent description of a system in sufficient detail for the system to be constructed?

1-10. Define analysis and synthesis. How are these concepts related to aggregation, decomposition, and system modeling?

EXERCISES AND DISCUSSION

1-1. Draw a diagram depicting and labeling the following: system, subsystem, component, system boundary, environment, interface.

1-2. Based on Figure 1-13 below, draw a more abstract diagram showing state, county, and city as a nested hierarchy of subsystems.

1-3. In Figure 1-7, which shows a highway network, identify the components of the system. Which are nodes, which are arcs? What is flowing through the

FIGURE 1-13. A state map.

system? How can the flow be measured? What does the length of each arc represent? Can it be measured in more than one way?

1-4. Draw a context diagram for a system with which you are familiar, such as the educational system in which you are a student, the company for which you work, a piece of equipment, or the solar system. (Hint: It will be similar to the diagram for Exercise 1-1 but will show specific features of the system you have selected. Also see Figure 1-8.)

1-5. Express Kepler's Laws in mathematical notation.

1-6. Describe a series of moves in a chess game in both descriptive and algebraic notation. You may wish to consult a book on chess for an explanation of the traditional descriptive notation. An explanation of algebraic notation with an example appears below.*

In algebraic notation, the files are designated "a" through "h," with "a" at White's left, and the ranks "1" through "8," with "1" closest to White. Each square has one name, consisting of the letter of the file followed by the number of the rank. To describe a pawn move, one simply writes the name of the square to which the pawn moves. For a move by a piece, one uses the customary abbreviation for a piece ("N" for knight, "B" for bishop, etc.) followed by the name of the square to which the piece moves.

Captures are indicated by an "x" between the name of the piece and the name of the square. When two knights or two rooks can capture the same square, another letter or number is added, preceding the "x," to show the file or rank from which the piece moved. Captures by pawns are always described by the file from which the pawn moves, an "x," and the square on which the capture takes place. Checks are shown by "+."

FIGURE 1-14. Alternative notations for squares on a chess board.

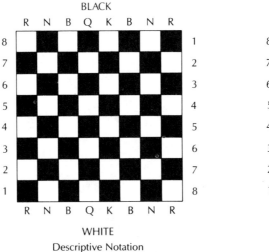

BLACK

Descriptive Notation

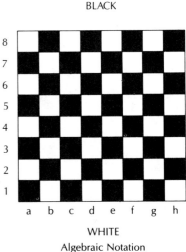

BLACK

WHITE

Algebraic Notation

*Jack Peters' chess column in the **Los Angeles Times**, Sunday, January 9, Part II, p. 2. Quoted by permission of the author.

Here are the identical opening moves in algebraic and descriptive:

1.	e4	e6		1.	P-K4	P-K3
2.	d4	d5		2.	P-Q4	P-Q4
3.	Nd2	Nf6		3.	N-Q2	N-KB3
4.	e5	Nfd7		4.	P-K5	KN-Q2
5.	Bd3	c5		5.	B-Q3	P-QB4
6.	c3	cxd4		6.	P-QB3	PxP
7.	cxd4	Nc6		7.	PxP	N-QB3
8.	Ngf3	f5		8.	KN-B3	P-B4
9.	dxf6	Nxf6		9.	PxPep	NxBP

What are the advantages and disadvantages of each notation?

1-7. Consider the games of chess and checkers as systems. What components do they share? What components differ? Could the descriptive chess notation be easily used for checkers? The algebraic? Is a king identical in both games? Why?

1-8. Frequency (pitch) and duration are important aspects of musical sound. Explain the conventional notation used for each in a musical score.

1-9. a. The meter (rhythmic structure) of poetry is described in terms of a primitive called a **foot.** Shakespeare's plays are written in **iambic pentameter.** Explain what this means and what approach to system description has been taken.

  b. Other poetic forms incorporate both meter and rhyme. Find an example of a **sonnet** and explain its structure.

2 Computer Information Systems for Business

The general systems concepts presented in Chapter 1 can be used to describe human organizations and activities as well the information systems that support them. This chapter takes a systems view of business organizations and summarizes their need for information as well as the business activities supported by computer information systems.

This chapter also introduces the principal functions of an information processing system and reviews the major components of a computer information system. It emphasizes the high degree of abstraction used in the description of computer information systems.

COMPUTER INFORMATION SYSTEMS FOR BUSINESS

Information systems are frequently classified by the kinds of applications they support. These broad areas of application include science, research, engineering, design, manufacturing, accounting, and finance. Clearly any or all of these functional areas can be found in a business organization.

Traditionally each of these areas has been associated with characteristic data and computations. Scientific, research, and engineering applications were typified by small amounts of data and extensive, complex, floating-point computation. Design applications added the manipulation of complex geometry. Manufacturing applications used computers to guide the paths of cutting tools or to monitor and control production process equipment. Monitoring industrial processes or research experiments required real-time readings from instruments measuring the values of critical variables.

Accounting and financial applications, on the other hand, needed only simple calculations, exact to the penny, but with large files of data.

Early hardware and software were intended either for scientific or commercial applications, and the two types of systems were generally incompatible. As computers have become more compact, less expensive, less demanding in their physical environment, and more widespread, these traditional distinctions have begun to blur. Scientific and engineering applications use increasingly large data bases. Business applications incorporate extensive statistical analyses and continuous monitoring of inventory levels. Word-processing and document-editing systems, a curiosity 20 years ago, now give support to people working in all these activities. Microcomputers, distributed computing, and color graphics displays provide further capabilities for a variety of computer applications.

This book is oriented toward business systems and takes its examples from applications such as order entry, inventory, purchasing, payroll, billing, and accounts receivable and payable. However, the principles and techniques of systems analysis and description are applicable to all kinds of computer information systems.

A Systems View of Business

A business provides products and services to the public or to other businesses. There are two principal types of businesses—***industrial*** or ***manufacturing*** firms, which manufacture goods, and ***service*** firms, which sell or distribute products made by others or which provide other services of various kinds. Some businesses expect to make a profit; others, primarily engaged in research and services, are nonprofit. Government organizations are public rather than private but are similar to business organizations in many respects, especially with regard to their use of information.

Figure 2-1 shows a context diagram for a manufacturing business. Its environment includes the government, which establishes the legal and regulatory framework; other businesses, which supply materials, parts, and services; competing

FIGURE 2-1. Context diagram for a manufacturing business.

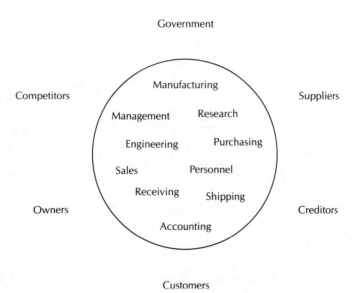

businesses; and customers. There are also the stockholders or owners, as well as those who have lent the company money, such as bondholders, banks, and insurance companies. The business itself comprises the management, the manufacturing functions (which are primary), and the secondary but necessary functions that support management and manufacturing, such as research, engineering, purchasing, marketing, personnel, receiving, shipping, and accounting.

Figure 2-2* shows the major functional units of a typical manufacturing business as well as the flows of material and information between these units.

The material required in the manufacturing process is purchased from suppliers. Deliveries from the suppliers are received and stored as inventory until used by manufacturing to make finished products. The goods produced are then shipped to customers, perhaps being stored awaiting sale or while in transit.

The major information flows help coordinate and control the flow of materials through the manufacturing process. Marketing receives orders for goods from customers. Information about those orders goes to Shipping, so that the goods can be shipped, and to Accounting, so that the customers can be billed. Information from Marketing about the level of sales goes to Manufacturing to specify the desired level of production and the product mix. Engineering may be required to prepare specifications to guide the manufacturing process for special orders. As material is needed for the manufacturing process, Manufacturing requisitions it from the stock on hand, or, if it is not available in inventory, asks Purchasing to buy it. Purchasing, in turn, orders material from suppliers as needed to maintain the desired inventory level or to fill special requests from Manufacturing. Copies of the Purchase orders go to Receiving to be compared against deliveries arriving from suppliers. Accounting sends bills to customers and pays the invoices from suppliers after they have been approved by Purchasing. Accounting also pays taxes, records and monitors the costs of doing business, and provides the company's management with information about its profitability. Research and development takes ideas for new products and turns them into prototypes for commercially viable products.

Figure 2-2 is deliberately simplified to present an overview. Thus it illustrates one of the benefits of an abstract model. It shows only major flows of material and information supporting the primary function of Manufacturing. A more detailed view would show other information flows that provide feedback within the system. Some of the important functions within the organization, notably Management and Personnel, do not appear at all.

A diagram presenting an overview of a service organization would show fewer functional units than Figure 2-2. There would be no Manufacturing function transforming material into finished products. Which other flows and functions would be eliminated would depend on the types of services performed.

Why Businesses Need and Use Information

Operating and managing a business requires information. Some information is required to carry out the operations on a continuing basis. What parts are required

---

*Figure 2-2 is based on Figure 3.3 of Robert J. Thierauf's and George W. Reynolds' **Systems Analysis and Design: A Case Study Approach**, Charles E. Merrill Publishing Company, Columbus, Ohio, 1980, p. 69. Adapted by permission.

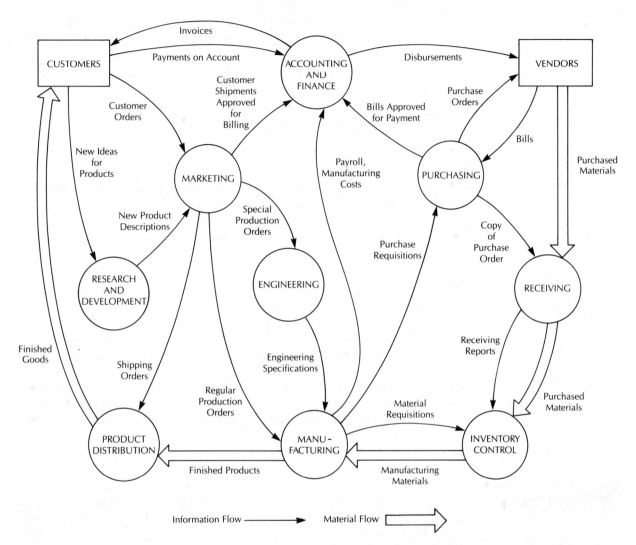

Information Flow ⟶   Material Flow ⟹

FIGURE 2-2. Major information and materials flows in a manufacturing business.

to make Product Y? Do we have them in stock? Where can we buy them? How much will they cost? How long does it take to get delivery? Is there a discount if we order in quantity? If we pay within five days?

Other information is needed to monitor the financial health of the business. What is our current level of profit? Is our debt too high? Are our production costs competitive? Are our customers paying promptly? Are our assets earning an adequate rate of return? Is our share of the market increasing?

Still other information is used to support long-range planning and decision making. What are the trends in the economy and in our markets? What kinds of capital investments, if any, should we make, and when? What will be the effect of introducing a new product? Of reorganizing our sales territories? Of an accelerated research and development program? Of revising our profit-sharing and retirement plan?

Similar questions, whether related to the immediate needs of day-to-day operation or the perspective of strategic management decisions, arise in every business organization. Information to answer them is essential to the production of goods and the provision of services as well as to the continuing vitality of the enterprise. Information is increasingly being regarded as a major corporate resource to be managed as wisely as people, materials, equipment, and money.

Information plays four major roles in a business:

1. It helps carry out the production and service functions of the organization.

2. It measures and monitors the performance of these primary business functions and the other functions that support them.

3. Based on these performance measurements, it helps direct the organization to meet performance targets.

4. It supports management decisions to improve the business by modifying the organization or changing its objectives.

In any business, these last three tasks—monitoring, controlling, and decision support—are almost exclusively information-processing activities. In many service organizations, such as an insurance company, a law firm, an advertising agency, or a market research group, even the primary function of the business is information processing. Only in such areas as manufacturing, distribution, and maintenance are the primary business functions material- and energy-intensive rather than information-intensive.

Problem Solving and Decision Making

Information is used to monitor, control, and support decisions at almost every level in a business organization, from the concerns of daily operation to the long-range strategic decisions of top managers. Information produced by monitoring the activities of the business can identify a discrepancy between what is happening and the expected performance of these activities. Such a discrepancy between actual and planned performance indicates that there is a problem. Indeed, the philosophy of management by exception provides for continuous monitoring but limits the reporting of information to those situations in which performance varies from what is planned or expected.

Rectifying a problem is a process that culminates in a decision. The purpose of the decision is to take action to eliminate the problem by making changes in the undesirable situation. Thus problem solving and decision making are closely related processes; problem solving leads to decisions, and effective decision making requires considering the underlying issues or problems.

The problem-solving process may be described as a sequence of steps for the purposes of presentation. In reality, however, there is considerable flexibility in the ordering of the steps and in the amount of effort devoted to each step, depending on the situation. One way of describing these steps is summarized here.*

1. Identify the problem. This step involves examining the situation to state clearly what the problem is. The problem statement implies the conditions that a solution must satisfy. These conditions are known as **constraints**.

2. Generate possible solutions. In general, problems may have many solutions. Considering several dissimilar solutions is likely to produce a better result than considering only a single proposal. In this step several possible solutions are proposed. Each solution is described in sufficient detail to carry out the remaining steps in the problem-solving process.

3. Eliminate the proposed solutions that do not solve the problem. Each proposal is checked to see whether it satisfies the constraints of the problem. If not, it is rejected or modified until it becomes acceptable.

4. Evaluate the expected performance or behavior of each proposed solution. The remaining proposals are evaluated qualitatively and quantitatively with respect to the most important factors affecting the choice of the best proposal. These factors are called evaluation **criteria.**

5. Compare the alternatives to select the best solution. As a result of the evaluation, the relative advantages and disadvantages of the proposals can be compared and the best one selected. At the end of this step, the best solution is known; the appropriate decision can be recommended. However, people in business make decisions in order to act—whether the action is immediate or requires considerable planning and preparation. Effecting the solution requires additional steps to be taken.

6. Plan how to implement the solution that was selected. This step includes everything that must be done to prepare for making the proposed solution a reality. If the solution involves extensive or complex changes to the organization, schedules and budgets must be prepared, resources allocated, and responsibilities assigned for implementing the solution.

7. Implement the solution. In this step the decisions and plans are effected.

8. Evaluate the performance of the solution after its implementation. Finally, the situation is monitored to determine whether the actions taken have solved the problem and whether the anticipated improvements have been achieved.

In the broadest sense, systems analysis is a discipline that analyzes problems, estimates the implications or consequences of various courses of action, and recommends what action to take to solve problems. In a business organization, the analysis of information systems is also directed toward problem solving and toward

---

*For other ways of describing problem solving, see Thomas Athey, **Systematic Systems Approach: An Integrated Method for Solving Systems Problems**, Prentice-Hall, Englewood Cliffs, N.J., 1982, or Joseph Allen and Bennet P. Lientz, **Systems in Action: A Managerial and Social Approach**, Goodyear Publishing Company, Inc., Santa Monica, Calif., 1978.

improved information systems that provide better support for business activities. All aspects of business rely increasingly on computer information systems to assure timely and accurate data displayed in a useful form.

FUNCTIONS OF AN INFORMATION SYSTEM

Information systems perform three principal functions—they transmit information, they store it, and they transform it.

1. Communication

Systems whose chief function is to transmit information are usually called communication systems. Communication systems move information from person to person or from place to place. Speech is the oldest and most fundamental form of human communication. Like all communication systems, it involves a transmitter (the vocal apparatus), which encodes information into elementary units (the sounds or phonemes); a signal or sequence of these encoded units (sound waves), sent via some information-carrying medium (the air); and a receiver (the ear), which decodes the signal (the sounds) and reconstructs the meaning of the message based upon a set of rules (the syntax and semantics of the language).

Communications problems arise through a faulty transmitter (a speech impediment), faulty transmission (perhaps due to noise or interference), a faulty receiver (lack of attention or partial deafness), or problems in encoding or decoding (neurological or linguistic disabilities).

Other familiar communication systems are the telephone, the radio, and television. Telephones encode sounds (primarily speech) into electromagnetic signals, which are transmitted through wires, through the air in the form of microwaves, and recently through glass filaments in the form of light. Radio transmits music as well as speech by means of airborne signals. Television, also transmitted by air or cable, adds pictures to music and speech.

2. Storage

Messages in a communication system are inherently transient; without a means of storage, there is no enduring record of the information. Carvings in stone, impressions in soft clay, ink on parchment, pencil on paper, phonograph records, photographic film, sound and video recordings on magnetic tape—all these are different ways of capturing information in a relatively permanent way.

In English and other Western languages, for example, the spoken language is recorded using an alphabet—a set of written symbols. A letter or combination of letters from the alphabet represents each sound of the language. The symbols can be recorded with pencil, pen, or other marking device on paper or some other surface. Thus a transient sequence of sounds has been converted into a permanent sequence of graphic symbols. The symbols are decoded visually into the sentences of the original utterance. But, surprisingly enough, when reading a book or newspaper, it is not necessary to utter the sounds. Thus (even without speed reading) information retrieval through reading is considerably faster than information transmittal through speech or information recording through typing or writing. Through techniques such as micrographics using microfilm or microfiche as the storage medium, or electronic storage using silicon chips, information storage today has become not only permanent, but extremely compact.

3. **Transformation**

As we have seen, even the transmission and storage of information often require a conversion from one information-carrying medium to another. However, when we speak of transforming information, we usually refer to a change in the content of the information rather than a change in the form of the information or in the medium that carries it.

From this point of view, we may say informally that a transformation is a computation or calculation: the result of the calculation is derived from known information by following a known procedure or set of rules. For example, given a list of purchases at a grocery store, the sales tax rate, and the rules for addition and multiplication, it is possible to calculate the total amount of the purchase. The price of each item on the list is summed; then the amount of the sales tax is computed by multiplying the tax rate by the sum of the items purchased; and, finally, the amount of the sales tax is added to the sum of the prices of the items purchased, giving the total amount of the purchase.

The list of purchases and the sales tax rates are called the **inputs** to the transformation. The total amount of the purchase is the **output** of the transformation. We say that the output is **derived** from the inputs. The rules that describe the transformation are known by a variety of names. **Procedure** is probably the most generally useful, although other terms are also used in more specialized contexts.

INFORMATION PROCESSING SYSTEMS

An **information processing system** is an information system that performs transformations. (Indeed, the word **process** is often used as a synonym for the word transformation.) Most information processing systems also need to store information and transport it from one part of the system to another. Thus, most information processing systems perform' all three of the functions discussed above—transmission, storage, and transformation.

But the essential function of an information processing system is to carry out one or more transformations of information—that is, to derive some desired information (the outputs) from some known information (the inputs). The specific information contained in these inputs and outputs depends, as discussed in Chapter 1, on the specific human objectives to be accomplished by the system and the specific context in which the information processing system occurs. We shall return to this topic later in the chapter.

For example, a simple payroll system will derive employee paychecks from time cards containing the number of hours per week each employee has worked and from the salary or hourly wages of each employee. The payroll system will also need information from which to calculate the amounts to be withheld for federal and state income tax and for contributions to the social security system. Other desired system outputs may include the annual statement of earnings and withholding

(Form W-2) for each employee and the summary of withholding for all employees sent by the employer to the federal and state governments periodically along with the amounts withheld.

Manual and Automated Systems

Thus far, everything that has been said about information processing systems has been independent of the technology used. We have given examples of manual systems as well as those assisted by sophisticated technology. Strictly speaking, every information processing system is based on a technology. The ancient abacus and the mechanical calculations of the late 19th and early 20th centuries were technological aids to increase the speed and accuracy of computation. But the invention of the electronic computer about 1950 and advances in the technology for information storage, transmission, and transformation over the past three decades have indeed revolutionized information processing systems. These advances have made a wide variety of applications of electronic computers economical.

The desire to use computers throughout a wide range of information processing tasks is what makes the study and understanding of information processing systems of great practical benefit and not merely of theoretical interest. (Although what is merely of theoretical interest today often turns out to be of great practical benefit tomorrow.) To be sure, the same concepts are useful in describing any information processing system, whether manual or automated, and the study of a manual system may lead to improvements in it. Nevertheless, the reason for carefully analyzing a manual information processing system usually turns out to be the hope that its effectiveness can be significantly enhanced by using a computer.

DESCRIBING A COMPUTER INFORMATION SYSTEM

When we apply the general system concepts of Chapter 1 to describing an information processing system, we expect to describe it in terms of its components, their arrangement or relationships, and their functions. Thus an information processing system consists of components that transport or transmit information, components that store it, and components that transform it. The system boundary separates the system from its environment and identifies the points of contact (the interfaces or connections) between the system and its environment.

Consider first a digital computer as an automated information processing system. Within the system boundary, information is transmitted, stored, and transformed in electronic form. At the system boundary are the interfaces between the computer and its environment—the human world in which information is stored in words, numbers, and pictures and transmitted orally and visually. These interfaces are devices for capturing data (such as keyboards, microphones, digitizers, touch-sensitive panels, card readers, optical scanners, and so on), as well as devices for displaying data (cathode ray tubes, voice synthesizers, plotters, printers, and the like). Within the system there are storage devices with a wide range of capacity and a wide range of access times. These include various levels of memory, magnetic disks, drums, tapes, data cells, and so on. And there are various devices for information transformation—logical and arithmetic processors in a variety of possible configurations. Connecting these components are channels, cables, wires, buses, and other paths through which information can flow from one component to another. Finally, there is a way to control and coordinate the various components so that they can work together to perform the information storage and transformation in the desired sequence (or, in some cases, in parallel).

Thus far, all the computer components named have been part of the hardware. But a computer information processing system also comprises software, the programs that direct the operation of the hardware. Some of these programs manage the hardware resources—naming and cataloguing information and allocating it to the storage devices, allocating time and space to various users and jobs in a multiuser or multijob system, and accounting for the usage of the system by the various users and jobs. These programs constitute the *operating system*.

Most computers are general-purpose information processing systems. The hardware provides a limited set of low-level operations for storing and transforming information. These operations must be combined in the proper sequence to perform the specific transformations of interest to a person or organization in the world outside the computer. That is, a general-purpose computer must be converted to a special-purpose computer. The computer software that adapts general-purpose hardware so that it carries out specific transformations or functions required by a user or group of users is called *application software* or *application programs*.

A Matter of Emphasis

Every description of a system, as we saw in Chapter 1, involves a selection of essential or significant characteristics. In our discussion of computer information systems, we have identified the transfer, storage, and transformation of information, as well as control of these functions, as the significant features. Each is essential. Which of these is most important?

Currently there are several points of view, each with a different emphasis. The most traditional view emphasizes the sequence or flow of control within a system. More recent approaches to describing computer information systems emphasize the structure of the information in the system or the flow and transformation of information. In many respects these approaches are complementary. The advanced student will do well to be familiar with them all. The first step is to understand one of them well and be able to use it. Because this is an introductory text, we will concentrate on one approach to describing information processing systems: the data flow approach.

Physical and Logical System Descriptions

System description also involves abstraction. The most concrete description of a computer information system would include the manufacturer, model, and serial number of each piece of hardware, as well as its capacity and speed, and perhaps even the color each cabinet is painted. It would also include a facsimile of each document containing input to the system and each report produced by the system. But we could imagine a specific set of application programs, such as the payroll system mentioned above, producing the same outputs from the same inputs, yet running on entirely different hardware—different in color, made by a different manufacturer, using different storage devices of different sizes and speeds. Or we could redesign the input documents and reformat the output without altering their information content or changing the kinds of required transformation. And yet, in spite of these obvious external differences, we intuitively feel that these are essentially the same system.

That is why we distinguish between a ***physical*** and a ***logical*** system description. In analyzing computer information systems, there is a need for both kinds of system description, yet we will most often be working with logical system descriptions.

In reality, a system description is almost never purely physical or purely logical; rather, there is a gradation between the two extremes. The conceptual distinctions we make apply at the extremes, but in practice our descriptions are more or less close to those extremes.

The distinction between a physical and a logical system description may be explained in terms of four pairs of opposites:

1. Nonessential versus essential
2. Form versus content
3. Concrete versus abstract
4. Implementation-dependent versus implementation-independent

Consider the time cards in a payroll system. The color of the cards is a physical detail. It doesn't matter whether the cards are printed on white or gray or cream stock. But suppose that the cards are color-coded: the hourly employees' cards are white, and the salaried employees' are blue. This may be a great convenience for those involved with the time cards, but the particular colors chosen are not essential. Any two easily distinguishable colors would do. The decision to color-code the time cards was a choice of the form in which the content (the classification of employees as hourly or salaried) would be shown. It was only one of many ways in which the distinction between the two types of employees could have been implemented, thus, it is implementation-dependent. Alternatively, the color-coding is one possible concrete realization of the abstract employee classification.

Similarly, the information on the time cards could be rearranged or formatted in a variety of ways without altering its content.

So a physical system decription incorporates implementation-dependent features; a logical system description is independent of the way in which the system is implemented. A physical system description includes the form in which information is stored or arranged; a logical system description focuses on the information content. A physical description includes accidental, superficial, or nonessential features; a logical description concentrates on what is essential. Thus a physical description views the concrete characteristics of a system; a logical description sees it abstractly.

WHAT IS INFORMATION?

The culmination of this movement from a physical to a logical point of view is the concept of ***information***. In its simplest, most essential and irreducible form, the minimum element of information is the ***bit***, represented and realized by 0/1, ON/

OFF, the presence or absence of a charge, etc. This is the level at which information is dealt with by digital computers. Bits are combined into patterns corresponding to **data**—numbers, words, and graphic symbols—as well as into other patterns, **instructions,** which control the storage, flow, and transformation of the data. Related data are further organized into larger-scale structures. At this level, **data** and **information** are used interchangeably. **Data** has the advantage of being shorter. There is a further level at which the aggregates of data or information acquire significance in relation to the human objectives of an information processing system. At this level, **information** is the term most widely used. Perhaps it is unfortunate that the word **information** has such a wide range of usage from the elementary context of a single bit to very comprehensive relationships of meaning in a large system. But that is how it is. Yet in each of these contexts the idea of information pushes beyond the concrete physical form or medium in which we find it to its abstract, logical content.

SUMMARY

Both manufacturing and service businesses make extensive use of information to carry out and support their activities. Information is increasingly being regarded as a major corporate resource to be managed as wisely as people, materials, equipment, and money. Information plays four major roles in a business:

1. It helps carry out the production and service functions of the organization.
2. It measures and monitors the performance of these primary business functions and the other functions that support them.
3. Based on these performance measurements, it helps direct the organization to meet performance targets.
4. It supports management decisions to improve the business by modifying the organization or changing its objectives.

Monitoring, controlling, and decision support are almost exclusively information-processing activities. Monitoring the operation of a business often leads to the recognition of problems to be solved. Problem solving and decision making are closely related activities leading to improved ways of doing business. Systems analysis can assist businesses in solving problems and making decisions. In particular, the analysis of information systems can provide improved monitoring, controlling, and decision support.

Information systems perform three principal functions—they transmit information, they store it, and they transform it. The essential function of an information processing system is to carry out one or more transformations of information—to derive some desired outputs from known inputs. Every information processing system can be described abstractly in terms of components that move, store, and transform information.

The degree of abstraction is expressed as a range from a highly physical to a highly logical system description. A physical system decription incorporates implementation-dependent features; a logical system description is independent of the way in which the system is implemented. A physical system description includes the form in which information is stored or arranged; a logical system description focuses on the information content. A physical description includes accidental, superficial, or nonessential features; a logical description concentrates on what is essential. Thus a physical description includes the concrete features of a system; a logical description sees it more abstractly.

REVIEW

2-1. Name the two types of outputs of a business.

2-2. Name the functions included in a typical manufacturing business. Which of these are primary and which are secondary?

2-3. List four major roles of information in a business.

2-4. State and explain briefly the steps in the problem-solving process.

2-5. What are the three principal functions performed by information systems?

2-6. How does an information processing system differ from a communication system or a filing (information retrieval) system?

2-7. How is a transform related to its input and output?

2-8. What are the principal components of a computer information processing system? Name some input and output devices. Which components of a computer information processing system carry out transformations? Name several types of computer software and briefly describe the purpose of each.

2-9. What approach to modeling information processing systems is presented in this book?

2-10. Explain the differences between physical and logical system descriptions. Do we ever find information in a purely logical form? Why?

2-11. Identify and explain three usages for the word *information.*

EXERCISES AND DISCUSSION

2-1. State some differences in function between a banking business and a manufacturing business. That is, which functions shown in Figure 2-2 are not part of a bank, and which functions in a bank are not part of a manufacturing business? List some specific products and services of a bank.

2-2. For each of the four major roles played by information in a business:
   a. Select an organization with which you are familiar.
   b. Identify a document or other collection of information.
   c. Explain how the document is used by the organization to fulfill its role.

2-3. In the light of businesses' need for information, what does *information resource management* mean, and what purpose does it serve?

2-4. Figure 2-3 shows a schematic diagram of a computer. Explain the purpose of each component shown there in terms of information storage, transmission, and transformation.

2-5. Figure 2-4 shows a system controlled by *feedback*. How does feedback affect the relationship between input, output, and transformation? Interpret Figure 2-4 as a diagram for a heating system that includes a thermostat.

FIGURE 2-3. Schematic diagram of a computer.

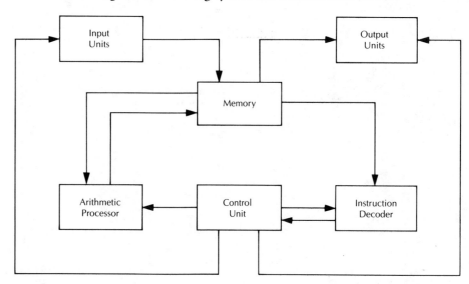

FIGURE 2-4. System controlled by feedback.

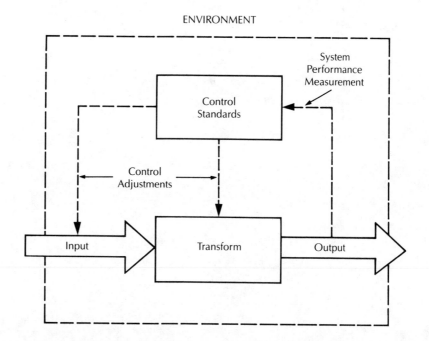

2-6. In Figure 2-2, identify some of the points at which feedback could be used to control the system. At each point, state what information could be obtained and how it could be used to control the system.

3 The Information System Life Cycle

In the previous chapter, we discussed computer information systems—in general and in relation to the information needs of a business organization. We also enumerated some of the kinds of applications software for business. But every system of application programs has to be developed—beginning with the identification of a need for a computer information system and ending with the system operating routinely and effectively to support a business organization. The ***information system life cycle*** provides an conceptual framework for presenting and understanding the activities involved throughout the system development process. The system life cycle also provides a management framework for scheduling and coordinating information system development and for monitoring its progress.

Among writers on information system development, there is little uniformity in defining and describing the life cycle. There is broad general agreement on the overall process but differences in the terms used to describe it. There is a reasonable consensus about the very detailed tasks but considerable variation in how these tasks are related within the life cycle. Unlike older fields, such as the sciences and engineering, information systems analysis has no universally accepted technical language. Currently the best we can do in this chapter is to present yet another information system life cycle. But the student is cautioned that others may view or describe the life cycle differently, and is encouraged to look behind differences in terminology to find the conceptual similarities and differences.

OVERVIEW OF THE INFORMATION SYSTEM LIFE CYCLE

FIGURE 3-1. The information system life cycle.

Figure 3-1 depicts a simplified overview of the information system life cycle. It shows only the major phases in the process and the technical information used or produced at each phase. It does not show the information used to manage the process or any of the more detailed activities within each phase. It is based on the use of structured methods of analysis, design, and programming. (In keeping with

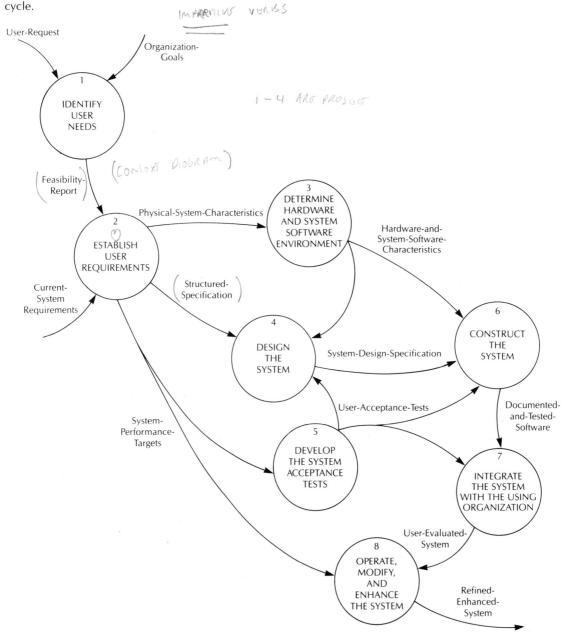

the best practices of structured analysis, as presented in Parts II and III, the phase names are imperative verb phrases instead of nouns. Though longer, these names are clearer and avoid some of the confusion due to the wide current variation in the choice of nouns for the phases of the life cycle.)

Figure 3-2 presents this life cycle in tabular form. It shows the phases in the life cycle, the participants in each phase, the products resulting from each phase, and the decisions required at the end of each phase. The participants with the most intensive involvement in each phase are shown in capital letters. The decisions are presented as questions to be answered affirmatively before proceeding to the next phase. If the answer to any question is "No," then parts of previous phases may have to be done over, or the development or use of the system may be discontinued.

FIGURE 3-2. Phases, participants, products, and decisions in the information life cycle.

ALL MUST BE YES

PHASE	PARTICIPANTS	PRODUCTS	DECISIONS
1. Identify User Needs	USERS ANALYSTS	Feasibility Report	Is the proposed system desirable to the users? Is it technically and organizationally feasible? Are the estimated benefits worth the expected costs?
2. Establish User Requirements *THIS BOOK*	USERS ANALYSTS	Structured Specification (includes: Physical System Characteristics and System Performance Targets)	Have the user' requirements been stated completely and accurately?
3. Determine Hardware and System Software Environment	DESIGNERS analysts	Hardware Characteristics System Software Characteristics	Are the proposed hardware and system software environment adequate to support the specified user requirements?
4. Design the System	DESIGNERS programmers	System Design Specification with Preliminary User Manuals	Does the proposed system design meet the specified requirements? Is it still technically, operationally, and economically feasible?
5. Develop the System Acceptance Tests	QUALITY ASSURANCE analysts	User Acceptance Tests	Do the acceptance tests measure compliance with all the stated performance targets with sufficient precision?
6. Construct the System	PROGRAMMERS designers analysts evaluators users	Documented and Tested System with User and Operations Manuals Report of Completed User Acceptance Tests	Are the documentation and manuals for the completed system adequate? Has the system passed all the user acceptance tests?
7. Integrate the System with the Using Organization	USERS analysts designers evaluators	Operational System Postimplementation Evaluation Report	Has the new system been smoothly integrated with the operations of the organization? Is the system still meeting the specified performance targets?
8. Operate, Modify and Enhance the System	USERS ANALYSTS DESIGNERS PROGRAMMERS EVALUATORS	Debugged, Refined, Enhanced System Performance Requirements for System Modifications and Enhancements	Have all the known bugs in the system been corrected? Is there a need to change parts of the system? Is there a need to enhance the capabilities of the system?

There are eight phases in the system life cycle:

1. Identify User Needs.   17
2. Establish User Requirements. 5-16
3. Determine the Hardware and System Software Environment.
4. Design the System.   19
5. Develop the System Acceptance Tests.
6. Construct the System.
7. Integrate the System with the Using Organization.
8. Operate, Modify, and Enhance the System.

Each of these is described further in the next section. Chapters 5 through 16 are devoted to a more detailed description of establishing user requirements. Chapter 17 discusses identifying user needs; Chapter 19 provides further detail about designing the system; Chapter 20 discusses the subsequent phases of the life cycle. Note that a slightly modified version of this life cycle could be used to describe the process of creating many other new systems, such as a house or an automobile. (See Exercise 3-5.)

MAJOR PHASES IN THE LIFE CYCLE

1. Identify User Needs

Computer information systems, like children, often begin as a "gleam in someone's eye." That someone, perhaps a potential user or perhaps a farsighted entrepreneur, hopes or dreams or imagines that a business (or parts of it) could be improved with the aid of a computer. Maybe some calculations are so complex that it is easy to make mistakes. Maybe the work is so tedious and repetitive that morale is low and personnel turnover high. Maybe the present way of gathering and processing information is so slow that it is too late for a manager to act by the time reports reach his desk. Maybe report deadlines are so tight that the quality of the information is unreliable. These desires for a better way of doing business need to be articulated and made explicit so that they can be communicated to others and evaluated. This is usually difficult, especially for one not trained in systems thinking.

The objectives and desired overall results may be keenly felt by the proponent of an improved system, but stating them concretely, completely, and clearly is rarely easy. The specifics can be elusive. "Oh, you know what I mean," usually means, "I wish you knew what I want to say because I'm having a hard time telling you."

Some people can envision a proposed system in great detail, but unfortunately its realization is beyond the current state of the art. Or it may be technically feasible but so expensive that its benefits are disproportionate to its costs.

Phase 1, Identify User Needs, takes a user's request for a new system; prepares an explicit statement of the system objectives, scope, and functions; and evaluates the request. The evaluation determines compatibility with the goals of the using organization, technical feasibility, and economic feasibility. In the process, user desires are sorted out into what is essential, what would be nice if not too expensive, and what is impractical or too costly. At this stage in the life cycle, the system description is necessarily so general and sketchy that economic evaluation is extremely imprecise. What is really addressed here is whether the proposed system promises sufficient potential benefit to invest the additional resources necessary to establish the user requirements in greater detail.

2. Establish User Requirements

The next phase establishes the users' requirements for the proposed system. It involves analyzing how things are currently being done and then describing the new system explicitly and in detail—what information enters and leaves the system, the flow of information through the system, the information stored in the system, and the transformations that occur. The result is the structured specification—a document containing the new system requirements and defining what parts of it are to be automated. With a definition of system requirements on hand, it is now possible to make more realistic estimates of the cost and effort to complete the system design and development.

3. Determine the Hardware and System Software Environment

The statement of user requirements has defined the scope of automation, the generic types of input/output devices, the modes of interaction between users and the computer, the required response times for various system functions, the amount of data to be stored and transformed, and other performance targets for the new system. The next phase determines the hardware and system software environment for the system. The system software environment may include more than the operating system. It may also encompass software for data base management; data base inquiry and reporting; creating and supporting computer graphics displays and other aspects of user-computer interaction; and any other software that facilitates the development and operation of the application software. The hardware selection may be made in terms of equivalent configurations from several manufacturers, to permit competition, or the decision may be narrowed to a single manufacturer.

4. Design the System

The requirements presented in the structured specification are also the basis for system design. This includes designing the application software for an automated system that will operate within the hardware and system software environment as well as the related manual procedures. The software design is specified as a system of program modules. At the system level, all the modules and the interfaces between them are defined. At the detailed level, the logic of the transformations performed by each module is specified. The design is also adapted to fit the hardware selected for implementing the system. In addition, details of the dialogue or interaction between users and the automated part of the system are defined, and the final form of inputs and outputs is determined. Draft versions of critical user documentation are also prepared as part of the system design process. The completed system design is also used to prepare plans, budgets, and schedules for programming, testing, and debugging the system.

5. Develop the System Acceptance Tests

The performance targets established in the structured specification are used to develop detailed and specific tests of acceptable system performance. These tests determine whether the system as constructed by its developers satisfies the users' requirements. The acceptance tests should be developed by people who are not involved in the design or construction of the system.

6. Construct the System

Constructing a computer information system from a design specification involves coding (using structured programming), debugging, and testing each of the modules. It means final choice of a programming language and detailed decisions as to how the modules of the system will be realized within the operating system and data base environment. It requires incremental addition of completed modules to the system until the entire system has been tested. At the end of this phase, there is an acceptance test to determine whether the performance requirements defined in the structured specification have been met. This test is performed by users and others who did not write and debug the programs. Final versions of the user manuals are developed in parallel with the software. Review, use, and evaluation of this documentation is part of the acceptance test.

7. Integrate the System with the Using Organization

After the documented and tested software has been delivered by its developers and accepted by the users, it must be integrated into the using organization. This requires the training of the users, the delivery and installation of any additional hardware, the conversion or creation of the files or data base for the system, and possibly a period of parallel operation of both the old and the new systems. User training aids, such as manuals, examples, videotapes, computer-aided instruction, and demonstrations or simulations, required at this stage will have been prepared in advance. Some refinement or tuning of the software and user procedures may be needed once the system is operational in its intended organizational setting. After the period of transition and integration is over, a postimplementation review should be held to evaluate whether the original system objectives as well as the users' requirements and performance targets continue to be met.

8. Operate, Modify, and Enhance the System

In this phase, development and installation of the system have been completed. The start-up and learning period is over, and users are familiar with the system. Routine operation can continue.

If the postimplementation review finds a discrepancy between the system's performance and the specified system requirements, it will be necessary to correct the deficiencies (or relax the performance requirements). But, in many cases, even though the evaluation is satisfactory and the system is operating effectively, there may be a need for subsequent modifications. Minor changes may become necessary to accommodate changes in hardware or operating sytems. Other modifications may be required as a result of changes in the business environment, such as new government regulations, increased customer expectations, or actions of competitors. Users seldom remain content, especially if they are being exposed for the first time to the power of an effective computer information system. Early successes usually create pressure to enhance the usefulness and augment the scope of a system.

THE LIFE CYCLE—LIMITATIONS AND FALLACIES

In understanding information system development concepts, tools, and techniques, it is always important to be aware of their limitations. So it is with the concept of the information system life cycle.

The idea of a system life cycle is based on an analogy—on similarities between the development of an information system and the development of an organism. Both have a beginning in time; their development passes through a sequence of stages; both must fit their environment; and eventually their existence terminates. As it evolves, an information system develops an existence that is objective and independent of its creators and developers.

But we must be careful not to carry the analogy too far, lest we commit what philosophers call the "biological (or organic) fallacy." A computer information system is not a frog, evolving from a tadpole, or a butterfly, evolving naturally from larva to pupa to chrysalis and eventually dying. An information system is an artifact—a purely human creation, designed and constructed to carry out human objectives by means of people and technology. There is no single, inherent, inevitable logic controlling the process. Each step along the way is the result of human decision and action and thus is subject to change. The entire project may be discontinued. And no single, irreversible sequence of activities is appropriate for every project. Some steps may have to be iterated. Sometimes, to make decisions early in the development process, it is necessary to assume or determine their implications for later stages. In practice, it is desirable to tailor the detailed activities of system development to the unique characteristics of each project. Some of the possibilities are discussed in Chapter 16.

Nevertheless, if we respect its limitations, the life cycle is a useful framework for thinking about the process of information system development, of orienting oneself in relation to this process, and of managing and monitoring progress throughout system development.

DON'T CARRY ANALOGY TO FAR! (LIFE CYCLE)

SUMMARY

The information system life cycle provides a conceptual framework for understanding and managing the process of information system development and use. Currently there is considerable variation in the names of the phases and the tasks allocated to each phase.

The system life cycle presented in this book consists of eight phases:

1. **Identify User Needs.** In this phase a user's request for a new or changed information processing system is evaluated for compatibility with the organization's goals, desirability, and feasibility.

2. **Establish User Requirements.** The current system is studied, and a structured specification is prepared for the new system to define the scope of automation and to state the users' requirements completely and accurately.

3. **Determine the Hardware and System Software Environment.** This includes selection of the generic hardware components and configuration, the operating sytem environment, the data base management software, and the system software to support the proposed application software.

4. **Design the System.** A realizable automated system that satisfies the users' requirements is specified, including its overall structure, its constituent program units, the user interface, and drafts of critical user manuals, as well as the related manual procedures.

5. **Develop the System Acceptance Tests.** This phase prescribes the tests used to determine whether the system constructed by the developers satisfies the requirements stated in the Structured Specification.

6. **Construct the System.** Each program unit is coded, debugged, and tested. Completed modules are added incrementally, and the interfaces are tested until the entire system is complete.

7. **Integrate the System with the Using Organization.** After the acceptance test is passed, the system is delivered and installed. Any necessary conversion from the old system takes place, user training is completed, and the system is used in its operational setting.

8. **Operate, Modify, and Enhance the System.** The system operates routinely. A postimplementation review may lead to modifications, and continuing use may identify desirable enhancements.

At the end of each phase there is a management decision as to whether to proceed to the next phase. Eventually the system may be superseded by a new one.

Though the information processing system life cycle is a very valuable abstraction, it does not correspond precisely to the realities of system development and use. In practice, the phases and activities within it are often overlapped and iterated. This idealized model must be tailored to the specific needs of each computer information system development project.

REVIEW

3-1. Name the eight phases in the information system life cycle.

3-2. Who are the major participants during each phase?

3-3. What are the products of each phase?

3-4. What is the major decision to be made at the end of each phase?

EXERCISES AND DISCUSSION

3-1. Why are there different major participants in the life cycle during the different stages?

3-2. Who decides whether the test at the end of each phase has been passed?

3-3. Which of the products developed during the life cycle presented in this chapter are primarily information? Which are not primarily information, and what are they then?

3-4. The description of the information system life cycle in this chapter is an abstraction from a very complex process. What components of the process have we chosen to abstract? What aspects of the process have been omitted? What features of the process are included in Figure 3-2 but are not shown in Figure 3-1? What features are discussed in the text of the chapter but not shown in Figure 3-1?

3-5. How would you modify the life cycle shown in Figure 3-1 to make it fit the development and use of a house rather than an information processing system?

3-6. Based on your answer to question 3-5, consider the following:

The processes of defining requirements for, designing, and constructing a house and a computer information processing system are similar because they deal with the creation of complex artifacts; they differ because they are directed toward producing different kinds of systems. Explain the similarities in developing a house and an information processing system in terms of the activities needed to specify and construct complex systems, and explain the differences in terms of the difference between a house and an information processing system.

4 Participants in Systems Analysis

Chapter 3 described the setting for systems analysis within the system life cycle. This chapter examines the participants in the process of systems analysis and the roles they play, looks more closely at the activities in the analysis portion of the system life cycle, and identifies the skills needed by a successful systems analyst.

In general (see Chapters 1 and 2), systems analysis is a process of studying a system—separating the system from its environment as an object for study by identifying the system boundary, determining the components that make up the system, and describing the rules that govern the interactions among the components. In a practical situation, when the system is not being studied for its own sake, analysts study a system to detect and correct deficiencies, to improve system performance, or to forecast the effects of changes to the system or its environment. Thus the purpose of systems analysis becomes to investigate systems problems, determine explicitly what constitutes a solution, and recommend a course of action that will lead to system change.

In the context of computer information systems for business, this means investigating problems related to an organization's use of information, stating precisely the users' requirements for information processing systems (including what portions should be automated), determining explicitly the system inputs, outputs, and transformations as well as the tests for satisfactory system performance, and recommending to management what action to take to change the system.

Although other system development people are occasionally involved, there are two principal groups of participants in this process—users and analysts. What are the major roles and functions of each group and how do the members of both groups interact?

ROLES AND FUNCTIONS OF THE USER

"The user" is a simple, convenient abstraction that often distorts or suppresses important perspectives on system development. It is better to recognize at the outset that there are several groups of users of a computer information system with different needs and concerns. Each group, in its own way, is able to stop, delay, sabotage, or ignore the completed system. Ignoring their differences is a frequent cause of failure.

Types of Users

DeMarco* identifies three types of users—the hands-on user, the responsible user, and the system owner—to which we will add a fourth, the beneficial user.

The *system owner* is a high-level manager and decision maker for the business area supported by the system, usually with profit-and-loss responsibility. In some organizations the cost of information system development is centrally funded. In others, development or operations costs are recovered from those supported by the system. Or the system owner may allocate funds for system development, as well as its continuing operation. An example of a system owner for a sales analysis and forecasting system might be a vice president of marketing.

The system owner's principal concerns are

1. The wise use of company resources in system development and operation
2. Top management's understanding of and support for the system
3. Compatibility of the system's objectives and functions with the goals of the organization as well as those of the part that the system owner manages
4. A working system delivered on schedule and within the budget
5. Satisfaction with the system on the part of all the other types of users
6. Protection from liability due to use or misuse of the system through auditability, adequate controls, and appropriate security

The *responsible user* is a low- to middle-level manager with direct day-to-day operational responsibility for the business functions supported by the system. An example of a responsible user for an accounts receivable system that records customer invoices and payments, produces monthly statements, and identifies overdue accounts would be the manager of an accounts receivable section.

The responsible user's principal concerns are

1. A system of appropriate scope for correct and detailed support of the operations that he supervises
2. Smooth integration of manual and automated procedures
3. Timely and understandable reports and other system output

*Tom DeMarco, **Structured Analysis and System Specification,** Yourdon, Inc., New York, N.Y., 1979, p. 14.

4. Adequate training and documentation for hands-on users
5. A reliable and dependable system
6. Enhanced productivity at an acceptable cost
7. Satisfaction on the part of the hands-on and beneficial users

The ***hands-on user*** is a person who interacts directly with the data capture and data display devices for the system. In a batch system, these may be part of the data-processing or computer operations staff rather than part of a group responsible for the business-related functions supported by the system. In an on-line or partly on-line environment, those who enter data and receive output at a terminal are directly supported by the system. An example of a hands-on user at a savings and loan association would be a teller who uses a terminal to enter deposits and withdrawals and to find out customer account balances.

The principal concerns of the hands-on user are

1. Ease of system use: procedures that are easy to use and well documented; effective error detection and correction features
2. A reliable and dependable system
3. Well-designed equipment and workstations to reduce the level of fatigue and stress
4. Increased personal satisfaction with working conditions as a result of the system

The ***beneficial user**** is one who has no direct contact with the automated system but who receives output from the system or provides input to the system. An example of a beneficial user of a banking system is a depositor who receives monthly statements and canceled checks. Note that a depositor using an automated teller machine becomes a hands-on user.

The principal concerns of the beneficial user are

1. The quality of service provided by the system
2. Understandable and usable reports and other system output
3. Security and privacy for confidential personal information
4. Appropriate error detection and correction features with effective response to problems by hands-on and responsible users and system owners who do not blame the computer for system inadequacies

It is also clear that a single individual may belong to more than one of these four categories, especially in small organizations.

---

*"Beneficial" is used here in the sense of "one who receives the benefits from," as in the "beneficial owner" of stock, in contrast to the owner of record.

Functions and Responsibilities of the Users

The users have—or share—the following responsibilities and functions in system analysis and development:

1. They decide to incorporate the automated system into their way of doing business.

   This is a decision that must eventually be made by all types of users whether the initiative comes from the beneficial users or the system owner. In many situations, however, as a beneficial user, an employee or customer may have little leverage beyond going elsewhere.

2. They relate the need for the system and its functions to the goals, policies, and objectives of the organization.

   This is ultimately the responsibility of the system owner but may also require the participation of the responsible user.

3. They know and understand the business functions supported by the information processing system.

   In large organizations it is rare for a single user or type of user to have a complete understanding of the entire system. An understanding of the place of the major system functions within the organization as a whole can be supplied by the system owner. Detailed knowledge and understanding must come from the beneficial users within the organization, as well as the responsible user.

4. They establish priorities and resolve conflicts among system objectives and among requirements.

   Corporate goals are not always consistent; desires or needs of various users may be incompatible. Sometimes it is possible to tailor the data collection or presentation of information to different groups. But fundamental conflicts must be resolved by the responsible owner if possible. Otherwise, they will require a decision from the system owner.

5. They review, understand, and approve the system development documents that define users' objectives and requirements.

   In particular these include the general statement of system objectives and requirements incorporated in the feasibility study and the more detailed statement contained in the structured system specification. In later phases of system development, users may also review user-oriented manuals and documentation.

6. They allocate the necessary resources to the system development process.

   This includes making key people available to the process as well as timely interaction with the analysts. It also includes the funds for system analysis and development.

7. They serve as a reliable information source.

   The user has a responsibility to communicate—to give thoughtful, considered, honest responses to the analysts' requests for information. Evasion, concealment, and even well-intentioned erroneous information all will work against the overall best interests of the organization.

8. They make decisions among alternatives, make trade-offs, and consider relative costs and benefits.

9. They review at milestones and make abort or continue decisions.

10. They provide support for desired change and continuing pressure for change.

Successful projects have strong, well-communicated psychological commitment on the part of management, especially the system owner. Continuing accessibility of system owners and responsible users to the analysts for communication, involvement in decision making, and interaction throughout the process are vitally important.

These last several responsibilities of the user are in management areas—motivation for change, decision making, allocation of resources, and support.

ROLES AND FUNCTIONS OF THE ANALYST

Analysts also have many roles and functions in information system development. Many of these are technical—requiring a knowledge of system concepts and expertise in system development techniques. Others are nontechnical—especially involving the business dimensions of the system or the interpersonal and human aspects of the development process. All the tasks of the analysts are directed toward an informed and adequately supported user decision to proceed with system development based on a precise statement of requirements.

A Scenario

A brief scenario for a day's activities involving a senior analyst with project leadership responsibility will illustrate some of these tasks and concerns.

In the morning she meets with the clerical staff, consisting of hands-on users of a proposed automated system, to gather information for a study of their department. At lunch she meets with her supervisor, the manager of information system development, to review progress on the project. After lunch she attends a meeting with top management to assist her manager with a presentation of budgets and schedules for projects during the next fiscal year. Then she participates in reviewing the design of a system for which her team of analysts produced the specification three months earlier. There she discovers some problems with the original acceptance test plan. Before leaving for the day, she answers questions from two junior analysts on the current project and then organizes her notes from the design review.

Functions and Responsibilities of the Analyst

Analysts are responsible for:

1. The technical quality of the products and procedures of systems analysis, as described in Parts II and III.

2. Determining (with the aid of other participants in the process) the implications—technical, economic, psychological, and organizational—of decisions about the scope and kind of automation.

3. Facilitating communication and understanding among the other participants in the process.

4. Providing effectively organized information to support user decisions leading to the development of the best computer information system for the organization.

5. Acting as an advocate or ombudsman for the users, especially during design and acceptance testing. This activity involves seeing that user requirements stated in the system specifications are satisfied in the later stages of system development.

6. Acting as a conscience for the development effort. The analyst helps other participants recognize negative as well as positive impacts of the information processing system under development, especially as they affect people in the organization. The analyst helps everyone to keep the development effort in perspective—as a part of the business and as the prelude to ongoing system use.

Required Analysis Skills

To carry out these responsibilities, an analyst requires the following skills or attributes:

1. Facility with the tools and techniques of analysis. Technical competence is also a major basis of credibility with other participants in the process.

2. The ability to master complexity, to assimilate large amounts of information, separating fact, opinion, hearsay and conjecture, and identifying what is relevant.

3. Facility at abstract thinking, the ability to generalize from limited and incomplete information without losing precision and concreteness of detail.

4. Language and communication skills. Much of the work of analysts is interpreting the significance of information and translating each other's technical jargon for users and system developers. Effective written, oral, and graphic communication is essential.

5. Information-gathering skills. The ability to extract information from a wide variety of sources, assess its relevance and reliability, and organize it in useful structures.

6. Interpersonal skills. The ability to inspire confidence and develop effective working relationships among the participants in the system development process.

7. The ability to function effectively in new situations, to tolerate a high degree of ambiguity and uncertainty, to function with imperfect information—deficient in quantity and quality, incomplete, imprecise, or redundant.

8. Sound judgement and common sense.

Do these requirements seem impossibly demanding? Perhaps so. Few people are equally competent or effective in every skill area. That is why systems analysis is often a team effort, so that the strengths of the team members can complement and support each other. That is why the process of system development must be carefully organized and why peer review of work products is built into structured analysis methods.

USERS AND ANALYSTS—A SYNERGISM

As we have seen, during analysis users serve primarily as information sources and decision makers, while analysts serve primarily as decision facilitators through a clear statement of proposed system requirements and their implications for the organization. Each group has its own responsibilities in the process, but system development at its best is truly a joint effort. Neither group has a monopoly on imagination, innovation, or wisdom. Experienced analysts often acquire a wealth of business knowledge, and the keenest insights into better systems often come from dedicated and imaginative users. Working together in close interaction with effective communication, users and analysts can achieve superior-quality computer information systems for business.

SUMMARY

There are two principal groups of participants in the process of systems analysis—users and analysts.

The group of users includes four types of people: beneficial users, hands-on users, responsible users, and system owners. Each user type has somewhat different interests in a computer information system and a different perspective on the system and its development. Users are responsible for establishing policies for the conduct of their business and for making business decisions about the information processing systems that support their organization. Strong user support and involvement and wise user decisions are essential to successful computer information system development and use.

System analysts, often working in teams, carry out both technical and nontechnical roles in system development. They are responsible for a technically competent statement of users' requirements. They must also understand, interpret, and communicate these requirements to all the other participants in the process. Facility with abstractions, superior communication skills, the ability to act effectively in unstructured situations amid complexity and uncertainty, and sound judgment and common sense are the most important characteristics for systems analysts.

REVIEW

4-1. What two groups are the principal participants in systems analysis?

4-2. Name the four types of system users. For each of them list at least four concerns with regard to a computer information system.

4-3. List at least eight responsibilities of users during system development.

4-4. List at least five responsibilities of analysts during system development.

4-5. What skills does a systems analyst require?

EXERCISES AND DISCUSSION

4-1. Explain why the users' responsibilities described in this chapter are theirs rather than those of the analysts.

4-2. Explain why the analysts' responsibilities described in this chapter are theirs rather than those of the users.

4-3. Technical competence alone is sufficient for a successful computer information system. Do you agree or disagree? Why?

4-4. Discuss why each of the analyst's required skills is needed. How do these skills reinforce each other?

4-5. Discuss why successful system development is usually a team effort involving several analysts and the various types of users.

5 Specifying User Requirements

The major task of systems analysis in relation to information system development is specifying the users' requirements for an information processing system. In the previous chapter we identified the roles and responsibilities of the various participants in the process. In this chapter, we take a closer look at what is involved in the process of requirements specification, emphasizing the special characteristics of structured methods of systems analysis.

Structured systems analysis may be characterized in terms of its goals, its procedures, its products, and its tools and techniques.

GOALS OF STRUCTURED ANALYSIS

The principal goals of structured analysis include

1. First, and fundamentally: ***To state users' requirements for a new information processing system accurately.***

   Because it is fundamental, this goal is common to all methods of information systems analysis.

   An accurate statement of user requirements has the following characteristics:
   a. It is ***explicit.*** It is documented and open to all the participants in system development. No one has to read between the lines, and there are no hidden assumptions.
   b. It is ***complete*** in all the essentials of concern to the user. It covers all the system requirements and the entire scope of the transformations to be carried

out by the computer. All the information entered, stored, transformed, and output is specified.

c. It is ***unambiguous.*** It leaves nothing vague, fuzzy, or subject to more than one interpretation by the reader.

d. It is ***consistent.*** There are no conflicts and incompatibilities built into the requirements specification. These have all been resolved during the process of analysis.

e. It is ***precise and specific.*** Requirements are presented clearly and definitely. They are sufficiently detailed so that nothing crucial has to be filled in later. They are expressed in terms that are tailored to the specific tasks of a specific system for a specific group of users rather than in general terms that could address a variety of other situations.

Growing out of the primary goal of accurately specifying the users' requirements are several related secondary goals:

2. *To understand the users' requirements.*

Accurate specification is impossible unless requirements have first been understood. Systems analysis is a process in which users and analysts help each other to an understanding of the system requirements that is sufficient for their accurate specification. This implies an understanding of all those dimensions of the users' business that affect or interact with the information processing system. It usually requires a knowledge of how things are currently done before deciding how the system should be changed.

3. *To communicate the current understanding of the current or proposed system.*

Arriving at a mutual understanding of users' requirements is achieved through interaction and communication. Communication makes understanding explicit and shared. If I cannot express my requirements so that you can understand them, it usually means one of two things: I do not really understand them myself (in which case communication may lead us both to my requirements); or I alone am able to act on my understanding—no one else can help me until we are able to communicate. In practice, successful communication is a prerequisite to successful system development. It must occur among users, among analysts, and between users, analysts, and others involved in developing a system.

4. *To prevent expensive mistakes.*

In practice, the goal of a perfect statement of users' requirements is unlikely to be fully achieved. Nevertheless the methods of systems analysis attempt to reduce the number of omissions, inconsistencies, and undetected errors as well as to minimize their impact on information system development.

DeMarco* has characterized systems analysis as a defensive activity, offering the following maxim: "The overriding concern of analysis is not to achieve success, but to avoid failure."

---

*Tom DeMarco, ***Structured Analysis and System Specification,*** Yourdon, Inc., New York, N.Y., 1979, p. 9. Reprinted by permission.

There are no guarantees of success. But the penalties of failure can be minimized by preventing mistakes, detecting errors as soon as possible, and correcting them quickly. The later in the information system life cycle changes are made, the more expensive they become.

5. *To state a design problem.*

An accurate statement of user requirements is not an end in itself. It is the basis for designing an information-processing system that will satisfy those requirements. Thus, the result of systems analysis must also communicate user requirements to the system designers, providing all the information they need to produce a realizable description of an executable system of computer programs that carry out the required transformations.

6. *To state the conditions for system acceptance.*

How can users be assured that the information processing system, when delivered by its developers, will in fact meet the stated system requirements? Of course, an accurate requirements specification is necessary. But there should also be some explicit performance requirements for the new system. These performance requirements will be used to test the completed system—to determine whether or not it is acceptable. They may be quantitative or qualitative, but they must be measurable.

THE PROCESS OF STRUCTURED ANALYSIS

The major activities within the analysis phase of system development are introduced here and described further in Part III. First we wish to highlight the more important general characteristics of the process.

Characteristics of the Process

1. *Analysis is inherently iterative.* In part, this is a natural human response to unstructured complexity. There is a process of stepwise refinement, beginning with a comprehensive, high-level view of the system and continuing until all the necessary details have been filled in. There is also a focusing on specific parts of the system to limit the amount of complexity from moment to moment. There is a need for correction when errors are detected. There is a need for revision when user requirements change.

   Iteration is also necessary because achieving understanding through communication requires considerable interaction.

2. *Effective analysis is systematic.* It is based on an organized body of skills and techniques.

3. *Analysis is dynamic.* It must be able to respond to and accommodate change as well as complexity.

4. *The process of analysis is inseparable from its tools and techniques.* Conceptually we distinguish between the process and the tools, but in practice they go together. The tools and techniques are integral to the process. They are the

natural by-products of doing analysis. Documentation produced by structured methods is not created artificially or after the fact. Rather, it is essential to achieve the goals of stating requirements accurately, understanding, and communication during an inherently interactive process.

The Activities of Structured Analysis

As discussed in Chapters 3 and 4, systems analysts participate throughout the system life cycle, but their heaviest involvement is at the first two phases. Therefore, the second phase, Establish User Requirements, is often called Systems Analysis or *The* Analysis Phase of system development. (Structured analysis methods prefer to name an activity concretely and specifically in terms of its product or output, using a forceful imperative verb, to eliminate ambiguity and confusion.) Figure 5-1 shows in greater detail the activities necessary to establish the users' requirements for a computer information system. Each of these activities is summarized below to provide an overview of the process of structured analysis. The details are the subject of Part III.

FIGURE 5-1. The activities of establishing user requirements.

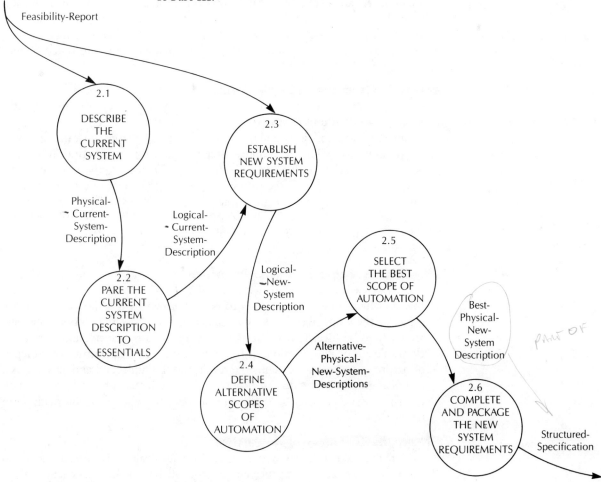

1. *Describe the Current System.*

   In most cases, a study of the current system is the best preparation for establishing users' requirements for a new system. This activity serves the technical purpose of identifying the information currently used, stored, and transformed by an organization. It also allows users and analysts to get to know each other and learn how to communicate with and understand each other. The result is an initial model of information flow in the current system. We try to make this description complete, correct, and consistent, yet it almost inevitably incorporates nonessential and implementation-dependent details that in Chapter 1 were called physical. Thus the initial current system description is usually, or at least in part, a *physical current system description.*

2. *Pare the Current System Description to its Essentials.*

   In the next activity, the current system description is reduced to its essentials, thereby transforming it into a *logical current system description.* Redundancy is eliminated, and details related to how information is transformed are removed, until the system description is minimal. Considerable effort is expended to describe the system data base in a comprehensive, unified, minimal, logical form. Logical data base description is discussed in Chapter 9 and illustrated further in Chapter 12.

3. *Establish New System Requirements.*

   Next the logical description of the current system is replaced by a *logical description of the new system.* The emphasis is on correctly specifying the required information flows and transformations; how these transformations will be accomplished will be determined later. The new system description is partly derived from the current system description by identifying the portions to be changed; it is partly invented in response to the objectives of the new system. The new system requirements must also be checked to see that they satisfy the constraints defined in the feasibility study. If the feasibility study allowed considerable latitude for the new system requirements, alternative new system descriptions may be produced.

4. *Define Alternative Scopes of Automation.*

   This activity designates which portion of the new system is to be automated— that is, which transformations are to be performed by a computer. The more extensive the scope of automation is, the greater the expected costs and benefits are likely to be. Because the decision to automate a transformation implies an electronic information medium, it has consequences for implementation in terms of hardware and software. Thus the result, though high-level and abstract, is a *physical new system description.* Several alternative physical new system descriptions are prepared so that the most appropriate scope of automation can be determined.

5. *Select the Best Scope of Automation.*

   Here the alternative new system descriptions are evaluated to determine which is the most desirable. The procedure is generally known as cost/benefit analysis but is not limited to those costs and benefits that are measurable in dollars. The first step is to define quantitative and qualitative measures of expected system performance. The overall performance of each alternative is estimated; then the estimates are compared to select the alternative with the greatest value to the organization. This ***physical new system description*** becomes the definitive statement of requirements for further system development.

6. ***Complete and Package the New System Requirements.***

Based on the selected alternative, specification of the new system requirements is completed. The information processing system model described in data flow diagrams, the system dictionary, and data base description is supplemented with any other essential user requirements and explicit targets for system performance. These are combined into the structured system specification—the basis for system design and the source of system acceptance tests.

PRODUCTS OF STRUCTURED ANALYSIS

The principal product of structured analysis is the structured specification—a document that explicitly defines the users' requirements for a computer information processing system. It consists of three major components—a set of data flow diagrams, a system dictionary, and a data base description. This chapter provides a brief overview of the structured specification and its components. Part II explains them in further detail.

Data Flow Diagrams   A data flow diagram depicts an information processing system. It shows the overall structure of the system in terms of the flow, storage, and transformation of information within the system. Each flow, storage, and transformation is named to identify its content in terms that are meaningful to the users.

System Dictionary   The system dictionary lists in an organized fashion every name appearing in a data flow diagram. It also defines the structure of every information flow and file in the system. Transformation descriptions, contained in the system dictionary, set forth the logic for a complete set of transformations necessary to produce the system outputs from the system inputs.

Data Base Description   The data base description provides a coordinated and coherent view of the users' requirements for access to information stored in the system.

The Complete Specification   In addition to the data flow diagrams, system dictionary, and data base description, the complete system specification contains any other specific essential user requirements, as well as an explicit list of performance targets that will be the basis for acceptance of the system by the users. See Chapter 14 for these aspects of the complete specification.

CHARACTERISTICS OF THE STRUCTURED SPECIFICATION

The characteristics of the structured specification serve to achieve the goals of the analysis process.

1. It is **graphic.** A data flow diagram presents the system structure visually—the most effective form of human communication.
2. It is **partitioned.** A data flow diagram decomposes the system into parts so that its structure can be understood and communicated.
3. It is **top-down.** A set of data flow diagrams presents the details of the system by partitioning the transformations hierarchically from the top down. Similarly, the system dictionary provides a decomposition of the data structures.
4. It is **nonredundant.** With "a place for everything and everything in its place," the structured specification is friendly to change. Minimum redundancy supports the many revisions to the specification necessary during an iterative process, accommodates change during system development, and facilitates modifications after the system is in production.
5. It is **accurate.** As discussed above, it is rigorous, precise, clear, consistent, and complete.
6. It is **minimal** in the sense that it eliminates nonessentials, omitting what can be decided later or left to the discretion of the system designers or implementers, but it must not omit details that will result in rejection of the system by the users.

TOOLS OF STRUCTURED ANALYSIS

The products described above are also the tools by which the process of structured analysis is carried out. The use of the tools in relation to the process is the subject of Part III. But to describe, understand, and communicate the structure of an information processing system means to draw pictures of it (data flow diagrams) and to redraw them until an accurate statement of the user requirements has been achieved. Giving precise content and meaning to the names with which a data flow diagram is labeled means developing the system dictionary until every data element in every collection of information in the system has been specified. Specifying the transformations that carry out the information processing functions of the system means making the correct logic for each transformation explicit. Defining the user requirements for access to a data base entails depicting the required paths into and through the data base by means of a data access diagram.

TECHNIQUES OF STRUCTURED ANALYSIS

It is clear that structured systems analysis must include techniques for producing and modifying the system specification. But other techniques in the analysis process are also important enough to be mentioned in this overview.

1. *Information-gathering Techniques*

   Basic techniques of gathering information, such as interviewing, observation, questionnaires, etc., are common to all approaches to systems analysis.

2. *Walkthroughs*

   A walkthrough is a technique specifically associated with structured analysis methods. A walkthrough (see Chapter 11) is a bug-identification session for an information processing system description. It is a peer review by systems analysts and users that focuses on finding errors. This serves the goal of detecting mistakes as soon as possible so that they can be eliminated.

3. *Data Base Description Techniques*

   These techniques are used to obtain a minimal, nonredundant, coherent, comprehensive view of the users' data base requirements. Expertise in data base description is an advanced skill in systems analysis and design, but the beginning student should at least have an understanding of its goals and results in relation to information systems analysis, as presented in Chapter 9.

SUMMARY

Structured systems analysis may be characterized in terms of its goals, its process, its products, and its tools and techniques.

The principal goal of structured analysis is to state users' requirements for a new information processing system accurately. Structured analysis is an iterative process, which is inseparable from its tools and techniques. It comprises six activities:

1. **Describe the Current System.** This activity produces a physical description of the current information processing system.

2. **Pare the Current System Description to Its Essentials.** The physical description of the current system, including the data base description, is transformed into a logical description by reducing it to a nonredundant, minimal, essential, consistent form.

3. **Establish New System Requirements.** This activity produces a logical description of the information flows, information storage, and transformations of information required in the new system, often by incorporating the desired changes in the current system description.

4. **Define Alternative Scopes of Automation.** In this activity, alternatives for automating the information processing are identified, leading to a highly abstract physical description of the new system for each alternative.

5. **Select the Best Scope of Automation.** The alternative new system descriptions are compared, and the best is selected as the basis for continued system development.

6. **Complete and Package the New System Description.** The description of the best alternative for the new system is completed, and all of the users' other critical requirements are incorporated in the structured specification—the principal product of structured analysis.

The structured specification is graphic, partitioned, top-down, nonredundant, and limited to the essentials. The system description contained in this document consists of three major parts: a set of data flow diagrams; a system dictionary, which includes transform descriptions; and a logical data base description. These system description techniques are also the principal tools for carrying out the process of structured analysis. They are supplemented by techniques for gathering information, by walkthroughs for peer review of system descriptions as they are produced, and by special techniques for data base description.

REVIEW

5-1. What are the goals of structured analysis?

5-2. List five characteristics of an accurate statement of user requirements for an information processing system.

5-3. List four characteristics of the process of structured analysis.

5-4. Establishing user requirements comprises six principal activities. List them and describe each briefly.

5-5. What are the three major components of the structured specification?

5-6. What else is included in a complete system specification besides the three major components?

5-7. List the characteristics of the structured specification.

EXERCISES AND DISCUSSION

5-1. Explain how each characteristic of the structured specification mentioned in this chapter helps achieve the goals of structured analysis.

5-2. How do information-gathering techniques and walkthroughs further the goals of structured analysis?

5-3. What kind of diagrams are Figure 3-1 and 5-1? What is the relationship of Figure 5-1 to Figure 3-1? Discuss these two figures in relation to the characteristics of a structured specification.

5-4. Discuss some of the impacts of lack of clarity, inconsistency, incompleteness, and lack of rigor on a statement of system requirements.

PART II

TOOLS AND TECHNIQUES OF STRUCTURED ANALYSIS

In the previous chapter we identified the major task of systems analysis as the accurate specification of the users' requirements for an information processing system. Further, we stated some desirable characteristics of the statement of user requirements; that is, a statement should be explicit, complete, unambiguous, consistent, precise, and specific. We also identified the following goals of systems analysis:

- to understand the users' requirements
- to communicate understanding of the current or proposed system
- to prevent expensive mistakes
- to state a design problem
- to state the conditions for system acceptance

Finally, we characterized the process of systems analysis as

- inherently interactive
- systematic
- dynamic
- inseparable from its analytical techniques

In Chapter 6 and the five that follow it we explore the major tools and techniques used in structured systems analysis—*the data flow diagram, system dictionary, transform descriptions,* and *data base description*—and the techniques which support them. Each has a purpose within the overall need to cope with complexity during systems analysis. Each plays a role in the major product of analysis—the *structured specification.*

Part II emphasizes the constituent parts of the structured specification as parts of an information processing system model. Chapter 6 presents the basic definitions, notation, and conventions used in sets of data flow diagrams and the concepts required to read and understand data flow models. It also introduces the FastFood Store system, which serves as an example throughout Parts II and III. Chapter 7 is devoted to the system dictionary—its organization and the conventions for defining data elements, data flows, and data stores. Chapter 8 describes techniques for specifying transform descriptions—Structured English, decision trees, and decision tables. Chapter 9 introduces the fundamentals of describing users' requirements for a data base. Chapter 10 explains how to draw data flow diagrams, stressing system descriptions that are not only complete, correct, and consistent but that also communicate effectively. Chapter 11 presents the criteria for evaluating and improving a system description; it also introduces walkthroughs, a quality control technique for system descriptions.

Part III emphasizes how to use these system description tools and techniques as a part of the systems analysis process.

6 Data Flow Diagrams

The data flow diagram (DFD) is the fundamental modeling tool for the systems analyst. It represents the essential functions of an information processing system in a highly abstract way using a minimum set of symbols. This chapter introduces data flow diagrams as tools for describing information processing systems. It defines each of the essential components and explains the symbols and conventions used in a leveled set of data flow diagrams. It also shows how data flow diagrams are related to the other parts of an information processing system description. The emphasis is upon how to read and understand data flow diagrams.

READING DATA FLOW DIAGRAMS: A DIALOGUE

Probably the best way to introduce DFDs is to let you observe how they are used as a communication aid between analyst and user. We present below a typical conversation between users and an analyst. It is important to note the setting; the following dialogue takes place at the second meeting between the analyst and the users, and is the users' first exposure to data flow diagrams.

At the first meeting the analyst was given a tour of the FastFood Store by the proprietors, Mr. and Mrs. Owner. At that meeting, Mr. Owner expressed an interest in automating some of his accounting practices. The analyst observed the daily operations for several hours, asked many questions about those operations, and had a conversation with Mrs. Owner, who does most of the bookkeeping. We now pick up the conversation between the analyst and Mr. and Mrs. Owner. (Paragraphs have been numbered to facilitate subsequent references to the dialogue.)

1   Mr. O: We're anxious to see what you've come up with.

2   Analyst: I hope you're not disappointed. I thought we should start with a picture of the existing operation as I understand it. I've brought along a copy of a set of diagrams for you and your wife to review with me. [Hands Mr. and Mrs. Owner each a modest stack of 8½ × 11 inch paper.] This is a picture of my understanding of your existing operation.

3   Mr. O: Hmmmm, I kind of expected some computer stuff. You know, flowcharts, report layouts, and the like. I mean, in the Introduction to Computing course I took at the community college, I was led to believe those were the kinds of products I should expect from an analyst. These don't look anything like that!

4   Analyst: It's true that these don't look much like the more traditional products of analysis. Maybe later on we'll produce some of those. But important new tools have been introduced to systems analysis over the last 10 years or so. For now, let's concentrate on your operation. Remember, you're the expert on how your business operates, and I'm the one who is trying to understand your operation.

5   Mrs. O: I'm not so sure that he's the expert. Oh, he knows how to cook and all that, but I'm the one who doesn't have any time with the children because I can't keep up with the paperwork here.

6   Analyst: Well, I hope we can do something about that. Let's start by reviewing these diagrams. This first one (Figure 6-1) represents the environment for your store. Here we see the various producers and consumers of paperwork for your system and the paperwork that flows between you and them.

7   Mr. O: Hey! We're in the fast-food business, not the paperwork business. Where's the raw meat, vegetables, and supplies coming in? And for that matter, where's the prepared food going out?

8   Mrs. O: Don't panic, dear. They're right here. Apparently all the supplies are lumped together as one shipment from one supplier and all the prepared food goes to one customer.

9   Analyst: That is a reasonable first interpretation, but not quite correct. This model would not be very useful if we had to show every kind of supplier or customer. We are going to try to suppress details like that for the time being. Really, how much difference is there between your suppliers? Based upon our conversation last time, every supplier requires a purchase order before delivering a shipment of goods. You don't pay a supplier until you have received an invoice for goods delivered. And presumably you pay your suppliers. I agree that you have many suppliers providing many different goods, but your interaction with them is highly stylized and follows a regular pattern independent of what they supply. That's what I've tried to capture here. A similar argument holds for customers. Sure they come in many sizes and shapes and order different things. But, again the interaction is the same. A person orders, pays, waits a short while for the food to be prepared, and then receives the prepared food along with a receipt.

CONTEXT DIAGRAM: FASTFOOD STORE

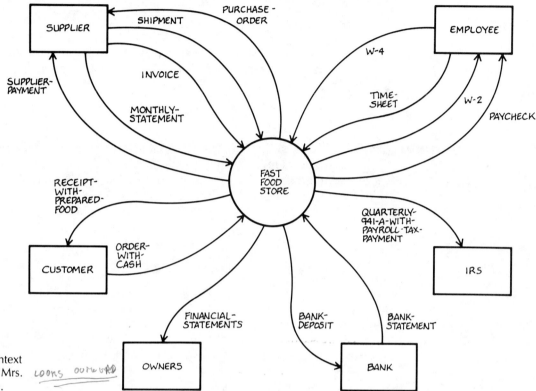

FIGURE 6-1. Context Diagram: Mr. and Mrs. O's FastFood Store.

10	Mr. O:	OK, but sometimes we make a mistake. You haven't shown that here.
11	Analyst:	True. How often does that happen? I assumed that it did not happen often. For the time being, I want to focus on the regular patterns of behavior in your system and deemphasize the errors. After all, we hope that your system behaves normally for the majority of the time it is in operation. After we get the big picture right, we'll concern ourselves with details such as errors.
12	Mrs. O:	Is there any reason you chose to show our store as just one circle? I mean there are many things happening within that circle that you haven't shown.
13	Analyst:	I agree, there certainly are many things that go on inside your system. What this diagram shows is the way your store interacts with its environment. So this diagram takes an outward-looking view. On succeeding diagrams we take an inward-looking view.
14	Mrs. O:	Before we go on, let me make sure I understand what this diagram represents. Each of these boxes is a person or business or institution that our business inter-

acts with. These lines with arrowheads indicate the path and direction of the paperwork or goods that flow between us and those persons, business, etc. And, at least at this level, we are not concerned with what goes on inside the circle that represents our store. Is this basically correct?

15	Analyst:	Yes. I think that is an excellent summary of this diagram. By the way, we call this the Context Diagram because it presents the context, or environment, for your business.
16	Mr. O:	OK, let's look at these other diagrams. I have the lunch-hour rush coming up soon.
17	Analyst:	Fine. This next diagram, labeled Diagram 0 (Figure 6-2), presents the major organizational breakdown of your store. I realize that you probably do not use these terms on a day-to-day basis. However, what we're trying to present here are the functional parts of your store.

FIGURE 6-2. Diagram 0: FastFood Store.

DIAGRAM 0: FASTFOOD STORE

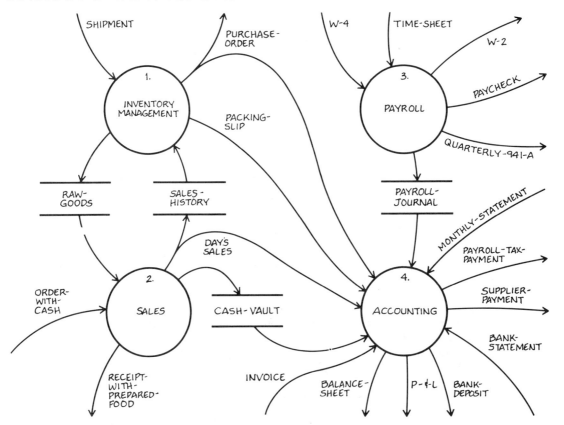

18   Mr. O: Where's the food prepared? Must be inside of Sales. Why don't we see our names anywhere? I mean, Mrs. O does the payroll and accounting, why not put her name in those bubbles?

19   Analyst: Hold on, one question at a time. Yes, the food preparation is included within Sales. If you look at Diagram 2 (Figure 6-3), which depicts what happens inside of Sales, you'll see food preparation. As to the names, if you feel that they would help you get your bearings, I'll add them. However, I'd like to encourage you to think in terms of the various tasks you do or the roles you play. I have tried to capture the function of a process rather than who does it. It will help later when we try to regroup and reassign tasks. For now, let's concentrate on logical functions or roles and ignore the individual who performs that role or function.

20   Mr. O: That seems reasonable because looking at Diagram 1 (Figure 6-4), I realize we both share in the inventory management. She prepares the purchase orders after

FIGURE 6-3.   Diagram 2: Sales.

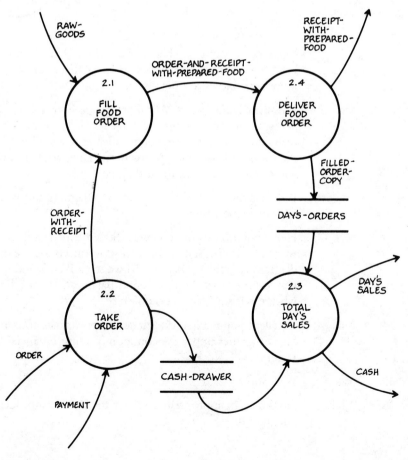

DIAGRAM 2: SALES

DIAGRAM 1: INVENTORY MANAGEMENT

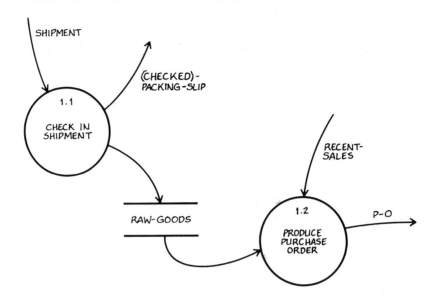

FIGURE 6-4. Diagram 1: Inventory Management.

I give her a count of what we need. I check in shipments. We would have to put both our names in that bubble. I agree that could be confusing. Good idea there, ignoring unnecessary details.

21 Mrs. O: This bubble number 4, Accounting (see Figure 6-2), looks awfully busy. You know, until you back off and look at things, you don't realize how much you really do. This gives me a better idea as to why I'm so busy. Hmmm, I think there's an inconsistency between this Diagram 0 and the previous one. What did you call it, the environment diagram?

22 Analyst: Well, actually I called it the Context Diagram, but you have the right idea. What appears inconsistent?

23 Mrs. O: Here [pointing to the Financial-Statements data flow on the Context Diagram] we see Financial-Statements being produced for the owners. However, on this diagram [points to Diagram 0] we see a P-&-L and a Balance-Sheet. Now I happen to know that the P-&-L and Balance-Sheet are Financial-Statements, but what about someone who doesn't?

24 Analyst: Excellent point. First, does it seem appropriate to join the P-&-L and Balance-Sheet together under the common heading Financial-Statements on the Context Diagram?

25 Mrs. O: Well, sure.

26 Analyst: Does it seem appropriate to show the P-&-L and Balance-Sheet separately on Diagram 0?

27   Mrs. O: I have no problem with that. But, what about someone who does not see the connection?

28   Mr. O: Yeah, like me. What's a P-&-L?

29   Mrs. O: It's a Profit-and-Loss-Statement. Sometimes it's called an Income-Statement. It shows net income from sales after deducting operating expenses and taxes. Usually it covers some accounting period like a month, quarter, or year. Shouldn't these terms be defined somewhere?

30   Analyst: I quite agree. In fact they're defined in this data dictionary [holds up several pages titled "Data Dictionary"]. Terms are filed in the data dictionary alphabetically. If we turn to the entry for P-&-L (Figure 6-5), we find a definition very similar to the one you gave, Mrs. O. The P-&-L is composed of sales, expenses, and net profit or loss.

31   Mrs. O: Hmmm, I see. But, this does not state how owners' equity and net profit or loss are derived. Isn't the method of calculation important too?

32   Analyst: Certainly it is! In order to keep the data dictionary simple, we only define the information content of data flowing in the system. We use another tool to describe how data flows are transformed within the system. Appropriately enough, these are called "transform descriptions." ← *METHOD OF CALCULATION*

33   Mrs. O: So if I wanted to see how a P-&-L is produced, I would just look for the transform description of the circle with P-&-L coming out of it. But how do I find it?

FIGURE 6-5. Excerpts from the data dictionary for Financial-Statements and its constituents.

FINANCIAL-STATEMENTS are composed of:
  BALANCE-SHEET and
  P-&-L
    .
    .
    .

BALANCE-SHEET is composed of:
  ASSETS and
  LIABILITIES and
  OWNERS'-EQUITY
    .
    .
    .

P-&-L, also known as INCOME-STATEMENT, is composed of:
  SALES and
  EXPENSES and
  NET-PROFIT-OR-LOSS

I mean, how are these transform descriptions filed? Oh, here's a section called "Transform Dictionary," but I don't see a detailed description for Bubble 4.

34   Analyst: Well, every bubble does not have a transform description. We keep breaking the bubbles apart into finer and finer detail until we reach processes that perform some straightforward calculations on relatively simple data flows. These are the only circles that have transform descriptions. Otherwise, we would be repeating the same calculation rules over and over.

35   Mrs. O: Let me see if I can figure this out on my own. Hmmm, the P-&-L is produced in Accounting [points to Bubble 4 on Diagram 0]. If we look at Diagram 4 (Figure 6-6), there is a bubble named "Produce Financial Statements." Is there another diagram for this bubble [pointing to Bubble 4.5 on Diagram 4]?

FIGURE 6-6.   Diagram 4: Accounting.

DIAGRAM 4: ACCOUNTING

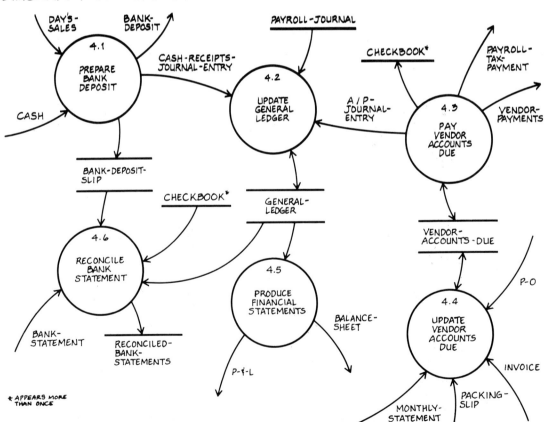

DIAGRAM 4.5 : PRODUCE FINANCIAL STATEMENTS

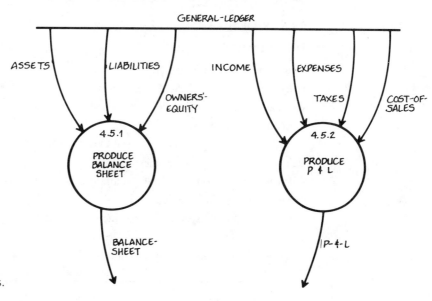

FIGURE 6-7. Diagram 4.5: Produce Financial Statements.

36	Analyst:	If there is, it should be named "Diagram 4.5: Produce Financial Statements." (See Figure 6-7.)
37	Mrs. O:	Ah yes, here it is. Now, I assume that each bubble here would be described by a transform description. How do I find them? By bubble number?
38	Analyst:	Excellent guess. What do you think the bubble number should be?
39	Mrs. O:	Well, if it's consistent with the way you numbered the diagrams, I would guess 4.5.2, since I see P-&-L coming out of it, and the bubble is labeled "Produce P & L." Is that right?
40	Analyst:	That's correct. Here is the transform description for Bubble 4.5.2: Produce P & L (Figure 6-8).
41	Mrs. O:	Let me read this. . . . Looks pretty good.
42	Mr. O:	Look, the lunch rush is coming up and I have to get ready. Thus stuff looks pretty good, but I'd like to have more time to look it over. Can we meet some time next week after we've had a chance to review these diagrams?
43	Analyst:	Sure. If you have any questions, you have my phone number. Don't hesitate to call.
44	Mrs. O:	We won't. Can we write on these diagrams?

FIGURE 6-8.   Transform description for Transform 4.5.2: Produce P & L.

TRANSFORM 4.5.2: Produce P & L

DESCRIPTION:

Determine Period-Ending-Date.

Calculate Cost-of-Goods-Sold = Beginning-Inventory + Purchase − Ending-Inventory.

Calculate Gross-Profit-on-Sales = Income-from-Sales − Cost-of-Goods-Sold.

For each Operating-Expense:

   Add the Operating-Expense to Total-Operating-Expenses.

Calculate Net-Operating-Income = Gross-Profit-on-Sales − Total-Operating-Expenses.

Calculate Gross-Income = Net-Operating-Income + Other-Interest-Income.

For each Interest-Expense:

   Add the Interest-Expense to Total-Interest-Expenses.

Calculate Net-Income-Before-Tax = Gross-Income − Total-Interest-Expenses.

Calculate Net-Profit-or-Loss = Net-Income-Before-Tax − State-Income-Tax −
                          Federal-Income-Tax.

---

45   Analyst: Absolutely! They're working documents. Feel free to mark them up in any way you see fit. Previous clients have written their questions right on the diagrams and saved their copies of the diagrams over the lifetime of the project. This gave them both a project history and an indication of progress. I look forward to our next meeting.

46   Mr. O: Until next week then.

47   Mrs. O: Thank you.

COMPONENTS OF THE DATA FLOW DIAGRAM

Now that we have seen an example of a set of data flow diagrams and selected portions of the data dictionary and transform descriptions which support it, we can explain explicitly and precisely the system modeling concepts illustrated in the dialogue above.

Recall that in Chapter 1 a system was defined as an interrelated set of components with a boundary distinguishing the system from its environment. Several general system structures were illustrated, including networks and hierarchical structures containing subsystems.

In Chapter 2 we described three principal functions of an information processing system: the transmission, storage, and transformation of information. Thus an appropriate abstraction for an information processing system would depict it in terms of just three types of components: one which shows the movement of information, one which shows the storage of information, and one which shows the transformation of information. These components are the data flow, data store, and transform, respectively. A data flow diagram models a specific information processing system by showing the relationships among components of these three types.

Basic Definitions

Thus *a data flow diagram is a network model of an information processing system. The arcs of the network represent data flows, and the nodes represent data stores, transforms, or selected elements of the environment.*

A *data flow* is a movement of information within the system or across the system boundary.

A data flow which crosses the system boundary to enter the system is called a *net system input,* or simply, *system input.* A data flow which crosses the boundary to leave the system is called a *net system output* or *system output.*

A *data store* is a time-delayed repository of information.

A *transformation,* or *transform,* is a process that changes incoming data flows into outgoing data flows.

To show the connection between a system and its environment, a fourth type of element is used. An *origin* is a person or organization or system outside the system that provides information to the system in the form of an incoming data flow. A *destination* is a person or organization or system outside the system that receives a system output. Origins and destinations are sometimes collectively called *external entities.*

Symbols

Data flow diagrams use the following symbols: Directed arcs represent data flows; parallel lines represent data stores; circles represent transformations (processes); and squares represent origins or destinations. (See Figure 6-9.)

Data Flow

We have defined a data flow as a movement of information from one point within the system to another. It may be thought of in several ways depending upon which aspect of the movement is emphasized. A data flow has a *direction.* If we view the data flow as connecting the points between which it moves, we see it as an *interface* between a transform and a file or another transform or between the system and an external entity. From the standpoint of a transform, a data flow is an *input* or *output.* A data flow entering or leaving a data store is *information stored or retrieved by an access* to that data store. These aspects of information flow are shown explicitly on a data flow diagram. For example, in Figure 6-1, the data flow Day's-Sales is an interface between Transform 2: Sales and the data store Sales-

FIGURE 6-9. Data flow diagram symbols.

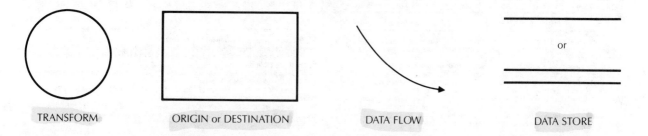

| TRANSFORM | ORIGIN or DESTINATION | DATA FLOW | DATA STORE |

History as well as between Sales and Transform 4: Accounting. It is an output of Sales, an input to Accounting, and represents information written to Sales-History. At other times we are primarily interested in the **information content** or data structure of a data flow; these are specified in the data dictionary.

Data Store

A data store is often referred to as a "file." Although **file** is a shorter word, the term **data store** is more general and abstract without the physical connotations of **file** with respect to the information storage medium or the organization of the data. Data stores arise if information must be retained within the system; if information must wait before being transformed; if a transformation requires an aggregation, or collection, of similar data flows to derive its output; or if different transformations occur at different time cycles. Examples of data stores in a nonautomated system include: a tax table used in computing allowable federal income tax deductions for state sales taxes (see Figure 6-10), a collection of invoices for customer purchases to be itemized when the monthly statement is produced, time cards which are filled out and collected weekly in a system in which paychecks are produced

FIGURE 6-10. Excerpt from a table for determining deductions for state sales taxes.

1983 Optional State Sales Tax Tables

(If you kept records that show you paid more sales tax than the table for your State indicates, you may claim the higher amount on Schedule A, line 10a.)

Your itemized deduction for general sales tax paid can be estimated from these tables plus any qualifying sales taxes paid on the items listed on page 19. To use the tables:

Step 1—Figure your total available income. (See note to the right).

Step 2—Count the number of exemptions for you and your family. Do not count exemptions claimed for being 65 or over or blind as part of your family size.

Step 3 A—If your total available income is not over $40,000, find the income line for your State on the tables and read across to find the amount of sales tax for your family size.

Step 3 B—If your income is over $40,000 but not over $100,000, find the deduction listed on the income line "$38,000-$40,000" for your family size and State. For each $5,000 (or part of $5,000) of income over $40,000, increase the deduction by the amount listed for the line "$40,001-$100,000."

Step 3 C—If your income is over $100,000, your sales tax deduction is limited to the deduction for income of $100,000. To figure your sales tax deduction, use Step 3 B but don't go over $100,000.

Note: Use the total of the amount on Form 1040, line 33, and nontaxable receipts such as social security, veterans', and railroad retirement benefits, workmen's compensation, untaxed portion of long-term capital gains or unemployment compensation, All-Savers interest exclusion, dividends exclusion, disability income exclusion, deduction for a married couple when both work, and public assistance payments.

Income	Alabama [1] Family size 1&2		3	4	5	Over 5	Arizona [2] Family size 1&2		3	4	5	Over 5	Arkansas [1] Family size 1	2	3	4	5	Over 5	California [3] Family size 1&2		3	4	5	Over 5	Colorado [2] Family size 1&2		3&4	5	Over 5	Connecticut Family size 1&2		5	Dist. of Columbia Family size 1	2	3	4	5	Over 5
$1-$8,000	91	113	120	130	141	160	98	113	113	120	126		78	97	102	109	116	132	125	147	155	164			50	59	62	65		126	139	146	94	112	125	125	132	140
$8,001-$10,000	107	129	140	151	164	185	115	133	133	142	148		91	111	118	127	135	152	147	173	183	193			60	70	74	77		150	167	175	110	129	145	146	155	164
$10,001-$12,000	121	144	159	171	185	207	131	152	154	162	169		103	123	134	143	153	170	167	198	208	219			68	81	85	88		172	194	203	125	145	164	166	177	186
$12,001-$14,000	135	158	176	190	204	227	146	170	173	181	188		115	134	148	158	169	187	186	220	232	243			76	91	95	99		194	220	229	139	159	182	186	197	206
$14,001-$16,000	148	170	192	207	223	246	160	186	191	199	207		125	145	161	173	184	202	204	242	255	266			84	101	105	110		214	244	254	152	173	198	205	216	225
$16,001-$18,000	160	182	208	224	240	265	173	202	209	216	224		135	154	174	186	198	217	222	263	276	288			92	110	115	120		234	268	279	165	186	213	223	234	244
$18,001-$20,000	172	193	222	240	257	282	186	218	226	233	241		145	163	186	199	212	231	238	282	297	309			99	119	125	129		253	291	302	177	198	228	240	252	261
$20,001-$22,000	183	204	236	255	273	298	198	233	242	249	257		154	172	198	212	225	245	254	301	317	330			106	128	134	138		271	313	325	189	210	242	256	268	278
$22,001-$24,000	194	214	250	269	288	314	210	247	258	264	273		163	181	209	224	238	258	270	320	336	349			113	136	143	147		289	335	347	200	221	256	272	284	294
$24,001-$26,000	204	224	263	283	303	329	222	261	274	279	288		172	189	220	235	250	270	285	338	355	368			119	144	152	156		306	357	369	211	232	269	287	300	310
$26,001-$28,000	214	234	276	297	317	344	233	274	289	294	303		180	197	230	246	262	282	299	355	373	386			125	152	160	165		323	378	390	222	242	282	302	315	325
$28,001-$30,000	224	243	289	310	331	358	244	287	304	308	317		188	204	240	257	273	294	313	372	391	404			131	160	168	173		340	398	411	232	252	295	317	330	340
$30,001-$32,000	234	252	301	323	344	372	255	300	319	322	331		196	211	250	268	284	305	327	389	408	422			137	168	176	181		356	418	432	242	262	307	332	345	354
$32,001-$34,000	244	261	313	336	357	385	266	313	333	335	345		204	218	260	278	295	316	341	405	425	439			143	176	184	189		372	438	452	252	272	319	346	359	368
$34,001-$36,000	253	269	324	348	370	398	276	325	347	348	359		211	225	270	288	306	326	354	421	441	455			149	183	192	197		388	458	472	262	281	330	360	373	382
$36,001-$38,000	262	277	335	360	383	411	286	337	361	361	372		218	232	279	298	316	336	367	436	457	471			155	190	200	205		403	477	491	271	290	341	373	387	396
$38,001-$40,000	271	285	346	372	395	423	296	349	374	374	385		225	239	288	308	326	346	380	451	473	487			160	197	207	213		418	496	510	280	299	352	386	400	409
$40,001-$100,000 (See Step 3B)	14	14	17	19	20	21	15	17	19	19	19		11	12	14	15	16	17	19	23	24	24			8	10	10	11		21	25	26	14	15	18	19	20	20

[1] Local sales taxes are not included. Add an amount based on the ratio between the local and State sales tax rates considering the number of months the taxes have been in effect.

[2] Local sales taxes are not included. Add the amount paid.

[3] If the State sales tax rate becomes 5¾ percent, November 1, 1983, taxpayers can add three percent to the table amounts. The 1¼ percent local sales tax is included. If a ½ of 1 percent local sales tax for transportation is paid all year (Alameda, Contra Costa, Los Angeles, San Francisco, San Mateo, Santa Clara and Santa Cruz counties) taxpayers can add 8 percent to the table amount.

[4] Local 1 percent sales taxes are included. If public transportation sales taxes are paid, compute the allowable deduction by the method in footnote 1.

[5] If your local sales tax applies to food for home consumption check your local newspaper during mid-January for the correct deduction. Otherwise see footnote 1.

[6] Sales tax paid on purchase of electricity of 750 KWH or more per month, can be added to the table amounts.

[7] Sales tax paid on the purchase of any single item of clothing for $175 or more can be added to the table amounts.

[8] Sales tax paid on purchases of natural gas or electricity can be added to the table amounts. For local sales tax see footnote 1.

[9] Local sales taxes are not included. If paid all year add 26 percent of the table amount for each 1 percent of local sales tax rate. Otherwise use a proportionate amount. For N.Y. City add 107 percent of the table amount.

FIGURE 6-11. A data store for food orders.

monthly, and customer orders in a restaurant that are waiting to be filled as soon as a cook is available. (See Figure 6-11.)

In a set of data flow diagrams there are some important conventions for where data stores would be shown. There are also rules for data flows to or from data stores. These are included in the discussion of leveling later in this chapter.

Transform

As discussed in Chapter 2, the ability to change the form or content of information permits an information processing system to support a business or other organization. Using the term *transform* for the system components which derive outputs from inputs emphasizes the act of changing one set of data flows into another. However, transforms are sometimes called by a variety of other names such as process or function. In this context *process* is synonymous with transformation, as exemplified in the familiar sequence input-process-output. The word "function" has several connotations. The first is a rather loose or general reference to a task, action, or activity within a system. Another refers to the role of a component within a system, as when we say that the function of a bicycle pedal is to transfer force from the foot to the drive chain. Another meaning is mathematical, expressing a related input-output pair. All these uses of function may be associated with the transforms in a data flow diagram, but the mathematical sense is the best way to think about a transform.

Transformations are the parts of the system that do the work. Associated with each transformation is a procedure or set of instructions for making its inputs into outputs. In information processing, the detailed procedure for a transformation is called an "algorithm," especially if the instructions are to be carried out by a computer. The data flow diagram merely shows which transforms are required in the system; the procedures themselves are specified by transform descriptions. But the transform descriptions, like the rest of the structured specification, are statements of user requirements. As such, they usually do not prescribe algorithms to the level of detail necessary for execution by a computer. Instead, the transform descriptions explicitly and unambiguously state the users' business policies for the transformations with sufficient rigor, completeness, and consistency for the system designers to supply the additional detail. This is discussed further in Chapter 8.

NAMING SYSTEM COMPONENTS

Names or labels on a data flow diagram serve two purposes. Formally, they identify the individual components; from a practical point of view, they also communicate as much specific information about the system as possible. Every constituent of a data flow diagram must have a unique name to distinguish it from the other components of the same type. Moreover, well-chosen names permit data flow diagrams to be read without constant reference to the data dictionary and transform descriptions, even though the meaning of the names is formally and rigorously defined only in the data dictionary and transform descriptions.

Appropriate names for the components of a data flow diagram are assigned as follows:

- Since data flows are moving packets of data, they take on the name of these packets.
- Since data stores are resting places for packets of data, they take on the name of the data that reside there.
- Since transformations change incoming data flows into outgoing data flows, their names state the activity involved in the transformation and incorporate the names of the inputs and outputs.
- Since origins and destinations are producers and consumers of data flows, respectively, they take on the name of the outside entity they represent—a person, an organization, or another system.

It is customary to hyphenate the names of data flows and data stores. Remember that transforms, in addition to being named, are identified uniquely by numbers.

SETS OF DATA FLOW DIAGRAMS

A single data flow diagram is inadequate to describe a complex system. As demonstrated by Figures 6-1 through 6-8, DFDs come in a hierarchically related set. The set depicts the decomposition of the system into successively finer detail. At the top is the Context Diagram, which presents the system being modeled as a single circle interacting via data flows with entities in its environment. These external entities are producers (origins or sources) of information needed by the system or consumers (destinations or sinks) of information from the system. The perspective of the Context Diagram is from the system outward toward the environment. At the next level down is Diagram 0. The perspective here turns inward. What was a single bubble on the Context Diagram is now partitioned into several bubbles representing the major functional areas or transformations composing the system being modeled and showing the flow of data between them.

Hierarchical Structure

Diagram 0 may be regarded as a subsystem of the Context Diagram. The transforms in the two diagrams are hierarchically related; therefore the transforms in Diagram 0 are the immediate subordinates of the single transform in the Context Diagram. For this reason, the Context Diagram and Diagram 0 are called *parent* and *child* diagrams, respectively, in relation to each other. *A data flow diagram showing the immediate subordinates of a transformation and the diagram containing that transformation are said to be child and parent diagrams, respectively.*

Balancing

Because the Context Diagram and Diagram 0 both represent the entire system, though in differing detail, the data flows entering the Context Diagram must be the same as the data flows entering Diagram 0, and the data flows leaving the Context Diagram must be the same as those leaving Diagram 0. This relationship among the data flows on the two diagrams is called *balanced.* There must be a balanced relationship between the data flows on every child diagram and those on its parent diagram for all the diagrams in the set. That is, if a data flow comes into or leaves the child diagram, then on the parent diagram the same data flow (or its equivalent) must come into or leave the bubble which corresponds to the child diagram. (See Figures 6-1 and 6-2.)

Numbering Conventions

The system is further partitioned into lower-level diagrams that are uniquely identified by number. The numbering scheme follows these conventions: Each child diagram takes the number of the corresponding bubble on its parent diagram. Bubbles on the child diagram are identified by appending a decimal point and a sequential number to the diagram number. (See Figures 6-3, 6-4, and 6-6.) Thus every bubble number shows its place in the hierarchy of transformations and permits its ancestry to be traced to the top (Figure 6-12). Because there is a transform for each diagram, the set of diagrams and the transforms have the same hierarchical structure.

In this manner we can provide a unique path from a diagram at the bottom of the hierarchy back up to Diagram 0. For example, suppose we want to see the diagrams along the path from Diagram 4.5.2 to Diagram 0. This path is depicted in the tree shown in Figure 6-13.

Because of its hierarchical structure, in which increasing detail is shown at the lower levels, the complete collection of diagrams describing an information processing system is often referred to as a *leveled set.*

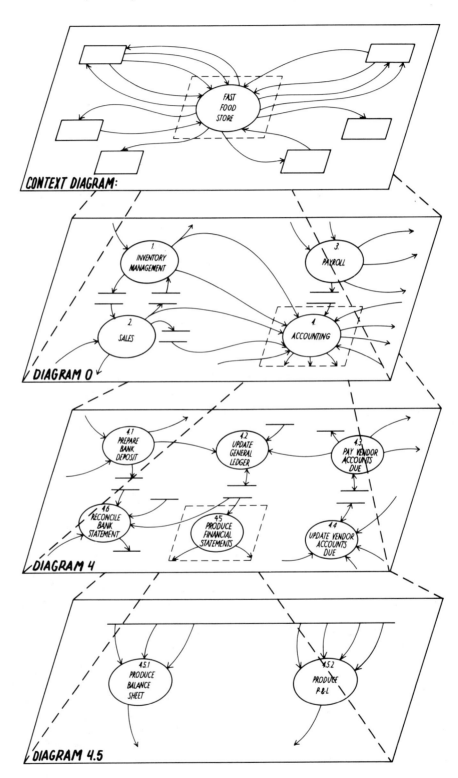

FIGURE 6-12. The hierarchy of data flow diagrams for the FastFood Store.

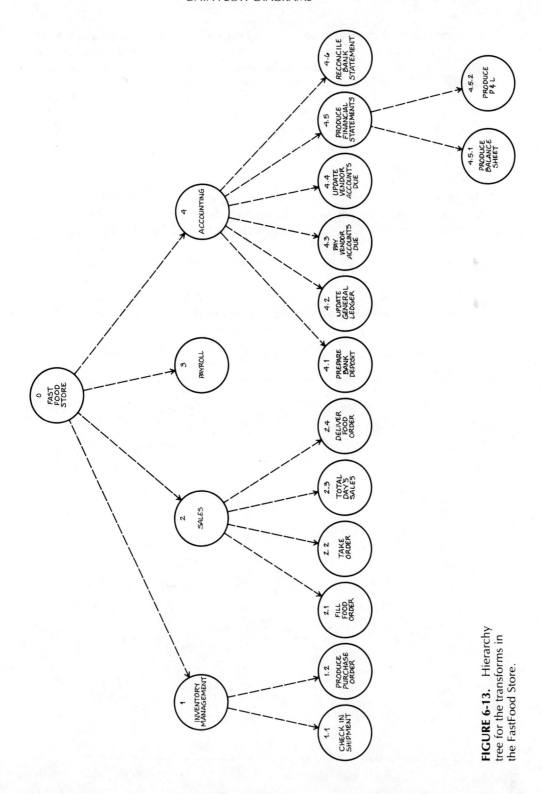

FIGURE 6-13. Hierarchy tree for the transforms in the FastFood Store.

**Primitive
Transformations**

At some point the decomposition of transforms stops, either because a transform cannot be partitioned further, or because we choose not to partition it further. These unpartitioned transforms are called ***primitive transforms*** or ***primitive functions***. By definition, a primitive transform has no child diagram.

When we encounter a bubble that is not further decomposed into child diagrams, we refer to the transform description that describes the policy governing that transformation. The relationship between a primitive transform and its transform description is illustrated in Figure 6-14.

**Conventions for
Data Stores**

The following conventions apply to data stores within a leveled set of data flow diagrams. These conventions have such a strong intuitive basis that diagrams containing data stores are easy to read. Therefore, they are merely listed here, with a discussion deferred to Chapter 10, which explains how to draw data flow diagrams.

1. Every data store in the system must appear at least once in the set of data flow diagrams.

2. A data store is always shown in a diagram if it is shared by (connected by data flows to) more than one transformation in that diagram. For example, each of

FIGURE 6-14. Relationship of a primitive transform to its transform description.

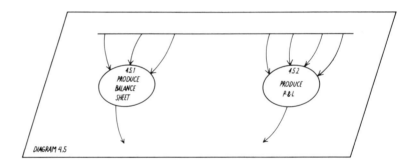

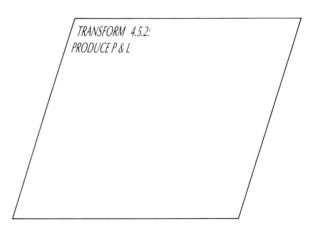

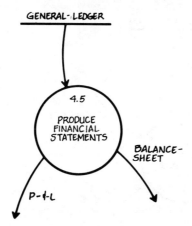

FIGURE 6-15. Access to a data store for reading.

the data stores Raw-Goods, Sales-History, Payroll-Journal, and Cash-Vault in Diagram 0 (Figure 6-2) is shared by transformations in the FastFood Store.

3. Data flows between a transformation and a data store distinguish three possibilities—access to read from the data store (Figure 6-15), access to write to the data store (Figure 6-16), and a two-way access (Figure 6-17) in which information from the data store is required as input to the transformation and must be read in order to produce an output which is then written to the data store. (Note that Figures 6-15, 6-16, and 6-17 have been extracted from Figure 6-6.) This distinction is sometimes expressed as follows: A data flow diagram shows only the *net flow* to or from a data store.

4. Data flows to and from a data store need not be labeled if the data store is shown. If the label is omitted, the implied name of the data flow is that of the data store. However, it is always proper to label an access to a data store to specify precisely what information is contained in the data flow.

FIGURE 6-16. Access to a data store for writing.

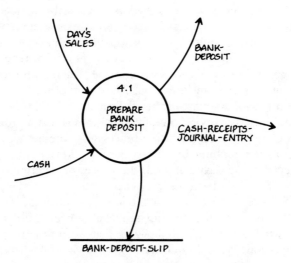

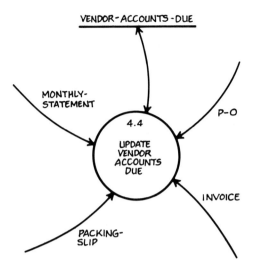

FIGURE 6-17. Access to a data store to read and write.

HOW TO VIEW A DATA FLOW DIAGRAM

The detailed conventions for data flow diagrams summarized above are straightforward and easy to understand with a bit of practice. More difficult for those first exposed to data flow diagrams is a general understanding of the kind of system model presented in a data flow diagram. This is due in part to the lack of previous experience with data flow diagrams; it is partly a result of being familiar with program flowcharts and trying to read a data flow diagram as if it were a program flowchart.

Data Flow Diagrams vs. Program Flowcharts

A **program flowchart** is a graphic representation of an algorithm. It emphasizes the step-by-step logic for a computational procedure. It shows the **flow of control,** that is, the order in which the various steps of the procedure will be carried out. A program flowchart depicts steps performed in **sequence** as well as **decisions** by which alternative computational sequences are chosen and **iterative** structures which control how many times a sequence of steps is to be repeated. Program flowcharts are appropriate tools for describing the detailed procedures within a computer program unit or transformation. Another tool for transform descriptions, more appropriate for systems analysis, is Structured English, discussed in Chapter 8.

This emphasis upon flow of control means that a program flowchart takes the point of view of an information processor—whether we think of that processor as a person or a machine. In each block of the flowchart the processor is carrying out instructions. At each diamond, the processor determines whether a condition is true or false and, based on the outcome, begins to execute one of the alternative sequences of instructions which follow. To control the iteration, the processor tests whether to repeat an instruction sequence or begin a different instruction sequence. In viewing a flowchart, we imagine that we are the processor, continually asking the question "What do I do **next**?"

This point of view, essential for a flowchart, is not applicable to data flow diagrams. Flowchart thinking will only get in the way of our understanding data flows diagrams and must be abandoned when we approach them.

A data flow diagram, on the other hand, deliberately suppresses these internal details of the transformations in order to focus on the large-scale structure of the system as a whole. It emphasizes partitioning, components, and interfaces. It shows the inputs and outputs of each transformation and the flow of information which connects transformations. Sometimes the connection is direct, so that an output of one transformation becomes an input to another; sometimes there is a delay, with some transformations placing output in a data store and some transformations taking input from a data store. A set of data flow diagrams shows the partitioning of a specific information processing system into transformations, data flows, and data stores and the specific connections among these components.

This emphasis on system structure and decomposition means that we must approach data flow diagrams with different images from those used to read flowcharts.

One helpful image is that of a map. For example, if we are planning a vacation in the United States and want to hike, ski, or swim, we can locate the mountains, the beaches, or the lakes. Perhaps we are especially interested in enjoying the outdoors in one of the national parks. Or, if we want to visit museums or historical sites, we may look for cities rather than the countryside. In each case the map will tell us in what part of the United States the places of particular interest to us are located.

FIGURE 6-18. A schematic map of the United States.

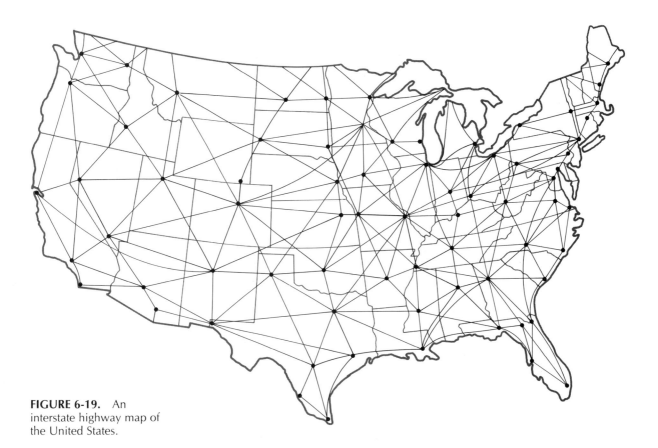

FIGURE 6-19. An interstate highway map of the United States.

In this sense, the map partitions the country into areas containing features of interest for planning a vacation. A highway map (Figure 6-19) would show us the movement patterns taking us from where we live to our selected vacation spot. There we can see the "lay of the land" and orient ourselves in relation to it. Similarly, data flow diagrams, especially at the upper levels, provide an overview of an information processing system.

In order to show more detail, an atlas of the United States is organized hierarchically. There will be one map showing the entire country with the major geographic features and the principal highway network. Then there will be a series of maps showing each state or contiguous group of states with the county and city boundaries and the state highway network. For the principal cities there will be maps showing the major streets and the highway connections and places of interest, and perhaps a detailed map of the city center depicting all the streets as well as individual buildings. A leveled set of data flow diagrams presents a similar hierarchical partitioning of an information processing system with increasing detail at the lower levels.

Another useful analogy is that of an industrial production line. Raw materials and components from outside suppliers enter the factory and are transformed and assembled until the finished product emerges. The entire process happens simultaneously with various stages of the transformation to the end product visible at every moment the factory is in continuous operation. We can focus attention on any portion of the process with its specific inputs and outputs. We can locate the part of the process in which each of the major subassemblies comes together. We can follow the movement of material from its entry into the factory until its emergence as part of the product. Or we can start with the product and trace its production backward through the subassemblies to the basic components. We can look at the stockpiles of materials, parts, and subassemblies as they wait to be incorporated into the next step of the process. All of these are valid and useful ways of viewing and understanding the production line.

A data flow diagram is like a snapshot of an information processing factory. It shows the factory in the steady state—in continuous production—with the various transformations carried out simultaneously. We can locate each transformation and data store in relation to the rest of the process. It is especially helpful to view the system from the standpoint of the data. We can imagine ourselves as an input, entering the information processing factory, moving along a conveyor belt to the next transformation, or waiting until needed by the next transformation. Or we can imagine ourselves as an output, and trace backward the series of transformations by which we were fabricated.

A final image for a set of data flow diagrams is that of a series of photographs at increasing magnification. Perhaps we begin with a landscape. A telephoto lens will enable us to isolate a tree. Further enlargement shows a leaf. A microscope will reveal the leaf structure. Increased magnification shows individual cells, and an electron microscope allows us to see the molecules. The area depicted in each of these photographs can be located on its parent within the hierarchy. Similarly, each data flow diagram within the leveled set has its place in the hierarchy. To see additional detail, we move to a lower level. As we move upward in the hierarchy, some of the lower level detail is no longer visible. However, in the data flow diagrams the range of magnification is limited, and the components of the information processing system at each level are always data flows, data stores, and transformations.

What a DFD Does Not Show

A data flow diagram, like other models, has limitations. It is important to be aware of these limitations. Otherwise, it is easy to misunderstand data flow diagrams by making invalid inferences or by trying to read into a data flow diagram meanings which were not intended.

A data flow diagram does not show flow of control. It is not a flowchart. Alternative outputs may result from decisions within a transformation; a data flow diagram shows only the alternative outputs, not the decisions. Only the incoming data flows appear as inputs to a transformation; if events or stimuli other than these inputs are necessary to activate the transformation, they are not shown. A transformation may be carried out many times while the system is in operation, but it is shown only once.

A data flow diagram does not show details linking inputs and outputs within a transformation. It shows only all possible inputs and outputs for each transformation in the system. The specific procedures linking the inputs and outputs are invisible. These procedures are defined by the primitive transform descriptions. Thus, in Figure 6-6 all we can tell is that Transformation 4.1: Prepare Bank Deposit requires Day's-Sales and Cash as inputs and produces Bank-Deposit, Cash-Receipts-Journal-Entry, and a data flow to the Bank-Deposit-Slip data store as outputs. What combinations of the inputs produce which outputs cannot be inferred from the data flow diagram. (However, your knowledge of the system, though not depicted in the data flow diagram, may include the relationships among these inputs and outputs.)*

   A data flow diagram does not show time. This is the best rule to guide a novice in understanding data flow diagrams. A data flow diagram does show relationships which are closely related to time, but it is better not to think of them in temporal terms. Although we have spoken of a data flow diagram as a snapshot of simultaneous transformations, it is better to think of it as a single picture of all the transformations in the system without regard to when they occur. A data flow diagram shows an ordering of transformations, but it is logical rather than temporal. The relationship is that of predecessors and successors and can be understood in either direction. Inputs are transformed into outputs; outputs are produced from inputs. A transformation producing an input for another transformation is a predecessor of that transformation. For a transformation to occur, all the required inputs must be available. Thus, a data flow diagram shows *information* **dependencies** rather than temporal sequence.

   A data flow diagram may show delay, but the duration is not shown. We may find it helpful to think of a data store as a time-delayed repository of data, and a data flow as transmitted instantaneously between transformations or between transformations and data stores. But we may also think of a data store as a collection of data with the same structure, each member of the collection being uniquely identifiable. This is discussed in more detail in Chapter 7.

RELATIONSHIP AMONG THE PARTS OF A SYSTEM DESCRIPTION

The relationship among the essential parts of an information processing system description is shown schematically in Figure 6-20. Balance-Sheet and P-&-L are data flows; General-Ledger is a data store. The name of each data flow and data store shown on a data flow diagram appears in the data dictionary portion of the system dictionary. Transforms 4.5.1: Produce Balance-Sheet and 4.5.2: Produce P-&-L are primitive transforms. Each primitive transform has a transform description in the transform dictionary portion of the system dictionary. In Figure 6-20, the data dictionary pages for Balance-Sheet, P-&-L, and General-Ledger are illustrated, as well as the transform descriptions for Transforms 4.5.1 and 4.5.2. Each is associated with the component of the same name in the data flow diagram.

---

*Some people annotate pairs of inputs and pairs of outputs to show that they are mutually exclusive or that they occur together, but the notation cannot always be unambiguously extended to more than two data flows. The symbols clutter up a data flow diagram. They are also redundant, duplicating the information in the transform description. It is better to omit such annotations and rely solely on the transform descriptions for details of the transformation.

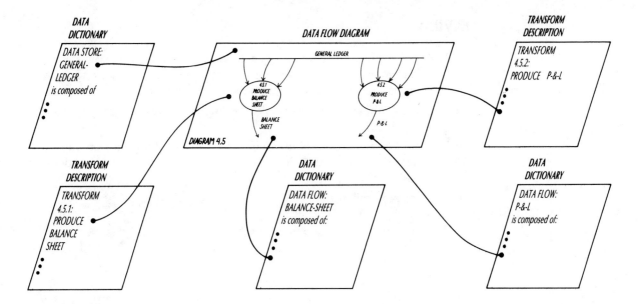

FIGURE 6-20. How the tools of structured analysis are related.

How PARTS AND TIED TOGETHER

DFD
DD
TD

SUMMARY

A data flow diagram is a graphic model of an information processing system. Movement of information is modeled by a data flow; storage of information by a data store; transformation of information by a transform, or process. The data flows are the interfaces connecting processes and data stores. The normal, or steady state, of the system is depicted; we are not concerned with how the system starts up (what happens when Mr. and Mrs. O open for business in the morning), nor are we concerned with how it shuts down (what happens at the end of the day).

A data dictionary is used to document the detailed information content of data flows. Transform descriptions are used to describe the policy for producing the outputs of a transform from its inputs.

In a set of data flow diagrams the system is decomposed into a hierarchy of networks and transforms. At the top level is the context diagram, which shows the system boundary and depicts the data flows which cross the boundary as interfaces between the system and those who supply its inputs and receive its outputs. When the process of decomposition stops, the undecomposed, or primitive, transforms and their interfaces provide a complete and detailed description of the entire system.

REVIEW

6-1. Name the three types of system components used in a data flow diagram to model an information processing system. How are they related to the three principal functions of an information processing system?

6-2. What symbol is used in a data flow diagram to represent each type of system component?

6-3. What symbol represents an origin or destination of information outside the system boundary? At which level in a set of data flow diagrams does this symbol appear?

6-4. Define the following:
a. data flow diagram
b. data flow
c. data store
d. system boundary
e. system input
f. system output
g. origin
h. destination

6-5. List 4 characteristics of a data flow that are of interest in modeling an information processing system.

6-6. How are the various components of an information processing system identified as distinct elements?

6-7. How are the names of data flow diagram components derived?

6-8. What is a context diagram?

6-9. What does it mean to say that data flows on parent and child diagrams are balanced?

6-10. Explain the numbering convention for transforms and data flow diagrams in a leveled set.

6-11. Define a primitive transform. Where are details of its logic specified?

6-12. Explain the difference between a flowchart and a data flow diagram.

6-13. List 3 things not shown on a data flow diagram.

6-14. Where is the description of the information contained in a data flow found?

EXERCISES AND DISCUSSION

6-1. In the set of data flow diagrams for the FastFood Store (Figures 6-1 through 6-4 and 6-6), which child diagram of Diagram 0 (Figure 6-2) is missing?

6-2. Demonstrate that Diagram 2: Sales (Figure 6-3) balances with its parent.

6-3. On which diagram would you expect to find a decomposition of the transform Update General Ledger in Figure 6-6?

6-4. Show how a set of data flow diagrams satisfies the following characteristics of a good system model stated in Chapter 1:
   a. an abstraction based on selected relevant system characteristics
   b. documentation that is easy to change

6-5. How does a set of data flow diagrams help the user and analyst cope with complexity in a computer information system?

6-6. Discuss the data flow diagram model of an information processing system in terms of the following concepts from general systems theory (see Chapter 1):
   a. system
   b. component
   c. system structure
   d. system functions
   e. system objectives
   f. network
   g. hierarchy
   h. environment
   i. interface
   j. abstraction
   k. decomposition
   l. subsystem

6-7. The following graphic elements are used in a program flowchart:

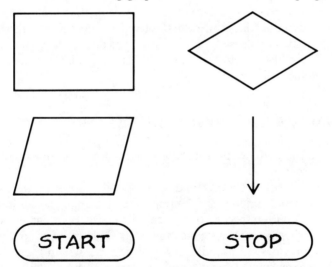

FIGURE 6-21. Program flowchart symbols.

What does each of these flowchart symbols represent? Is any of the graphic components of a flowchart also a graphic component of a data flow diagram? Is any of the elements represented in a flowchart also represented in a data flow diagram? If so, by what?

7 The System Dictionary

The previous chapter discussed one of the principal system modeling tools—the data flow diagram. We indicated that the data flow diagram is composed of several different types of components—data flows, data stores, transforms, and origins and destinations. In the system dictionary we list the name of each system component and define it.

GOALS OF THE ANALYSIS PHASE SYSTEM DICTIONARY

There are two principal goals for the system dictionary used in system requirements specification.

Goal 1: The system dictionary will be the sole reference for definitions and will contain a rigorous definition for each component in the system model.

With respect to the system dictionary, rigor entails being scrupulously accurate when defining system components. This attention to strict precision and exactness provides the foundation for testing the system model. It is important to have a system model that is complete, correct, and consistent. A rigorous system dictionary allows the analyst to check the system model for completeness, correctness, and consistency. Without the system dictionary, a set of diagrams lacks rigor; without the diagrams, the system dictionary has limited usefulness.

Goal 2: The system dictionary will be organized for ease of reference and maintenance.

A system model may be composed of thousands of components. During development and refinement of the system model, new components may be defined and existing components may be modified or eliminated. The effort to maintain the system dictionary increases with the number of components defined in it. There are several ways to make the system dictionary easy to use and modify. First, partition the system dictionary so that the terms referring to system components of the same type appear together. Second, sequence the names within each class according to some meaningful criterion. Third, use a concise, unambiguous, and straightforward notation for definitions. Fourth, minimize the analyst's maintenance burden by avoiding redundant information.

THE ORGANIZATION OF THE SYSTEM DICTIONARY

Typically when the system dictionary is consulted, both the type of component and its name are known. To facilitate quick reference, the system dictionary is organized to exploit this knowledge. The system dictionary is divided into subdictionaries—the data dictionary, transform dictionary, and origin/destination dictionary. Within each subdictionary, names are organized by component type and sequenced by component identifier. This hierarchical partitioning of the system dictionary and the sequencing of names within each subdictionary are illustrated in Figure 7-1.

For example, to find a specific data flow in the system dictionary, we would consult the data flow section of the data dictionary. We would search the names alphabetically until the desired data flow name was found.

Entries in the System Dictionary

The *entry* is the basic unit of storage in the system dictionary. An entry in the system dictionary contains everything we care to record about a named system component. The name of the entry is the name used for the component in the DFD, in the data dictionary, or in the data base description, as the case may be.

Entries in a natural language dictionary consist of more than a definition. We may see a word's pronunciation, synonyms, antonyms, etymology (the derivation of the word), grammatical usage, syntactic classification, etc. Similarly, an entry in the system dictionary may consist of: a component name, its aliases or synonyms,

FIGURE 7-1. Organization of the system dictionary.

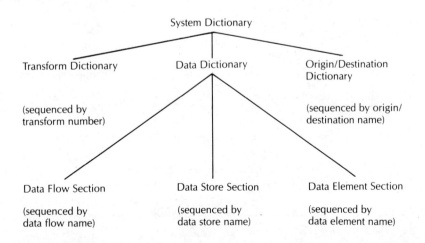

a component definition, and any necessary notes. The specific form of an entry depends upon the type of system component being described. This is illustrated in Figure 7-2. The definition part of an entry also varies by component type. For data flows and data stores we define their composition. For transforms we define the rules for the transformation.

Before we can embark on a detailed discussion of entries in the data dictionary, we need to explain how a data structure is described.

DESCRIBING DATA STRUCTURE

Data structure is a generic term referring to both data flows and data stores. Gane and Sarson state this concept as follows: "Data flows are data structures in motion; data stores are data structures at rest."*

It is useful to separate the discussion of data structure into two parts: one concerned with content—the constituent parts—and one concerned with structure—how the parts are arranged. The content and structure of a data structure are called its **composition.** To define a data structure, its composition is specified.

The Constituents of Data Structure

Data structures are composed of either of two types of constituents: subsidiary data structures or data elements.

*Chris Gane and Trish Sarson, **Structured Systems Analysis,** Prentice-Hall, Inc., Englewood Cliffs, N.J., © 1979, p. 50. Reprinted by permission of Prentice-Hall, Inc.

```
Data Element Name:    Job

        Alias(es):    Job-Code

        Values        Meanings
         10           Order Taker
         20           Grill Cook
         30           Fryer Cook
         40           Soda Fountain

Notes:
```

```
Data Store Name:    Employee

      Alias(es):

    Composition:    Employee-Number
                    AND
                    Employee-Name
                    AND
                    Employee-Payroll-Info

Notes:    Key is Employee-Number
```

```
Data Flow Name:    Employee-Name

      Alias(es):

    Composition:    First-Name
                    AND
                    Middle-Initial
                    AND
                    Surname

Notes:
```

```
Transform Number:    3.4.2

  Transform Name:    Determine-Net-Pay

     Description:    Net-Pay = Gross-Pay
                     − Taxes −
                     Other-Deductions

Notes:
```

Figure 7-2.   The different types of system dictionary entries.

A *data element* is an item of information that does not require any decomposition in the system of which it is a part. For the purpose of data definition, a data element is the lowest, or atomic, level. It is defined in terms of the values it may take on and the meaning associated with each value. In practice, a data element may be referred to by any of several terms—data item, data primitive, or field. We will use the term "data element."

Lack of decomposition is what characterizes data elements. Sometimes it makes no sense to further subdivide the data. At other times we choose not to further subdivide the data because we are not interested in its structure. Every data structure is ultimately defined in terms of data elements.

A *data structure* is a named collection of data elements or other data structures arranged in a specified manner. In contrast to a data element, a data structure is decomposable. Thus, a data structure is defined in terms of the arrangement of its components. Since a data structure is a collection of items, sometimes it is referred to as a "data aggregate" or "group." We will use the term "data structure."

The Data Structuring Operations

Three operations are used to specify how data elements or subsidiary data structures form a data structure. These operations are sequence, selection, and iteration.

Sequence. The sequence operation concatenates (joins together) other data structures or data elements. This is illustrated in Example (7.1). The order of the items joined is generally not important. The two definitions for Employee-Name in Example (7.1) are *structurally* equivalent to each other.

A sequential data structure IS COMPOSED OF
    item 1 AND
    item 2 AND
    item 3 AND
      .
      .
      .
    item N

where: item may be another data structure or a data element

Employee-Name IS COMPOSED OF
    Surname AND
    First-Name AND
    Middle-Initial

Employee-Name IS COMPOSED OF
    First-Name AND
    Middle-Initial AND
    Surname

Work-Assignment IS COMPOSED OF
    Job AND
    Shift

(7.1)

Selection. One of the components of a data structure may be a selection from alternative data structures or data elements. This is shown in Example (7.2). The alternatives are mutually exclusive—we must select one and only one of the alternatives. Each alternative appears on a separate line. Thus, Daily-Assignment is composed either of Day-Off-Date or of Work-Date and Work-Assignment.

A selection data structure IS COMPOSED OF
   SELECT ONE OF:
     item 1 OR
     item 2 OR
     item 3 OR
      .
      .
      .
     item N

   where: item may be another data structure or a data element

Daily-Assignment IS COMPOSED OF
   SELECT ONE OF:
   Day-Off-Date OR
   Work-Date AND Work-Assignment   (7.2)

Iteration.   One of the components of a data structure may be a repetition of another data structure or data element. The minimum and maximum numbers of iterations are always specified, either implicitly or explicitly. Unless explicitly indicated, the minimum number of iterations is assumed to be zero, and the maximum number of iterations is assumed to be infinity. Thus, in Example (7.3) there are at least 28 and no more than 31 iterations of Daily-Assignment.

An iterated data structure IS COMPOSED OF
   FROM lower-limit TO upper-limit ITERATIONS OF item
    or
   any number of ITERATIONS OF item

Monthly-Work-Schedule IS COMPOSED OF
   FROM 28 TO 31 ITERATIONS OF Daily-Assignment   (7.3)

The Notation for Describing Data Structure

The sample definitions of Examples (7.1) through (7.3) show the need for a more concise notation. Example (7.4) identifies the symbols for each of the data structuring operations. These are called *data structure operators.* The ampersand is used to indicate concatenation of a sequence of items.* Brackets are used to enclose a list of alternative items. Braces are used to enclose repeated items.

---

*Most authors use the symbol + to indicate concatenation. Because + is also a symbol for the arithmetic operation of addition, we have chosen to use the & to avoid any possible confusion. Usually the context shows whether + refers to addition or concatenation.

operation	expanded form	concise form	
composition	IS COMPOSED OF IS DEFINED AS	=	
sequence	AND	&	
selection	SELECT ONE OF	[ ]	
iteration	ITERATIONS OF	{ }	(7.4)

The notation in Example (7.4) is the minimum necessary for defining data structure. However, consider Example (7.5). This seems an awkward, though perfectly correct, notation for an optional data structure or data element.

$$0 \{ \text{Area-Code} \} 1 \qquad (7.5)$$

Since optional items occur so often, we introduce a simpler alternative by enclosing optional items in parentheses, as illustrated in Example (7.6).

$$(\text{Area-Code}) \qquad (7.6)$$

We also find it useful to separate items in a selection list with a vertical bar. This is illustrated in Example (7.7).

Daily-Assignment = [Day-Off-Date | Work-Date & Work-Assignment]   (7.7)

Though it may be inappropriate for communicating with most users, this concise notation reduces the effort of describing data structures. If we had an automated method of converting the concise notation to the more verbose notation, we could maintain the description of the data structures in the concise form and publish it for users in the expanded form.

Describing Nested Data

The composition operations of sequence, selection, and iteration may be combined, or nested, to describe a hierarchical data structure. Consider a work schedule for either of two kinds of employees—one scheduled daily by the month, the other scheduled weekly by the quarter. We can illustrate the hierarchical structure of the definition:

Employee-Work-Schedule = Employee-Name
                     & [Monthly-Schedule | Quarterly-Schedule]

as in Figure 7-3. In Figure 7-3, the structure of Employee-Work-Schedule is the sequence of Employee-Name and the choice of Monthly- or Quarterly-Schedule.

Adding the following definitions to the hierarchy of Figure 7-3 we get the hierarchy of Figure 7-4.

FIGURE 7-3. First-level decomposition of Employee-Work-Schedule.

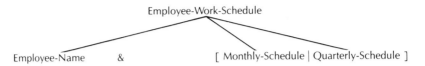

Employee-Name = First-Name & Middle-Initial & Surname
Monthly-Schedule = 28 { Daily-Assignment } 31
Quarterly-Schedule = 13 { Weekly-Assignment } 13

Here, Employee-Name is the concatenation of the data elements First-Name, Middle-Initial, and Surname. Monthly-Schedule is from 28 to 31 iterations of Daily- Assignment. This demonstrates that the lower and upper bounds on iterations may differ—in this case, depending on the number of days in a month. There are exactly 13 iterations of Weekly-Assignment for a Quarterly-Schedule.

FIGURE 7-4. Second-level decomposition of Employee-Work-Schedule.

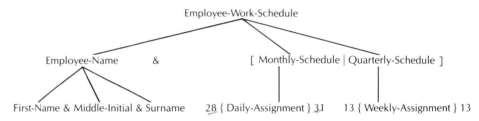

The components of Employee-Name are data elements, but Daily-Assignment and Weekly-Assignment require decomposition.

Daily-Assignment = [Work-Date & Work-Assignment | Day-Off-Date]
Weekly-Assignment = Work-Week-Start-Date & Work-Assignment
Work-Assignment = Shift & Job

Since Work-Date, Shift, Job, Day-Off-Date, and Work-Week-Start-Date are data elements, no further decomposition is necessary. Furthermore, Figure 7-5 shows that Employee-Work-Schedule is ultimately described in terms of data elements.

ENTRIES IN THE DATA DICTIONARY

All entries in the data dictionary have a similar format. Figure 7-6 shows the templates, or formats, for data dictionary entries for a data flow, a data store, and a data element. The similarities between the templates include: identification of the type of entry (data flow, data store, or data element), the name and aliases for the system component, and provision for notes. The principal difference lies in the definition part. For data flows and data stores, we define their composition. We also specify the key for a data store. For data elements, we define values and meanings. The following three sections discuss the details for each type of entry.

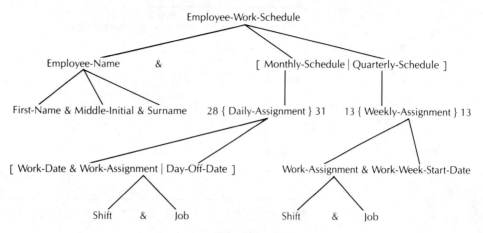

FIGURE 7-5. Decomposition of Employee-Work-Schedule to the data element level.

The Data Flow Entry   Figure 7-7 shows the data flow entries for Employee-Work-Schedule and its alias, Schedule. We use the notation introduced in the previous section to define a data flow by describing its composition.

Data Element Name:

    Alias(es):

        Values   Meanings

Notes:

Data Flow Name:

    Alias(es):

    Composition:

Notes:

Data Element Name:

    Alias(es):

  Range of Values:

Notes:

Data Store Name:

    Alias(es):

    Composition:

Notes:

FIGURE 7-6. Formats for data dictionary entries for data elements, data flows, and data stores.

```
Data Flow Name:     Employee-Work-Schedule

      Alias(es):    Schedule

   Composition:     Employee-Name &
                    [Monthly-Schedule |
                       Quarterly-Schedule]




Notes:
```

```
Data Flow Name:     Schedule

      Alias(es):    Employee-Work-Schedule

   Composition:




Notes:
```

FIGURE 7-7. Sample entries for a data flow and its alias.

There are no restrictions on what may appear in the Notes section. However, notes should pertain only to the entry in which they appear. Examples of the kind of information sometimes entered as notes include: volume, cyclicality, encoding, security, validation, and format.

The Composition section of the entry Schedule has been left blank because it is an alias. This is done to eliminate the burden of maintaining redundant definitions. To find the definition for Schedule, we simply refer to the composition of the principal data flow, Employee-Work-Schedule. The choice of which name is to be the principal data flow is a judgement made by the analyst. Even though the goal for aliases is total elimination, realistically we can hope only to minimize their occurrence. At the expense of some redundancy, we suggest that aliases be treated as in Figure 7-7—each referring to the other.

The Data Store Entry   The data dictionary entry for a data store is very similar to that for a data flow. The treatment of aliases and notes is the same, but there are some additional requirements for the definition of a data store.

Since a data store is usually a collection of records with the same data structure, we nearly **always** define the data store as being composed of repetitions of a data structure.* Thus, the description of the composition of a data store usually begins and ends with braces. When the lower or upper limits are known, they are specified. Otherwise, they are left implicit. Consider a data store for Employee-Work-Schedules. It could be defined as in Figure 7-8.

In addition to specifying the repetitive nature of a data store, we also need to indicate how specific records may be identified for the purposes of storage, update, and retrieval. A data element or concatenation of data elements that uniquely identifies a record in a data store is called a **key.** When defining the composition of the data store, the key is underlined. For example, Figure 7-8 shows Employee-Name as the key. This implies that to retrieve a schedule for a specific employee, all we need to know is the employee's name. Care must be exercised when selecting the key. A key **must be able to uniquely identify each record in a data store.** This may not be true for Employee-Name, in which case another key must be found, for example, Employee-Number.

Some analysts insist that the physical organization of a data store be included in its entry in the data dictionary. This may be appropriate when describing a data store in an existing system. For a new system, the choice of organization should be left to the system designer. The analyst should indicate to the designer that access to a particular data store is primarily random or sequential.

The Data Element Entry

The treatment of the name, aliases, and notes for a data element entry is the same as that for a data flow or data store. However, the definition of a data element is unlike that for a data flow or data store. A data element definition consists of two parts: the values that it may take on and the meanings of those values. These are usually presented in a table, as in Figure 7-9.

---

*An obvious exception is the case where there is **never** more than one record in the data store. The appropriate notation is either parentheses indicating that there may be no or one occurrence of the record or just the description of the structure for the case where there is **always** exactly one record. Since this is an unusual situation, a comment in the Notes section may be appropriate.

FIGURE 7-8. A sample entry for a data store with the key underlined.

```
Data Store Name:   Employee-Work-Schedules

      Alias(es):   None

   Composition:   { Employee-Name &
                   [ Monthly-Schedule |
                     Quarterly-Schedule ] }




        Notes:
```

```
Data Element:    Job

    Alias(es):   Job-Code

                 Value        Meaning

                   10         Order Taker
                   20         Grill Cook
                   30         Fryer Cook
                   40         Soda Fountain
```

```
Data Element:    Shift

    Alias(es):   Work-Shift

                 Value        Meaning

                   1          7 am - 4 pm
                   2          4 pm - 11 pm
                   3          11 pm - 7 am
```

FIGURE 7-9. Sample entries for discrete data elements.

There are two kinds of data elements—discrete and continuous. The distinction between the two is determined by the values for the data element. A data element is said to be **discrete** if each of its values is separate and distinct from the others and the values are discontinuous. Examples of different types of discrete data element values include: a set of integers, the alphabet, the set of English punctuation marks, and U.S. currency (dollars and cents). A data element is said to be **continuous** if for any two of its values, a third value can be found that lies between them. This implies that the number of values for a continuous data element is infinite. Examples of continuous data elements include: temperature, distance, and volume.

PERFORMING DATA DECOMPOSITION

As indicated in previous sections, data flows and data stores are frequently hierarchical structures. This section demonstrates some techniques of data decomposition used in developing the data dictionary.

A Comprehensive Example

Recall the Employee-Work-Schedule example from the section on describing nested data. There we performed a top-down decomposition. We perform similar decomposition in this section.

The example here deals with one of the principal financial statements in the FastFood Store system—the balance sheet.

A balance sheet for the FastFood Store is illustrated in Figure 7-10. This document is represented as a data flow in the system description, and is a system output.

This financial statement is based upon the fundamental accounting equation

Assets = liabilities + owners equity

A balance sheet reflects the assets of the business and the liabilities (the claims upon those assets) at a specific point in time, in this case, the year ended December 31, 198_. Claims upon assets include the claims of the owners and those of the creditors to whom the business has promised to pay money. The balance sheet is used to determine whether the assets of a business are adequate for the activities the business is carrying out, and whether the assets are adequate to satisfy all the claims upon them.

Where should we start to determine the data structure of this balance sheet? We observe that it is composed of a combination of letters and digits. A description such as that below is too detailed to be useful.

Balance-Sheet = {[Letter | Digit | Punctuation]}

Until the letters, digits, and punctuation marks are combined to form data element values, they have no meaning in relation to the system description.

The physical layout of the balance sheet in Figure 7-10 gives some clues to its hierarchical data structure. Note the use of centered titles to indicate the sections of the balance sheet

In Figure 7-11 we have circumscribed each major section to identify its boundary. Having made this initial partitioning, we can record the highest-level structure of the balance sheet. It is a sequence consisting of: the title block, the section for assets, and the section for liabilities and owners' equity. We could record this sequential data structure with the data dictionary notation in Example (7.8).

Balance-Sheet = Balance-Sheet-Title-Block & Assets-Section &
                 Liabilities-And-Owners'-Equity-Section   (7.8)

How should we continue from this initial partitioning? We could further subdivide each block or section in much the same manner. It probably makes little difference which section we decompose next. We continue with the decomposition of the Balance-Sheet-Title-Block.

Balance-Sheet-Title-Block = Balance-Sheet-Title & Business-Name &
                             Balance-Sheet-Date   (7.9)

The structure of Balance-Sheet-Title-Block could be described as shown in Example (7.9). But, suppose that the balance sheet were typed on a preprinted form where only the date needs to be supplied to complete the title block. Would it be important

FIGURE 7-10. A balance
sheet for the FastFood
Store.

The FastFood Store
Balance Sheet
on December 31, 198___

Assets

Current Assets				
Checking Account		1,000		
Passbook Savings Account		7,500		
Inventory		5,000		
Total Current Assets			13,500	
Fixed Assets				
Equipment	50,000			
Less: Accumulated Depreciation	22,000	28,000		
Building	10,000			
Less: Accumulated Depreciation	9,000	1,000		
Land		5,000		
Total Fixed Assets			34,000	
Total Assets				47,500

Liabilities & Owners' Equity

Current Liabilities				
Note Payable (New Refrigerator)	1,500			
Trade Accounts Payable	500			
Payroll Taxes Payable	0			
Total Current Liabilities		2,000		
Long-term Liabilities				
Mortgage: Land	600			
Mortgage: Building	1,200			
Total Long-term Liabilities		1,800		
Total Liabilities			3,800	
Owners' Equity				
Mr. & Mrs. O., Capital, Jan. 1		38,700		
Net Income for year ended Dec. 31	35,000			
Less: Withdrawals	<30,000>			
Change in Capital		5,000		
Mr. & Mrs. O., Capital, Dec. 31			43,700	
Total Liabilities & Owners' Equity				47,500

FIGURE 7-11. An initial partitioning of the balance sheet.

┌─────────────────────────────────┐
│   The FastFood Store   │
│   Balance Sheet   │
│   on December 31, 198___   │
└─────────────────────────────────┘

Assets

Current Assets			
Checking Account		1,000	
Passbook Savings Account		7,500	
Inventory		5,000	
Total Current Assets			13,500
Fixed Assets			
Equipment	50,000		
Less: Accumulated Depreciation	22,000	28,000	
Building	10,000		
Less: Accumulated Depreciation	9,000	1,000	
Land		5,000	
Total Fixed Assets			34,000
Total Assets			47,500

Liabilities & Owners' Equity

Current Liabilities			
Note Payable (New Refrigerator)	1,500		
Trade Accounts Payable	500		
Payroll Taxes Payable	0		
Total Current Liabilities		2,000	
Long-term Liabilities			
Mortgage: Land	600		
Mortgage: Building	1,200		
Total Long-term Liabilities		1,800	
Total Liabilities			3,800
Owners' Equity			
Mr. & Mrs. O., Capital, Jan. 1		38,700	
Net Income for year ended Dec. 31	35,000		
Less: Withdrawals	<30,000>		
Change in Capital		5,000	
Mr. & Mrs. O., Capital, Dec.31			43,700
Total Liabilities & Owners' Equity			47,500

to include the title and business name in the DD entry for the title block? What if a preprinted form is not used, but the complete balance sheet, headings and titles included, is printed by computer on a blank sheet of paper? The answers to these questions determine whether or not we need to include titles or identifying labels such as "Balance Sheet" or "The FastFood Store" in a DD definition. The working rule is: Do not include labels for the blanks or blocks on a preprinted form. These labels usually become the name of data flows or data elements and thus appear in the data dictionary. If in doubt about other titles, include them. You can always delete them later.

The question arises for the components of the Balance-Sheet-Title-Block, "Have we reached the data element level?" Balance-Sheet-Title and Balance-Sheet-Business-Name are obviously data elements. But what about Balance-Sheet-Date? Isn't it composed of the month, day, and year? Nevertheless, in this document it is not important that the date be subdivided; its components are always used together. Therefore, Balance-Sheet-Date is also a data element.

Figure 7-12 illustrated the decomposition of the assets section and also shows the data dictionary definition that corresponds to this partitioning. Total-Assets is a data element.

In the description of the current assets section, shown in Example (7.10), the account name is important. It distinguishes one current asset account from another.

Current-Assets-Section = { Current-Asset-Line } & Total-Current-Assets

Current-Asset-Line = General-Ledger-Current-Asset-Account-Name &
General-Ledger-Current-Asset-Account-Balance   (7.10)

FIGURE 7-12. Decomposition of the assets section of the balance sheet.

Assets

Current Assets		
Checking Account	1,000	
Passbook Savings Account	7,500	
Inventory	5,000	
Total Current Assets		13,500

Fixed Assets			
Equipment	50,000		
Less: Accumulated Depreciation	22,000	28,000	
Building	10,000		
Less: Accumulated Depreciation	9,000	1,000	
Land		5,000	
Total Fixed Assets			34,000

Total Assets	47,500

Assets-Section = Current-Assets-Section & Fixed-Assets-Section & Total-Assets

Note that all fixed assets are not reported in the balance sheet in the same manner. Some fixed assets are depreciable; some are not. In the data dictionary we can distinguish between depreciable and nondepreciable assets as depicted in Example (7.11).

Fixed-Assets-Section = { Fixed-Asset-Line } & Total-Fixed-Assets

Fixed-Asset-Line = General-Ledger-Fixed-Asset-Account-Name &
                    General-Ledger-Fixed-Asset-Account-Balance &
                    (General-Ledger-Fixed-Asset-Accumulated-Depreciation &
                    Balance-Sheet-Fixed-Asset-Balance)   (7.11)

This completes our decomposition of the assets section of the balance sheet. Decomposition of the liabilities and owners' equity section is left as an exercise (Exercise 7-4).

Decomposition and Data Elements

Recognizing data elements can be a problem. You might try the following guideline: When you make the transition from concern about data structure to concern for values of data items, you have reached the data element level.

Context quite often determines when something is considered a data element. For instance, consider a date. Under some circumstances we may consider this to be a data element. However, if we need to determine that the date lies within a particular season (Fall, Winter, Spring, Summer), then the date must be considered a data structure because we need to use the constituents Month and Day-Of-Month to determine the season.

Care must be exercised when assigning meanings to values. Consider Example (7.12).

Data Element: Age

Value	Meaning
1–11	Child
12–17	Adolescent
18–99	Adult

(7.12)

These ranges really refer to different age groups. The definition of Example (7.13) is more appropriate.

Data Element: Age

Values: 1 to 99

Data Element: Age-Group

Value	Meaning
1–11	Child
12–17	Adolescent
18–99	Adult

(7.13)

There is a subtle interaction between values and meanings. It has to do with the values or symbols used to convey meaning. This is illustrated in Example (7.14), where a meaning is associated with a single value.

Data Element Name: Sex

Value	Meaning
F	Female
M	Male
?	Unknown

(7.14)

We raise the point because a data element's set of values is more likely to change than its set of meanings. For example, we do not expect to find a new sex, but we can envision the substitution of new values for old ones or the assignment of existing meanings to new values, as in Example (7.15).

Data Element Name: Sex

Value	Meaning
F	Female
f	Female
M	Male
m	Male
?	Unknown
blank	Unknown

(7.15)

When defining a data element, an astute analyst makes a point of distinguishing the values (which are more subject to change) from the meanings.

Some Data Decomposition Guidelines

There are several concluding observations we would like to make with respect to top-down data decomposition.

Proceed in an orderly, incremental fashion. Concentrate on the data structure. Follow the boundaries indicated on the document, form, or report. Decompose hierarchically following these boundaries. For ease of reference, keep entries in the DD sorted alphabetically by name within each DD class.

Avoid overly nested structures. The example of complex nesting in Figure 7-13 does require fewer entries to describe part of a balance sheet. However, what has been sacrificed? It impedes understanding by disguising the hierarchical data structure. Furthermore, there are no named, intermediate, reusable data structures.

Lower-level structures should not be iterated, so that alternative higher-level structures may be composed from them.

A data structure that is not a data store is recorded using the data flow entry format. It is included in the data dictionary with the entries for data flows that appear somewhere on a DFD.

Proliferation of aliases can be a problem. They should be recorded only when

FIGURE 7-13. A deeply nested description of the composition of the balance sheet and a preferred decomposition.

Balance-Sheet = Title & Business-Name & Date &
{ Current-Asset-Account-Name &
Current-Asset-Account-Balance } &
{ Fixed-Asset-Account-Name &
Fixed-Asset-Account-Balance &
(Fixed-Asset-Accumulated-Depreciation &
Fixed-Asset-Net-Balance) } &
Total-Assets

A deeply nested description.

Balance-Sheet = Balance-Sheet-Title-Block & Assets-Section &
Liabilities-And-Owners'-Equity-Section

Balance-Sheet-Title-Block = Balance-Sheet-Title & Business-Name &
Balance-Sheet-Date

Assets-Section = Current-Assets-Section & Fixed-Assets-Section &
Total-Assets

Current-Assets-Section = { Current-Asset-Line } & Total-Current-Assets

Current-Asset-Line = General-Ledger-Current-Asset-Account-Name &
General-Ledger-Current-Asset-Account-Balance

Fixed-Assets-Section = { Fixed-Asset-Line } & Total-Fixed-Assets

Fixed-Asset-Line = General-Ledger-Fixed-Asset-Account-Name &
General-Ledger-Fixed-Asset-Account-Balance &
(General-Ledger-Fixed-Asset-Accumulated-Depreciation &
Balance-Sheet-Fixed-Asset-Balance)

A highly partitioned alternative.

the users cannot agree on a name. The analyst should discourage aliases. However, the analyst should not be pedantic about the issue.

Analysts should avoid the innocuous introduction of aliases. For example, consider the generic data structure

Date = Month & Day-Of-Month & Year

and its use in

Work-Date = Date
Day-Off-Date = Date
Work-Week-Start-Date = Date

While these definitions stress a structural similarity that **may** exist between Work-Date, Day-Off-Date, and Work-Week-Start-Date, they do not emphasize the semantic differences between them. Furthermore, some overzealous analyst might try to suggest that Work-Date, Day-Off-Date, and Work-Week-Start-Date are aliases of Date—clearly an erroneous suggestion. Work-Date, Day-Off-Date, and Work-Week-Start-Date are probably data elements. Hence, we do not care about their structure.

THE ORIGIN/DESTINATION DICTIONARY

An entry in the origin/destination dictionary is relatively simple. Figure 7-14 shows the format or template for an origin or destination as well as a completed example. At a minimum, the entry for an origin or destination must include its name. An origin or destination may have an alias. We may also wish to include some descriptive information about the role the external person or organization plays. If the external entity is an automated system, we may record other information for future use, such as the manufacturer of the computer hardware, the configuration of computer hardware, the type and version of the operating system, or specialized software packages such as data base management systems or accounting systems.

THE TRANSFORM DICTIONARY

An entry in the transform dictionary may consist of: the transform number and name, a specification of the rules for the transformation, and notes. Transforms do not have aliases. Figure 7-15 shows the template for an entry in the transform dictionary.

FIGURE 7-14. Format and example of the system dictionary entry for an origin or destination.

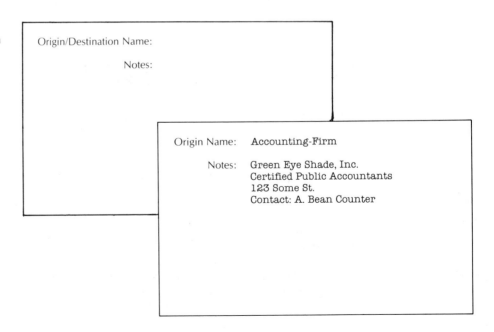

Origin/Destination Name:

    Notes:

Origin Name: Accounting-Firm

    Notes: Green Eye Shade, Inc.
    Certified Public Accountants
    123 Some St.
    Contact: A. Bean Counter

```
Transform Number:

  Transform Name:

     Description:




Notes:
```

FIGURE 7-15. Format of the system dictionary entry for a transform.

If a transform is a functional primitive, its definition will look like that of Figure 7-16. All nonprimitive transforms are ultimately decomposed into primitive transforms. The other transforms are simply aggregations of the primitive transforms. Thus a transform description for a nonprimitive transform is redundant, and the definition section is left blank. Some analysts prefer to omit entries for nonprimitive transforms. We prefer to include them for consistency and completeness checking.

The alternative forms for specifying transforms are subjects of Chapter 8.

SUMMARY

The system dictionary contains an entry for every named component in a set of data flow diagrams. Associated with each name is a definition appropriate to the specific type of component—data element, data flow, data store, origin, destination, or transform. The entries are organized into sections by component type. The data dictionary comprises the sections for data flows, data stores, and data elements; the origin/destination dictionary contains entries for external entities; and the transform dictionary contains entries for the transforms. Entries for data elements, data flows, data stores, origins, and destinations are in alphabetical order; entries for transforms are in sequence by bubble number.

FIGURE 7-16. An example of a system dictionary entry for a primitive transform.

```
Transform Number:   3.4.2

  Transform Name:   Determine Net Pay

     Description:   Net-Pay = Gross-Pay − Taxes −
                    Other-Deductions




Notes:
```

A data element is the primitive for data flows and data stores. The definition of a data element specifies its values and may include the meaning of those values. A data element is either discrete or continuous depending on the nature of its values.

A data flow is defined in terms of its composition. The constituents of a data flow are either subsidiary data structures or data elements. Sequence, selection, and iteration operations are used to combine these constituents. The data structuring operators provide a concise, symbolic notation for the operations.

The definition of a data store has two parts. It states the composition of a typical record of the data store and shows the key by which a specific record is uniquely identified and accessed.

The data dictionary should be developed from the top down, by decomposition of the data flows and data stores, usually in parallel with the data flow diagrams.

The entry for an origin or destination includes its name and perhaps a brief description.

The entry for a transform includes its number and name. Primitive transforms also have the rules for the transformation specified in their definition section. The other transforms are aggregations of the primitives.

REVIEW

7-1. State the two principal goals for the system dictionary.

7-2. Distinguish between a system dictionary entry and a system dictionary definition.

7-3. Name two things that should be included in a data dictionary entry for a data element.

7-4. Identify the three basic operations for expressing data structure. Give the concise and expanded notation for each.

7-5. What is the purpose of the braces and the underscore in a data store description?

7-6. Briefly state the guidelines for hierarchical data decomposition.

EXERCISES AND DISCUSSION

7-1. Identify which of the following are data elements and supply data element definitions for them:
   a. Marital-Status
   b. Primary-Color
   c. U.S.-Currency-Denomination
   d. Military-Time
   e. Latitude-And-Longitude

7-2. Given the definition of Date and each of its data elements below, identify which of the following instances are unacceptable.

$$Date = Month \& Day\text{-}Of\text{-}Month \& Year$$

Month =	Value	Meaning
	1	January
	2	February
	3	March
	. . .	

$$Day\text{-}Of\text{-}Month = Values: 1 \text{ through } 31$$
$$Year = Values: 1900 \text{ through } 1999$$

a. April 24, 1953
b. 4/24/53
c. 24 IV 53
d. 53-4-24
e. 53113 (where 113 indicates the 113th day of the year)

Expand the definition as necessary to accommodate the instances you deemed unacceptable.

7-3. Figure 7-17 is abstracted from a Request for Classes document that students use for pre-enrolling. Write a DD definition for this block of information. (**Hint:** Decide whether this is a data flow or a data element.)

7-4. Complete the decomposition of the liabilities and owners' equity section of the balance sheet of Figure 7-10.

7-5. Give a DD definition for a data store containing balance sheets such as that of Figure 7-10. Be sure to identify the key.

7-6. Indicate what is wrong with the following DD notation:

a. [Class]
b. [Class & Professor]

FIGURE 7-17. An excerpt from a request for classes document.

SCHEDULE ME AS FOLLOWS:	MARK ONLY ONE
DAY CLASS ONLY (All classes ending before 6:00 P.M.)	⓪
NIGHT CLASS ONLY (All classes starting 4:00 P.M. or later)	①
DAY AND NIGHT CLASS	②

7-7. Are the following equivalent? Write a one-sentence literal interpretation of each.

    a. 1 { Menu-Item } 2
    b. [Menu-Item | Menu-Item & Menu-Item]
    c. Menu-Item & (Menu-Item)

7-8. Explain the meaning of the following and suggest a simpler notation for each:

    a. (1 { Stock-Item } 10)
    b. 0 { Middle-Initial } 1

7-9. Having to maintain redundant information in a system model can compromise completeness, consistency, and correctness. We can avoid maintenance difficulties caused by redundancy by recalling the purpose of the system modeling tools:

- The ***routing*** of data is depicted on the DFD.
- The ***composition*** of data flows and data stores is described in the data dictionary.
- The ***rules*** for transforming data are prescribed in the transform dictionary.
- The ***names*** of producers and consumers of data are listed in the origin/destination dictionary.

Identify any violations of the above guidelines that appear in Figure 7-18. Suggest how each violation may be remedied.

Data Element Name: Shift

	Value	Meaning
	1	7 am to 4 pm
	2	4 pm to 11 pm.
	3	11 pm to 7 am

Notes: part of Work-Assignment.

Data Flow Name: Financial-Statement

Composition: [Balance-Sheet | Income-Statement]

Notes: input from Origin: Accounting-Firm
output to Destination: Owners

Data Element Name: Net-Pay

Value: $0.00 to $999.99

Notes: calculates as: Net-Pay =
Gross-Pay − Taxes − Other-Deductions

FIGURE 7-18. Some redundant notes in data dictionary entries.

8 Transform Description

Chapters 6 and 7 described two major parts of the structured specification—data flow diagrams and the system dictionary. Data flow diagrams specify the partitioning of the system being modeled, depicting all the system components. Entries in the system dictionary define or specify all these components. Chapter 7 discussed the organization of the system dictionary and presented the rules for the decompositon and definition of the data flows and data stores as well as the entries for data elements, origins, and destinations.

In a computer information system, the real work of transforming incoming data flows into outgoing data flows takes place in the transforms. Within the set of data flow diagrams, the primitive transforms shown on the bottom level are not further decomposed. These primitive transforms require specification of their processing logic. How we specify the processing logic for transforms is the subject of this chapter.

Before you say to yourself, "Finally, we get to look at some code!" remember that we are describing system requirements, not solutions to those requirements. Therefore we will not use code to represent processing logic. Rather, we will explore several alternative forms for expressing and presenting processing logic. These include textual, tabular, graphical, and hybrid forms.

GOALS AND OBJECTIVES FOR TRANSFORM DESCRIPTIONS

Transform descriptions, like the rest of the structured specification, state users' requirements to be met by the system designers and builders. In order for users to

118

be able to agree to the adequacy of a transform description as a requirements statement, it must be expressed understandably. In an automated information processing system, a transformation will be implemented as a group of instructions in a computer program.

From the point of view of the system designers and builders, a transform description provides the essentials of an **algorithm**—the step-by-step specification of the operations or actions which accomplish the transformation. The users of a business information system are accustomed to thinking about an algorithm in a somewhat different way and talk about it in different language.

In normal business parlance, the processing logic for a transformation involves both **procedure** and **policy.** In some organizations procedures and policies are formal and written; in others they are informal and unwritten. In most organizations, there is likely to be a mixture of formal and informal, written and unwritten. Business procedures define the details of actions taken to accomplish the tasks of operating the business. The procedures are developed from and guided by business policies—relatively high-level statements of the way management wishes to operate the business. These policies may specify the results desired in given situations or the rules for decision-making and action under a variety of circumstances expected in the normal course of business.

As logical descriptions of algorithms, transform descriptions incorporate features of both business procedure and policy. The steps in the algorithm must be complete and detailed enough to be the basis for writing a computer program. Yet business procedures exist in a specific organization with specific resources and technologies, and therefore include physical details of how the procedures are implemented. Analysts must capture the essentials of the algorithm in a way which is independent of the specific means by which the algorithm is implemented. A transform description must also capture the essentials of the business policy underlying the procedure. Thus the important rules for decision and action will be incorporated in the statement of the algorithm in a way which is consistent with the management's desires. Few businesses can afford to have important decisions made by default by a computer programmer and built into an information system because the rules governing a transformation were inadequately defined in the requirements specification.

The following objectives guide the development of transform descriptions.

- There is one transform description for each primitive transform.
- The transform should be described in a page or less.
- The transform description should emphasize what the transform is to do. This description should be independent of how the procedure or algorithm is to be implemented—the specific means or technique. It should also be independent of who is to carry out the procedure—the agent.
- The transform description should provide a complete, precise, unambiguous statement of the algorithm.
- The logic for the transformation should be stated in terms of a few minimal types of structural elements or building blocks.
- The transform description should not introduce any redundancy into the structured specification.

The sections that follow present the alternative forms for describing transformations. In the review at the end of the chapter we ask you to evaluate each description technique with respect to the objectives of transform description.

SOME TYPICAL QUESTIONS

Before proceeding, we would like to address some questions that typically arise here.

"Why don't you produce transform descriptions for all transforms?"

In the first place, it would require a lot of effort to prepare transform descriptions for each transform in all but the simplest system. After all, the systems you will model may require hundreds if not thousands of transforms. However, avoiding work is not sound justification for producing descriptions for only the primitive transforms.

Perhaps your question meant "Aren't you leaving something out?" To produce transform descriptions for all bubbles would be redundant. After all, any intermediate or high-level bubble merely represents the combination of all its descendants. Completely describing the primitive transforms completely specifies all the transformations performed in the entire system.

"Why don't you include the method or technique for implementing the transformation, or for that matter the agent responsible for implementing the transformation?"

Remember that the structured specification states a design problem; it defines the requirements that the designer's solution must satisfy. By specifying a method or technique, we encroach on the system design. It is the responsibility of the designer to decide how to implement a particular algorithm for a business procedure. Moreover, the method or technique for implementing a business procedure is more likely to change than either the essentials of the procedure or the underlying policy itself. The agent is also subject to change when there is a change of personnel, a reorganization of the business, or the introduction of automation. By capturing the essential policy basis for the algorithm, we maximize the flexibility of that portion of the structured specification containing the transform descriptions.

"Who's going to read these transform descriptions?"

We have already mentioned the system designer. Certainly, other analysts on a development team will read them. However, the most important readers of the transform descriptions will be the users who assess their adequacy and correctness. Recall the user-analyst dialogue of Chapter 6. There the point was made repeatedly that the tools of structured analysis are principally communication aids. This is true for transform descriptions.

ALTERNATIVES FOR POLICY DESCRIPTION

The alternative transform description techniques range from textual to tabular to graphical. In the sections that follow we describe four specific techniques and the details of each:

1. Structured English
2. Decision tables
3. The graphical equivalent of decision tables, decision trees
4. Hybrids of the previous three

STRUCTURED ENGLISH

Unrestricted natural language is not an appropriate tool for transform specification. DeMarco summarizes its inherent weakness as a specification tool:

> It is imprecise, wordy, redundant, and full of implications, connotations, and innuendo. . . .
> These very "weaknesses" are what give English its remarkable depth and vitality as a medium for artistic communication. But for specification, they only get in the way.*

Instead, we use Structured English—a subset of the English language that is pruned to minimize ambiguity and misinterpretation. Because Structured English is a form of natural language, it can communicate procedures to both nontechnical and technical people in a familiar medium. Because Structured English is restricted, it is better suited to the objectives of transform description, including clarity, precision, and lack of ambiguity.

What is Structured English?

We answer this question in three steps. First, what is structured about this form of transform description? Second, what are the English parts of this form of transform description? Third, we examine some examples of Structured English.

What is Structured about Structured English?

Around 1966 the work of Boehm and Jacopini laid the foundation for a "structure theorem."[†] This theorem guarantees that **any** flowchartable logic can be represented using **only three** types of control structures. This is in sharp contrast to the earlier practice of flowcharting with unrestricted control branching operations. The three control structures—sequence, selection (decision), and iteration (repetition)—and their flowcharts are illustrated in Figure 8-1.

*Tom DeMarco, **Structured Analysis and System Specification**, Yourdon Press, New York, N.Y., © 1978, p. 177. Reprinted by permission of Yourdon Press.

[†]C. Boehm and G. Jacopini, "Flow Diagrams, Turing Machines and Languages with only Two Formation Rules," **Communications of the Association for Computing Machinery**, 9, 5, May 1966, pp. 366–371.

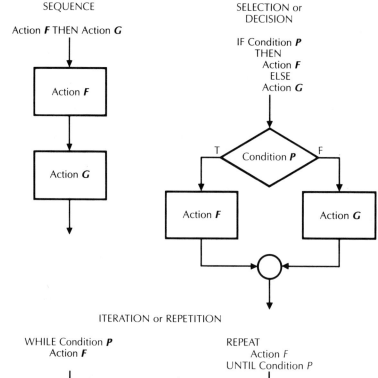

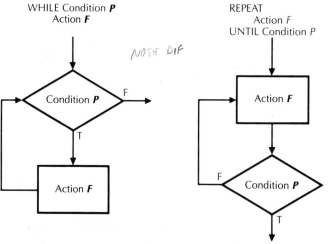

FIGURE 8-1. The three basic control structures.

F and *G* are actions or flowcharts with one entry point and one exit. *P* is a test.

Sequence. In the sequence control structure of Figure 8-1, first action *F* is performed, followed by action *G.* In principle, there could be an arbitrary number of actions performed in sequence. This is illustrated in Figure 8-2.

Selection. In the selection control structure, a condition is tested. The outcome of the test determines what action is to be performed next.

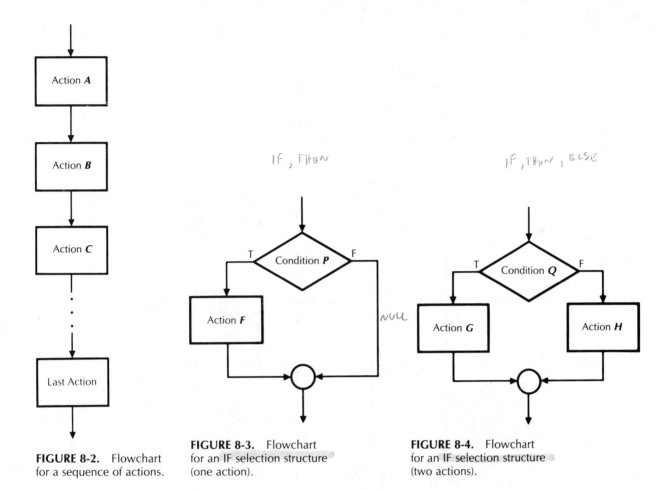

IF, THEN

IF, THEN, ELSE

NULL

FIGURE 8-2. Flowchart for a sequence of actions.

FIGURE 8-3. Flowchart for an IF selection structure (one action).

FIGURE 8-4. Flowchart for an IF selection structure (two actions).

With the decision structure, sometimes it is necessary to indicate that only one action is performed as a result of a decision; the other action is null. Thus, in Figure 8-3 the True path of condition **P** indicates that action **F** is performed. However, the False path indicates that no action is performed. This form of decision structure is called *if-then.* The structure ilustrated in Figure 8-4 is called *if-then-else*. When condition **Q** is True, action **G** is performed. When condition **Q** is False, action **H** is performed. Note that in the if-then-else structure the alternative paths are mutually exclusive. That is, either action **G** or **H** is performed, but not both.

Sometimes we have a number of alternative conditions and actions, as indicated in Figure 8-5.

In Figure 8-6 we have rearranged an equivalent layout of the flowchart symbols to illustrate condition-action pairs. For example: condition **P** is paired with action **F**; condition **Q** is paired with action **G**. Note that action **J** is done only if each of the conditions **P** through **S** is False. This situation arises often enough that we are

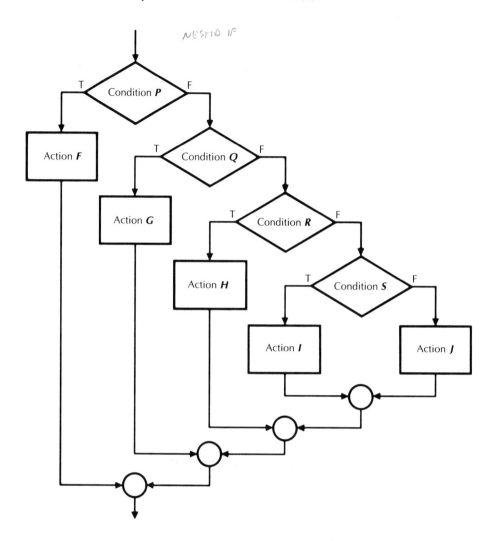

FIGURE 8-5. Flowchart for nested IFs.

justified in suggesting a concise form for what would otherwise be an awkward nested decision structure. We call this alternative the **case** structure. This term captures the essence of the decision structure—the existence of many possibilities, only one of which is applicable in a given situation.

Iteration. The iteration control structure causes an action to be repeated. How many times the action is repeated depends upon a condition which is tested at each iteration. This control structure is often called a **loop**; the action to be repeated is called the **body** of the loop. As Figure 8-1 shows, there are two forms depending on whether the test for terminating the iteration comes before or after the action is performed. The **while** structure tests the condition first, whereas the **repeat-until** structure tests the condition afterwards.

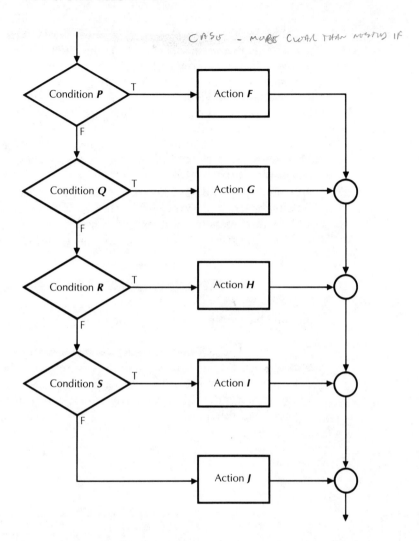

CASE — MORE CLEAR THAN NESTED IF

FIGURE 8-6. Flowchart for the CASE selection structure.

A natural consequence of the arrangement of the decision with respect to the action in the **repeat-until** structure is that the action will always be performed at least once. This is because the decision whether to terminate is placed after the action.

A natural consequence of the arrangement of the decision with respect to the action in the **while** structure (see Figure 8-1) is that the action may not even be performed once. This is because the condition for termination is evaluated before the action is performed. If the condition for termination is satisfied upon initial entry to the control structure, the exit path will be taken without activation of the action path.

In Figure 8-1, note the restrictions on the control structures involving the actions or flowcharts **F** and **G**. They may have only a single point of entry and a single exit. The flowchart for each control structure satisfies the restrictions on the number of entry and exit points. These restrictions and the fact that each control structure observes them have powerful consequences. They allow the control structures to be nested as demonstrated in Figure 8-7.

The action *F* in Figure 8-7 is composed of the sequence of actions *A, B, C,* and *D*. The action *G* is composed of the repetition of action *H*. Action *H* is itself a repetition of action *I*.

Note the boundaries indicated in Figure 8-7. Each boundary marks the **scope** of a control structure. The scope of a control structure consists of the actions and decisions included in the structure. It may be delimited by indicating where the structure begins and where it ends. As we will see later, it is important to clearly indicate the scope of each control structure when describing transform logic.

Thus three basic control structures and the ability to nest them provide the necessary constructs to express the flow of control for the logic of any transform description.

Expressing the Control Structures

Actions. An action is expressed in structured English by an ***imperative sentence***, or command. As illustrated by flowcharts earlier in this chapter, the control structures are built from actions and conditions.

Sequence. The sequence construct is composed of two or more actions performed in order. Figure 8-8 illustrates the simplest format for the sequence control structure and shows an example—two imperative sentences. The rectangle enclosing the actions indicates the scope of the construct. Scope is indicated in the same way in Figures 8-9 through 8-17.

FIGURE 8-7. Flowchart showing substitution of a control structure for an action.

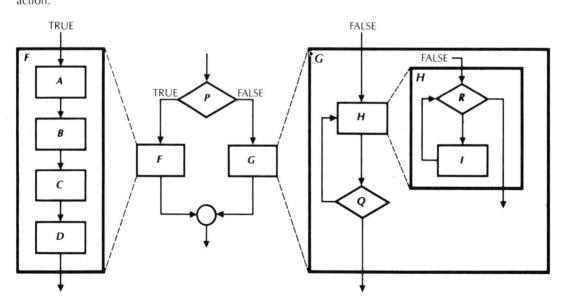

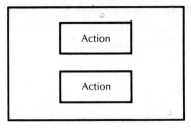

FIGURE 8-8. A sequence of actions with a structured English example.

Compute Sales-Tax-On-Item = Item-Cost × Sales-Tax-Rate.
Add Sales-Tax-On-Item to Sales-Tax-Subtotal

Selection. The formats for the *if-then* and the *if-then-else* decision structures and examples of their application are given in Figures 8-9 and 8-10. The condition to be tested is expressed in a *conditional clause*; the action is expressed by an imperative sentence. The use of a parenthetical comment in Figure 8-10 to explicitly

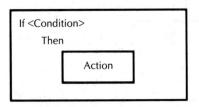

FIGURE 8-9. IF-THEN selection structure with a structured English example.

If the Item is taxable.
 Then
  Compute Sales-Tax-On-Item = Item-Cost × Sales-Tax-Rate.

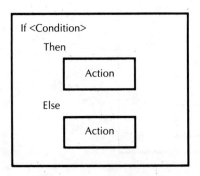

FIGURE 8-10. IF-THEN-ELSE selection structure with a Structured English example.

If the Fixed-Asset is depreciable.
 Then
  Calculate Balance-Sheet-Fixed-Asset-Value =
   General-Ledger-Fixed-Asset-Account-Balance −
   General-Ledger-Fixed-Asset-Accumulated-Depreciation.
 Else (the Fixed-Asset is not depreciable)
  Use General-Ledger-Fixed-Asset-Account-Balance for
   Balance-Sheet-Fixed-Asset-Value.

state the else condition is a good practice. Remember, we want to unambiguously communicate to the reader the rules governing the transform.

Conditional expressions are constructed using the relational operators and the Boolean operators summarized in Figure 8-11.

The format for the *case* selection structure is stated in Figure 8-12 along with

FIGURE 8-11.   The relational and Boolean operators with examples of conditional expressions.

Relational Operators	Boolean Operators	NOT T , F
> greater than	AND	
< less than	OR	
= equal to	NOT	

Previous-Quantity 0
Change-In-Quantity NOT = 0
New-Quantity = 0
Previous-Quantity 0 AND Change-In-Quantity = 0
Previous-Quantity NOT New-Quantity OR Change-In-Quantity = 0

FIGURE 8-12.   CASE selection structure with a Structured English example.

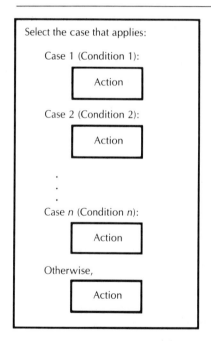

Select the case that applies:
    Case 1 (Asset is an Auto or Truck)
        Use 3 years for Asset-Useful-Life.
    Case 2 (Asset is Office-Furnishings or Office-Equipment)
        Use 5 years for Asset-Useful-Life.
    Case 3 (Asset is a Building)
        Use 10 years for Asset-Useful-Life.
    Otherwise.
        Use 5 years for Asset-Useful-Life.

an example of its application. Note how the pairing of conditions and actions demonstrated in Figure 8-6 is carried over and clearly indicated here. While it is not always necessary to provide the "Otherwise" alternative for understanding, it is a good practice to include it for completeness. It should be included even if the action is "Do nothing."

Iteration.   The format for the ***repeat-until*** iterative control structure and an example of its application are ilustrated in Figure 8-13.

The ***while*** iterative control structure is shown in Figure 8-14. Note how each of these control structures accurately reflects the position of the condition for termination with respect to the action. The condition for termination is tested ***after*** the body of the actions in the repeat-until structure; the termination condition is tested ***before*** the body in the while structure. (Compare Figures 8-13 and 8-14 with Figure 8-1.)

Quite often we are not concerned with explicitly stating the condition for termination or the position of the condition for termination (before or after the action). Figure 8-15 demonstrates the ***for-each*** version of the iteration construct. In this example it is implicit that the repetition stops when there are no more current accounts in the general ledger. This form of an iteration construct is preferred when we choose not to explicitly specify the condition for termination. Another variation of the iteration construct uses ***For all*** instead of ***For each***.

FIGURE 8-13.   REPEAT-UNTIL iteration structure with a Structured English example.

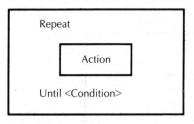

Repeat:
    Add General-Ledger-Current-Asset-Balance to
        Balance-Sheet-Total-Current-Assets.
Until there are no more General-Ledger-Current-Assets.

FIGURE 8-14.   WHILE iteration structure with a Structured English example.

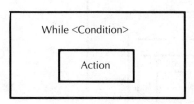

While there are more General-Ledger-Current-Assets in the General Ledger:
    Add General-Ledger-Current-Asset-Balance to
        Balance-Sheet-Total-Current-Assets.

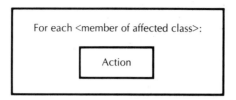

FIGURE 8-15. FOR EACH iteration structure with a Structured English example.

For each Current-Asset in the General Ledger:
    Add the Current-Asset-Account-Balance
        to the Balance-Sheet-Total-Current-Assets.

Combining the Control Structures

Control structures may be combined in the following way. Wherever an action appears in the formats shown in figures 8-8 through 8-15, it may be replaced by any one of the three control structures. The example of nesting shown as flowcharts in Figure 8-7 is shown in Structured English format in Figure 8-16.

Figure 8-17 presents a Structured English description of the computation of total cost of an order as well as a diagram of the Structured English constructs of which it is composed.

Indicating the Scope of the Control Structures

When writing transform descriptions in Structured English, it is especially important to make the scope of the constituent constructs clear. The examples thus far have used indentation to show subordination when constructs are combined. Figure 8-18 shows what Figure 8-17 would look like without indentation as well as several possible interpretations, each of which describes a different procedure. Avoiding

FIGURE 8-16. The structure of Structured English for the example of Figure 8-7.

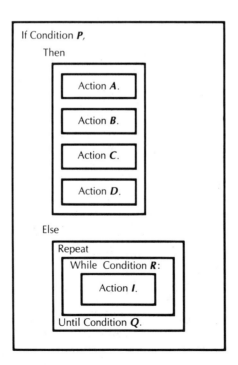

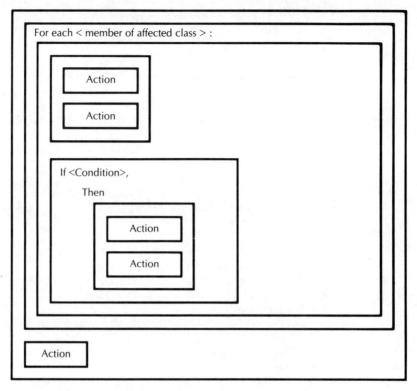

For each < member of affected class > :

> Action
>
> Action
>
> If <Condition>,
>> Then
>>> Action
>>>
>>> Action

Action

For each Item in an Order:

    Compute Item-Cost = Item-Quantity × Item-Unit-Price.
    Add Item-Cost to Order-Subtotal.
    If the Item is Taxable;
        Then
                Compute Sales-Tax-On-Item = Item-Cost × Sales-Tax-Rate.
                Add Sales-Tax-On-Item to Sales-Tax-Subtotal.
Compute Order-Total = Order-Subtotal + Sales-Tax-Subtotal.

FIGURE 8-17. A Structured English example of sequence, selection, and iteration.

confusion, ambiguity, and misinterpretation is the reason for making the structure of a transform description explicit by clearly stating the scope of each constituent construct.

There are several alternatives for explicitly indicating the scope of each control structure. They are stated below in order of preference.

1. Use indentation to show subordination.

2. Use a hierarchical numbering scheme similar to that used for numbering diagrams and transforms.

3. Use labels to mark the end of the control structures: ***end-if, end-case, end-for, end-while, end-repeat***.

4. Draw a box around each control structure. Figure 8-17 shows the use of indentation; Figure 8-19 shows the other alternatives for the same example.

FIGURE 8-18.   Alternative interpretations of an ambiguous transform description.

For each Item in an Order:
Compute Item-Cost = Item-Quantity × Item-Unit-Price.
Add Item-Cost to Order-Subtotal.
If the Item is Taxable,
Then
Compute Sales-Tax-On-Item = Item-Cost × Sales-Tax-Rate.
Add Sales-Tax-On-Item to Sales-Tax-Subtotal.
Compute Order-Total = Order-Subtotal + Sales-Tax-Subtotal.

For each Item in an Order:
   Compute Item-Cost = Item-Quantity × Item-Unit-Price.
Add Item-Cost to Order-Subtotal.
If the Item is Taxable,
   Then
      Compute Sales-Tax-On-Item = Item-Cost × Sales-Tax-Rate.
Add Sales-Tax-On-Item to Sales-Tax-Subtotal.
Compute Order-Total = Order-Subtotal + Sales-Tax-Subtotal.

For each Item in an Order:
   Compute Item-Cost = Item-Quantity × Item-Unit-Price.
   Add Item-Cost to Order-Subtotal.
If the Item is Taxable,
   Then
      Compute Sales-Tax-On-Item = Item-Cost × Sales-Tax-Rate.
      Add Sales-Tax-On-Item to Sales-Tax-Subtotal.
      Compute Order-Total = Order-Subtotal + Sales-Tax-Subtotal.

For each Item in an Order:
   Compute Item-Cost = Item-Quantity × Item-Unit-Price.
   Add Item-Cost to Order-Subtotal.
   If the Item is Taxable,
   Then
      Compute Sales-Tax-On-Item = Item-Cost × Sales-Tax-Rate.
      Add Sales-Tax-On-Item to Sales-Tax-Subtotal.
      Compute Order-Total = Order-Subtotal + Sales-Tax-Subtotal.

Indentation is preferred because it is a natural and obvious visual indication. Furthermore, it is easy to implement and maintain. Hierarchical numbering is only slightly less preferable because of its tendency to clutter the presentation. The use of labels or boxes is much less desirable than either indentation or numbering, though for different reasons. Labels make Structured English look like something it is not—a computer program. Drawing boxes indicates the scope clearly, but is difficult to maintain, and the boxes may interfere with reading the transform description.

1. For each Item in an Order:
   1.1 Compute Item-Cost = Item-Quantity × Item-Unit-Price.
   1.2 Add Item-Cost to Order-Subtotal.
   1.3 If the Item is Taxable,
         Then
                 1.3.1 Compute Sales-Tax-On-Item = Item-Cost × Sales-Tax-Rate.
                 1.3.2 Add Sales-Tax-On-Item to Sales-Tax-Subtotal.
2. Compute Order-Total = Order-Subtotal + Sales-Tax-Subtotal.

   For each Item in an Order:
         Compute Item-Cost = Item-Quantity × Item-Unit-Price.
         Add Item-Cost to Order-Subtotal.
         If the Item is Taxable,
            Then
                 Compute Sales-Tax-On-Item = Item-Cost × Sales-Tax-Rate.
                 Add Sales-Tax-On-Item to Sales-Tax-Subtotal.
         End-If
   End-For (Item)
   Compute Order-Total = Order-Subtotal + Sales-Tax-Subtotal.

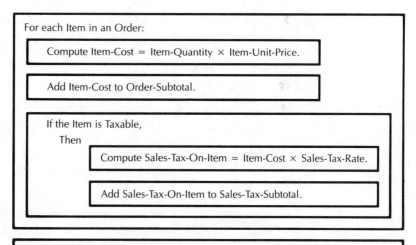

FIGURE 8-19. Alternatives for clear specification of the scope of the control structures.

The English Parts of Structured English

The English parts of Structured English can be partitioned into two groups—those that are system-related and those related to flow of control. The system-related vocabulary includes the names of data elements, data flows, data stores, origins, and destinations and the imperative verbs used in describing actions. The control-related vocabulary consists of words used to identify the control structures, the relational operators, and the Boolean operators. If you choose to use scope labels to mark the end of control structures, they are also included in the control-related vocabulary.

Those parts of the English language missing from Structured English are

- most qualifiers (adjectives, adverbs)
- compound sentence structures
- all verbs except transitive active imperatives
- expletives, articles, prepositions, and most conjuctions

Thus, Structured English is a terse form of natural language. It consists of

- a small group of words used consistently for stating the control structures—sequence, selection (decision), and iteration (repetition). These are: **If, Then, Else, Select, Case, For Each, Repeat, Until,** and **While.** Some practitioners prefer to capitalize every letter in these words.
- your choice of transitive active verbs for stating actions in transformations, such as **Compute, Sum, Type, Print, Write, Copy, Read, Update, Retrieve.**
- nouns taken from the data dictionary. These nouns are the names of data flows, data stores, and data elements, and perhaps meanings of the values of some discrete data elements.
- adjectives taken from the DFD. These adjectives are likely to be meanings of the values of some discrete data elements. For example, **New, In Good Standing, Overdue, Delinquent,** and **Closed** are adjectives describing **Customer-Account-Status.**
- the relational operators $>$, $<$, and $=$.
- the Boolean operators **And, Or,** and **Not.**

Some Structured English Examples

This section gives a detailed example of Structured English for the Transform 1.2.2: Determine Order Quantity. This primitive transform is shown in Figure 8-20 with some relevant data dictionary entries. A sample purchase order form is given in Figure 8-21. Figure 8-22 shows Structured English for Transform 1.2.2.

First, note the data flows entering and leaving Bubble 1.2.2. We expect to see them (or their data elements) manipulated somewhere in the transform description. This expectation is borne out by the description in Figure 8-22, which refers to Change-In-Quantity and Order-Quantity.

An analyst should develop the ability to write Structured English in a variety of styles as necessary to communicate better with users. We conclude this discussion of Structured English examples with an illustration of a more relaxed style, less like a computer program. In Figure 8-23 we have dropped the use of hyphens and capital letters in DD names and made the sequential statements flow more naturally by replacing the relational operators by English phrases, by adding some articles ("an" and "the"), and by referring to increases and decreases in quantity. However, we have retained the indentation to indicate the major control structures. This transform description has the same structure and the same rigor as that shown in Figure 8-22, but will be easier for many users to understand. It may require some slight extra effort on the part of the system designers and constructors to correlate with the data dictionary.

DIAGRAM 1.2: PRODUCE PURCHASE ORDER

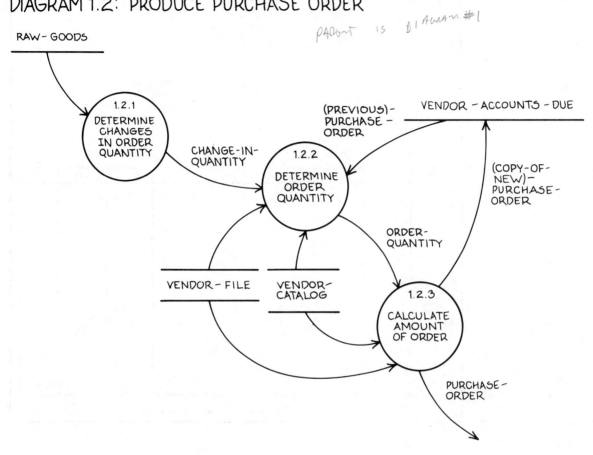

(handwritten annotation: PARENT IS DIAGRAM #1)

Vendor-Catalog = { <u>Vendor-Name</u> & { <u>Vendor-Stock-Number</u> &
    Description & Unit-Of-Measure &
    Minimum-Order-Quantity & Unit-Price } }

Vendor-File = { <u>Vendor-Name</u> & Vendor-Address & Vendor-Telephone &
    Vendor-Representative }

Purchase-Order = P-O-Date & P-O-Vendor-Block & { P-O-Line } & P-O-Total *136*

P-O-Vendor-Block = P-O-Vendor-Name & P-O-Vendor-Address & P-O-Vendor-Representative

P-O-Line = P-O-Line-Quantity & P-O-Line-Stock-Number &
    P-O-Line-Description & P-O-Line-Unit-Price & P-O-Line-Total

FIGURE 8-20. Diagram
1.2: Produce Purchase
Order with the
compositions of selected
data flows and data stores.

> PURCHASE ORDER
> The FastFood Store
> 123 Some St.
> Anytown, Ca.
> Tel. 714-555-1212
>
> *TITLE BLOCK FORGET IT*
>
> To:
>
> PO Number: 4321
> Date:
>
> Attn:
>
> Please enter our order for the following:

Quantity	Stock No.	Description	Units	Unit Price	Total

Authorized Signature _____

FIGURE 8-21. A purchase order form for the FastFood Store.

Guidelines for Structured English

We summarize the guidelines for Structured English as follows:

- With respect to vocabulary,

  Use names from the data dictionary.
  Except for the most relaxed structured English, retain the hyphens between words and the capital letter for each word of a data element name or data flow name.

- With respect to the control structures,

  The logic for any transformation should be described using a combination of only the sequence, selection (decision), and repetition (iteration) control structures.
  Use a consistent set of words to state these control structures.
  Keep punctuation to a minimum as illustrated in the formats for each control structure.

FIGURE 8-22. A transform description for Transform 1.2.2: Determine Order Quantity.

Description for Transform 1.2.2: Determine Order Quantity

For each Vendor in the Vendor-File: KEY
   Retrieve Previous-Purchase-Order from Vendor-Accounts-Due by Vendor-Name.
   For each item supplied by a Vendor:
     Select the case that applies:
       Case 1 (Previous-Quantity = 0 and Change-In-Quantity = 0):
         Skip this item.
       Case 2 (Previous-Quantity > 0 and Change-In-Quantity = 0):
         Order-Quantity is Previous-Quantity.
       Case 3 (Previous-Quantity = 0 and Change-In-Quantity > 0):
         Access Vendor-Catalog by Vendor-Name and Vendor-Stock-Number.
         Order-Quantity is Minimum-Order-Quantity.
       Case 4 (Previous-Quantity > 0 and Change-In-Quantity > 0):
         Compute Order-Quantity = Previous-Quantity + Change-In-Quantity.
       Case 5 (Previous-Quantity = 0 and Change-In-Quantity < 0):
         Skip this item.
       Case 6 (Previous-Quantity > 0 and Change-In-Quantity < 0):
         Compute New-Quantity = Previous-Quantity + Change-In-Quantity.
         Access Vendor-Catalog by Vendor-Name and Vendor-Stock-Number.
         If New-Quantity < Minimum-Order-Quantity,
           Then
             Skip this item.
           Else (New-Quantity not < Minimum-Order-Quantity)
             Order-Quantity is New-Quantity.

Indicate scope and nesting by indentation, hierarchical numbering, labels, boxes, or a combination of these.

- With respect to conditional expressions,

 Use parenthetical expressions to clarify the condition associated with the else part of an if-then-else structure and to state the condition for each case in a case structure.

 As Victor Weinberg suggests: "Do not use non-negative logic wherever possible."[*] Positive logic is simpler.

 Use nested decision structures instead of complex conditional expressions. Complex conditional expressions should be avoided. Break compound expressions into combinations of simpler structures. (Or use a decision table or decision tree.)

[*]Victor Weinberg, **Structured Analysis,** Yourdon Press, New York, © 1979, p. 115. Reprinted by permission of Yourdon Press.

FIGURE 8-23.   An alternative description for Transform 1.2.2: Determine Order Quantity.

Description for Transform 1.2.2: Determine Order Quantity

For each vendor in the vendor file:
   Retrieve the previous purchase order from the vendor accounts due by the vendor's name.
   For each item that is supplied by that vendor:
    Select the case that applies:
      Case 1 (The previous quantity was 0 and the change in quantity is 0):
        Skip this item.
      Case 2 (The previous quantity was greater than 0 and the change in quantity is 0):
        The order quantity is the same as the previous quantity.
      Case 3 (The previous quantity was 0 and there is an increase in the quantity):
        Access the vendor catalog by the vendor's name and the stock number of the item.
        The order quantity is the minimum order quantity.
      Case 4 (The previous quantity was greater than 0 and there is an increase in the quantity):
        The order quantity is the previous quantity plus the change in quantity.
      Case 5 (The previous quantity was 0 and there is a decrease in the quantity):
        Skip this item.
      Case 6 (The previous quantity was greater than 0 and there is a decrease in quantity):
        Compute the new quantity by adding the previous quantity and the change in quantity.
        Access the vendor catalog by the vendor's name and the stock number of the item.
        If the new quantity is less than the minimum order quantity,
          then skip this item.
        Otherwise, the order quantity is the new quantity.

In summary, Structured English should read like sensible English; transform descriptions should be revised until they communicate effectively.

DECISION TABLES

Decision tables (and their counterparts, decision trees) are special-purpose means of transform description. This is in contrast to Structured English, which is capable of describing the logic for any transform because it includes a way of stating all three of the control structures used in transform description. Decision tables, as their name implies, show only decision constructs. A **decision table** presents a decision control structure in tabular form. It is especially well suited to specifying what actions to take when complex combinations of conditions arise in a decision-making situation.

The decision table in Figure 8-24 shows the rules for paying a claim covered by a medical insurance policy. After a deductible amount (specified in the policy) has been satisfied, the insured is reimbursed a stated percentage of the medical expenses covered under the policy. This percentage varies depending on whether the expense is for an office visit, hospital services, or laboratory tests. This example will be used to help explain the organization of a decision table.

	1	2	3	4	5	6
Deductible Satisfied?	X		X		X	
Type of Covered Expense						
Office Visit	X	X				
Hospital Visit			X	X		
Laboratory Work					X	X
Percentage Paid						
0%		X		X		X
50%	X					
80%			X			
100%					X	

FIGURE 8-24. Example of a limited-entry decision table.

The structure of a decision table is shown in Figure 8-25. It consists of four major parts. The upper left quadrant is called the **condition stub.** In it are listed the conditions or decision variables for the policy.

Condition Stub	Condition Matrix
Action Stub	Action Matrix

FIGURE 8-25. The structure of a decision table.

The upper right quadrant is called the **condition matrix.** It contains entries for the values associated with each condition or decision variable. The lower left quadrant is the **action stub**. It contains a list of the independent actions for the decision policy. Entries in the **action matrix**, the lower right quadrant, link the actions to combinations of conditions shown in the condition matrix.

A **decision rule** is a combination of conditions and the actions to be taken when that combination of conditions is True. A decision rule is represented as a column of a decision table. For example, rule 3 of Figure 8-24 states that when the deductible has been satisfied and the covered expense was incurred in the hospital, the percentage of reimbursement is 80 percent.

Sometimes all conditions are stated as questions requiring a YES or NO answer. In this case, the entries in the condition matrix are limited to the values YES and NO and the decision table is called a **limited-entry decision table**. Figure 8-24 is an example of a limited-entry decision table.

At other times, each condition is stated in terms of values of a ***decision variable.*** This permits each decision variable and the corresponding entries in the condition matrix to take on more than two values. Therefore, the resulting table is called an ***extended-entry decision table***. As Figure 8-26 shows, an extended-entry decision table is more compact than a limited-entry decision table.

	1	2	3	4	5	6
Deductible Status	S	N	S	N	S	N
Type of Covered Expense	O	O	H	H	L	L
Percentage Paid	50	0	80	0	100	0

S   Deductible Satisfied
N   Deductible Not Satisfied

O   Office Visit
H   Hospital Visit
L   Laboratory Work

FIGURE 8-26. An extended-entry decision table for the example of Figure 8-24.

A decision table shows what actions are to be taken for all possible combinations of conditions. Thus there are as many decision rules as there are distinct combinations of conditions. How many is that? Each combination of conditions includes one value for each decision variable. Thus the number of combinations and the number of rules is the product of the number of values each variable can take on, for all of the decision variables.

For example, if there are three decision variables, there will be three terms in the product. If the first variable has three possible values, the second has two, and the third has four values, there will be

$$3 \times 2 \times 4 = 24$$

combinations and 24 decision rules, requiring 24 columns in the decision table.

Relation of Decision Variables to the Data Dictionary

How are the decision variables in a decision table related to the data flow diagrams and the data dictionary? Decision variables or names used in stating conditions are data element names and appear in the data dictionary. We expect them, like other information required by a transform, to be contained among the data flows coming into the transform.

Note that a decision table is based upon conditions which take on a limited set of values. This means that the decision variables in a decision table are expressed as discrete data elements. When a data element used in a decision table takes on continuous values, the range of values must be divided into segments with clearly defined boundaries. Each segment then behaves as if it were a single, discrete value. For example, in Figure 8-27 the decision variable Age is used to determine whether a person is a minor or an adult. Its range of values, from 0 to perhaps 120, has been divided into two segments or intervals.

FIGURE 8-27. A continuous data element used as a decision variable.

Constructing a Decision Table

The structure of a decision table leads to the following orderly procedure for its construction:

1. Identify all the decision variables and the possible values for each.

In some cases, a decision is based upon a condition with a YES or NO answer. In other cases, the decision variable, as mentioned before, is related to a data element. The values for a discrete data element are relatively easy to identify. For a data element with continuous values, it is necessary to divide the range of values into the segments upon which the decision is based. Be sure that a value at a segment boundary is assigned to one of the adjacent segments, so that there will be no ambiguity in specifying the policy. In Figure 8-27 the value 21 is such a boundary point.

Is someone 21 years old a minor or an adult? If we specify the two intervals above as

$$\text{Age} < 21 \quad \text{and} \quad \text{Age} > 21$$

the value 21 is not accounted for. If we specify

$$\text{Age} \le 21 \quad \text{and} \quad \text{Age} \ge 21$$

there is an inconsistency—an ambiguity as to which policy applies to a person 21 years old. In this case the appropriate choice is

$$\text{Age} < 21 \quad \text{and} \quad \text{Age} \ge 21.$$
$$\quad\text{Minor} \qquad\qquad \text{Adult}$$

For the medical insurance claim example, there are two decision variables: Deductible-Status (with values Satisfied and Not Satisfied) and Type-Of-Covered-Expense (with values Office Visit, Hospital Visit, and Laboratory Work).

Now fill out the condition stub.

Each condition or decision variable is listed in a separate row.

2. Determine the number of rules.

Compute the product of the number of values for each of the decision variables. This determines the number of columns in the table.

There are two values of Deductible-Status and three of Type-of-Covered-Expense. Thus there are 6 decision rules—the product of 2 and 3.

3. *Systematically generate all possible combinations of values of the decision variables.*

Any systematic procedure will do. The most straightforward approach takes one variable at a time and replicates patterns of combinations of values until all the combinations have been generated. The procedure is best explained by an example. We begin with Deductible Status. Its values are Satisfied and Not Satisfied (S and N). Example (8.1) shows the initial pattern.

Deductible Status   S   N   (8.1)

The next decision variable, Type-of-Covered-Expense, can have values Office Visit, Hospital Visit, and Laboratory Work. Since there are three values, we make three copies of the pattern SN, Example (8.2). If there were more decision variables, we would continue the processs of repeating the pattern.

Deductible Status   S   N   S   N   S   N   (8.2)

The value O is placed underneath the first copy; the value H under the second; and the value L under the third (see Example (8.3)).

Deductible Status   S   N   S   N   S   N
Type of Covered Expense   O   O   H   H   L   L   (8.3)

Now fill out the condition matrix. The procedure above has just generated it.

4. *Identify the independent actions and fill out the action stub.*

Requiring the actions to be independent keeps the action stub small. Other actions can then be expressed as combinations of the independent actions.

5. *Identify the action or combination of independent actions to be taken for each combination of conditions or decision variables. Fill in the action matrix one column at a time.*

Note that in the limited entry decision table in Figure 8-24, three actions are shown. In Figure 8-26, however, the action stub is expressed as a single variable, Percentage Paid, and its values entered in the action matrix to produce a more compact table.

In the process of producing a decision table, you may discover problems with the underlying user policy or with your understanding of it. That is one of the reasons for specifying user policies explicitly. Actions may be missing for some of the rules. Users may specify several different or even conflicting actions for other rules. There may also be areas of ambiguity. This is an important characteristic of a decision table—it requires explicit consideration of what to do for every condition that may occur. Closer examination may determine that in some combinations of

	1	2	3	4	5	6
Deductible Satisfied?	Y	N	Y	N	Y	N
Type of Covered Expense						
Office Visit	Y	—	N	—	N	—
Hospital Visit	N	—	Y	—	N	—
Laboratory Work	N	—	N	—	Y	—
Percentage Paid						
0%		X		X		X
50%	X					
80%			X			
100%					X	

FIGURE 8-28. The medical payments decision table with indifference shown.

conditions, some of the decision variables do not affect the action to be taken. For example, if the deductible has not been satisfied, no payment is made regardless of the type of expense. We say that the percentage of reimbursement is **indifferent** to the value of Type of Expense. To show this in the decision table, we may use the hyphen as an indifference symbol, as shown in Figure 8-28.

6. *Formulate questions for the users so that they can resolve omissions, inconsistencies, or ambiguities in the decision policy.*

7. *Revise and complete the decision table based upon the users' answers to your questions.*

Another Decision Table Example

Figure 8-29 shows another example of a decision table. It presents the FastFood Store's policy for deciding how much to change the previous order quantity when reordering food or supplies. For the FastFood example, there are three decision variables: Recent-Sales-Activity (with values Fast, Normal, and Slow), Inventory-Level (High, OK, or Low), and Perishable (Yes or No). There are three values of Recent-Sales-Activity, three of Inventory-Level, and two of Perishable. Thus there are 18 decision rules—the product of 3, 3, and 2. The action stub contains the variable Percent-Change, the values of which are entered in the action matrix as shown.

FIGURE 8-29. Extended-entry decision table for changing the order quantity.

	1	2	3	4	5	6	7	8	9	10	11	12	13	14	15	16	17	18
Recent Sales Activity	F	N	S	F	N	S	F	N	S	F	N	S	F	N	S	F	N	S
Inventory Level	H	H	H	O	O	O	L	L	L	H	H	H	O	O	O	L	L	L
Perishable	Y	Y	Y	Y	Y	Y	Y	Y	Y	N	N	N	N	N	N	N	N	N
% Change	0	−10	−15	0	0	−5	+5	+5	0	+5	−5	−10	+10	0	−10	+15	+5	0

DECISION TREES

A *decision tree* presents a decision control structure in the form of a tree. It is most appropriate for situations in which each rule has a single action. It represents a policy as a *sequence of decisions*.

The tree appears horizontally with the root at the left. The root represents the first decision and corresponds to the first decision variable. The outcomes of the decision (values of the variable) are shown as branches from the root. Each node of the tree represents a decision point, with each branch out of the node corresponding to one possible value of the associated decision variable. The leaves of the tree represent the actions. A path from the root to a leaf is associated with a specific sequence of decisions, corresponding to a combination of values of all the decision variables. Thus, *each path represents a decision rule*.

Figure 8-30 shows the decision tree for the medical insurance reimbursement rules in Figures 8-24 and 8-25.

Constructing a Decision Tree

Constructing a decision tree is also a straightforward process.

1. Identify all the decision variables and the possible values for each.

This is the same as the first step for constructing a decision table.

2. Generate the tree systematically from left to right.

   a. Take the first decision variable, draw the root, and label it with the name of the first decision variable.
   b. Draw one branch out of the root for each value of the first decision variable, and label each branch with the associated value. (See Figure 8-31.)
   c. The right end of each branch is now a decision point. Take the next decision variable. At each decision point, draw one branch for each possible value of the decision variable, and label it with the associated value.
   d. Repeat step c until there are no more decision variables.

The tree is now complete.

FIGURE 8-30. The medical payments decision tree.

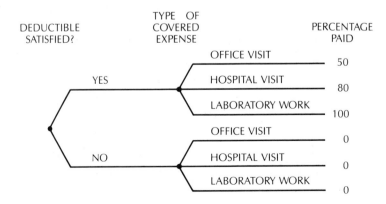

DEDUCTIBLE SATISFIED?	TYPE OF COVERED EXPENSE	PERCENTAGE PAID
YES	OFFICE VISIT	50
	HOSPITAL VISIT	80
	LABORATORY WORK	100
NO	OFFICE VISIT	0
	HOSPITAL VISIT	0
	LABORATORY WORK	0

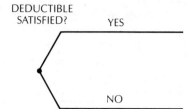

FIGURE 8-31. Initial step in generating the medical payments decision tree.

3. *Determine the number of decision rules and check to see if they are all shown in the tree.*

This step is the same as step 2 for a decision table.

4. *Identify the independent actions, and label each leaf with the actions to be taken.*

If there are more than one action, the actions must be listed at the end of each leaf. Many actions per rule will make a decision tree hard to read.

The next two steps are the same as those for producing a decision table.

5. *Formulate questions that will enable the users to resolve omissions, inconsistencies, and ambiguities in the decision policy.*

6. *Revise and complete the decision tree based on the users' answers to your questions.*

7. *Consider pruning or simplifying the tree.*

Figure 8-32 shows a decision tree for the FastFood Store policy for determining the change in the reorder quantity. This policy was shown as a decision table in Figure 8-29.

Once we have determined that Recent-Sales-Activity has been Normal and that the Inventory-Level is OK, it makes no difference whether the item to be ordered is Perishable or not. In either case, the required action is No Change in the order quantity. The case of Slow Recent-Sales-Activity and Low Inventory-Level is similar, as shown in Figure 8-30. These irrelevant decisions can be eliminated to produce the simplified tree in Figure 8-33. Note that we could not tell whether these decisions made a difference until we had looked at all the combinations of conditions.

Decision tables may also be reduced in a similar way. For example, the fourth and sixth columns in Figure 8-28 could be eliminated, since they are identical to the second column.

An inherent danger in this kind of simplification is that, if the policy is changed, it is easy to overlook the consequences of the decisions eliminated from the original tree or table.

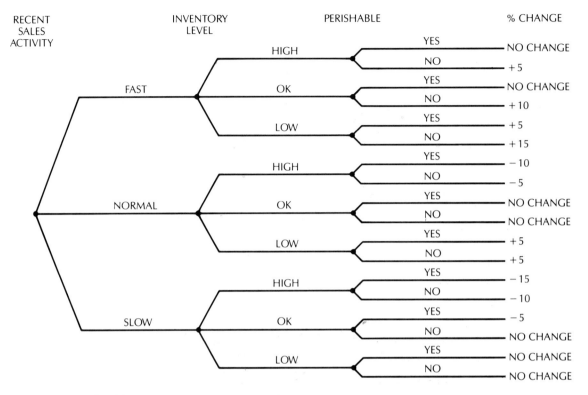

FIGURE 8-32. Decision tree for changing the order quantity.

When to Use a Table and When to Use a Tree

It should be clear from the preceding discussion that the major advantage of a decision table is the systematic consideration of all the conditions and associated actions involved in a decision. For most users, however, the sequential and graphic character of a decision tree makes it a more effective way to present a decision policy.

These observations lead to the following guidelines for deciding when to use a decision table and when to use a decision tree.*

Guideline 1: *Use a decision table when there are many decision variables* to ensure that all the combinations of conditions have been considered.

Guideline 2: *Use a decision table when the decision rules require combinations of actions.*

More than a few actions for a single rule are difficult to depict clearly in a decision tree.

Guideline 3: *Use a decision tree when there are few decision variables and only one action per rule.*

*See Gane and Sarson, *Structured System Analysis: Tools and Techniques,* Prentice-Hall, Inc., Englewood Cliffs, N.J., © 1979, p. 95. Adapted by permission of Prentice-Hall, Inc.

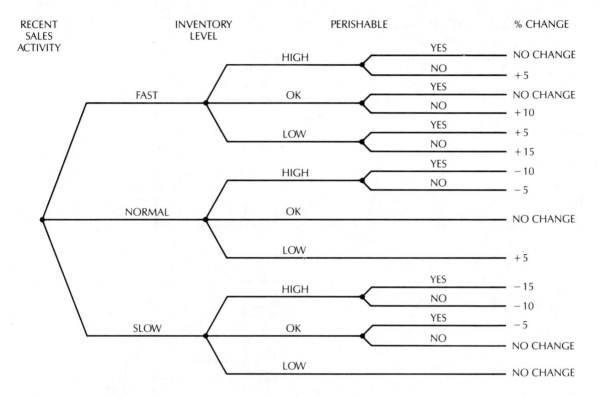

FIGURE 8-33. Simplified decision tree for changing the order quantity.

Guideline 4: *Use a decision table as a working tool whenever you need it to check all the combinations of conditions, but present the policy as a decision tree when it conforms to Guideline 3.*

If there are many conditions, so that a decision table or decision tree becomes too complex, decompose. First, try to decompose the transform. If that is impossible, the decision table itself can be partitioned.

Remember that the final criteria for all transform descriptions are precision of specification and ease of understanding.

HYBRID TRANSFORM DESCRIPTIONS

Hybrid transform descriptions combine Structured English with decision tables or decision trees.

Figure 8-34 gives a transform description that is equivalent to that in Figure 8-29. In this example a decision table is substituted for the case structure. A decision tree embedded within Structured English forms another kind of hybrid transform

Transform Description for; 1.2.2: Determine Order Quantity

For Each Vendor in the Vendor-File:

Retrieve Previous-Purchase-Order from Vendor-Accounts-Due by Vendor-Name.

For each item supplied by a Vendor:

	1	2	3	4	5	6
PREVIOUS QUANTITY	0	>0	0	>0	0	>0
CHANGE IN QUANTITY (N: NO CHANGE, I: INCREASE, D: DECREASE	N	N	I	I	D	D
SKIP THIS ITEM	X				X	
ORDER QUANTITY IS PREVIOUS QUANTITY		X				
ORDER QUANTITY = PREVIOUS QUANTITY + CHANGE IN QUANTITY				X		
ORDER QUANTITY IS MINIMUM ORDER QUANTITY			X			
IF PREVIOUS QUANTITY + CHANGE IN QUANTITY IS LESS THAN MINIMUM ORDER QUANTITY THEN SKIP THIS ITEM						X
ELSE ORDER QUANTITY IS MINIMUM ORDER QUANTITY						X

FIGURE 8-34. Hybrid transform description for determining the order quantity.

description. Alternatively, imperative sentences may be embedded in a decision tree or table to state the actions.

Hybrid transform descriptions provide additional flexibility to specify policy clearly and effectively.

SUMMARY

Each primitive transform in a set of data flow diagrams requires specification of its processing logic, incorporating the users' governing business policies. The transform description states what must be done to transform the inputs into the outputs in a form which is independent of how the transformation is implemented.

The transform descriptions are expressed in terms of the three control structures of structured programming—sequence, selection (decision), and iteration (repetition). These are characterized by a single entry point and a single exit point and are combined to be read sequentially.

The principal techniques for transform description are Structured English, decision tables, and decision trees.

Structured English incorporates all three control structures. In Structured English a simple sequence construct is expressed as an imperative sentence. A decision structure associates one or more conditions with the corresponding actions. It is expressed as a two-way decision (if-then-else) or as a multiway decision using the case selection structure. An iteration structure is expressed using repeat-until or while.

Decision tables and decision trees are specialized forms for expressing decision structures only. They are most useful when there are complex combinations of conditions. In a decision table the combinations of conditions are shown in a condition matrix, and the associated actions in an action matrix. Each column of the table represents a decision rule. A decision tree is a graphic representation of a decision structure. Each branch represents a decision, and a path through the tree shows a decision rule—a combination of decisions and the associated actions.

Structured English, decision trees, and decision tables may be combined to present a transform description in a hybrid form.

REVIEW

8-1. State the objectives for the development of transform descriptions.

8-2. Name the alternatives for transform description. Identify the form of each. Evaluate each alternative with respect to the objectives for developing transform descriptions.

8-3. Name the three control structures.

8-4. Give an example of a decision structure with: a null path; two paths; many paths.

8-5. Under what circumstances is the for-each repetition structure preferred over the repeat-until or while?

8-6. Why do we use indentation in Structured English? What are the alternatives to indentation?

8-7. Name the four major parts of a decision table.

8-8. Define the following:
   a. decision rule
   b. decision variable
   c. limited-entry decision table
   d. extended-entry decision table

8-9. Define the folowing parts of a decision tree:
   a. root
   b. branch
   c. leaf
   d. decision rule

8-10. State the guidelines for deciding when to use a decision table and when to use a decision tree.

EXERCISES

8-1. Identify every control structure in Figure 8-22 and state which of the three basic types it is. Also show the scope of each control structure by explicitly marking its beginning and its end. This may be done in one of the following ways: Use pairs of symbols, such as {}, [], (), or <>, each pair denoting a different type of construct. Draw boxes around each construct. Insert end-if, end-for, end-select, end-repeat, or end-while markers, as appropriate.

8-2. The following narrative describes what must be done to produce monthly checking account statements. Using Structured English, express the procedure for preparing a group of these statements.

Our bank sends a statement each month to each customer with a personal checking account unless there is no activity in the account. Personal checking accounts are divided into different groups based on the customer's last name so that we don't have to send out all the statements at the end of the month. This spreads out the workload.

To produce a customer account statement we must show the account balance as of the date of the last statement and the balance as of the end of the period covered by the statement. We must also list each deposit and each withdrawal or debit (such as charges for printing checks or fees for bounced checks). We also compute and subtract service charges as follows: Special checking accounts have a service charge of 25 cents per check. Regular checking accounts have no service charge if the minimum balance during the month was $500 or more or the average balance was $1000 or more; otherwise the service charge is $2 plus 18 cents per check.

8-3. Produce a decision table which is equivalent to the decision tree in Figure 8-35.

8-4. Produce a decision tree which is equivalent to the decision table in Figure 8-36.

8-5. Produce a decision table that is equivalent to the Structured English below.

Policy for Termination and Retirement Benefits

If the employee has less than 5 years' service,
  Then
    The employee's contributions are returned as a termination benefit.
  Else (Employee has 5 years or more of service)
    If the employee is 50 years old or older,
      Then
        If the employee was hired before 1965,
          Then
            Retirement benefits are payable at the rate of 1 percent of the final year's salary per year of service plus 0.1 percent of the final year's salary per year of service for each year of age over 50.
          Else (Employee hired in 1965 or after)
            Retirement benefits are calculated as for employees hired before 1965 and then reduced $150 per month to coordinate with Social Security.
    Else (Employee is less than 50 years old)
      The contributions of the employee and employer are returned as a termination benefit.

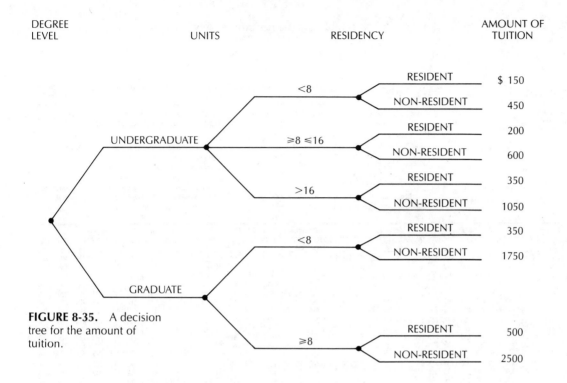

DEGREE
LEVEL   UNITS   RESIDENCY   AMOUNT OF TUITION

<8   RESIDENT $ 150

NON-RESIDENT   450

UNDERGRADUATE   ≥8 ≤16   RESIDENT   200

NON-RESIDENT   600

>16   RESIDENT   350

NON-RESIDENT   1050

<8   RESIDENT   350

NON-RESIDENT   1750

GRADUATE

≥8   RESIDENT   500

NON-RESIDENT   2500

FIGURE 8-35. A decision tree for the amount of tuition.

8-6. Simplify the Structured English developed for Exercise 8-2 by incorporating either a decision table or a decision tree to state the procedure for calculating service charges.

FIGURE 8-36. A decision table for airport parking rates.

TYPE OF PARKING	V	V	V	S	S	S	L	L	L
BASIS OF RATE	H	D	W	H (0-12 HOURS)	D (12-24 HOURS)	W (≥6 DAYS)	H (0-4 HOURS)	D (1-5 DAYS)	W >5 DAYS)
RATE	$3/ HOUR	$3/ HOUR	$3/ HOUR	$1/ HOUR	$12/ DAY	$72/ WEEK	$.75/ HOUR	$3/ DAY	$15/ WEEK

V: VALET   H: HOURLY

S: SHORT TERM   D: DAILY

L: LONG TERM   W: WEEKLY

FIGURE 8-37. Excerpt from table for determining deduction for state sales taxes.

8-7. Derive a decision table for the tax table shown in Figure 8-37.

1983 Optional State Sales Tax Tables

(If you kept records that show you paid more sales tax than the table for your State indicates, you may claim the higher amount on Schedule A, line 10a.)

Your itemized deduction for general sales tax paid can be estimated from these tables plus any qualifying sales taxes paid on the items listed on page 19. To use the tables:

Step 1—Figure your total available income. (See note to the right).

Step 2—Count the number of exemptions for you and your family. Do not count exemptions claimed for being 65 or over or blind as part of your family size.

Step 3 A—If your total available income is not over $40,000, find the income line for your State on the

tables and read across to find the amount of sales tax for your family size.

Step 3 B—If your income is over $40,000 but not over $100,000, find the deduction listed on the income line "$38,000-$40,000" for your family size and State. For each $5,000 (or part of $5,000) of income over $40,000, increase the deduction by the amount listed for the line "$40,001-$100,000."

Step 3 C—If your income is over $100,000, your sales tax deduction is limited to the deduction for

income of $100,000. To figure your sales tax deduction, use Step 3 B but don't go over $100,000.

Note: Use the total of the amount on Form 1040, line 33, and nontaxable receipts such as social security, veterans', and railroad retirement benefits, workmen's compensation, untaxed portion of long-term capital gains or unemployment compensation, All-Savers interest exclusion, dividends exclusion, disability income exclusion, deduction for a married couple when both work, and public assistance payments.

Income	Alabama [1]						Arizona [2]					Arkansas [1]						California [3]				Colorado [2]				Connecticut			Dist. of Columbia					
	1	2	3	4	5	Over 5	1&2	3	4	5	Over 5	1	2	3	4	5	Over 5	1&2	3&4	5	Over 5	1&2	3&4	5	Over 5	1&2	3,4&5	Over 5	1	2	3	4	5	Over 5
$1-$8,000	91	113	120	130	141	160	98	113	113	120	126	78	97	102	109	116	132	125	147	155	164	50	59	62	65	126	139	146	94	112	125	125	132	140
$8,001-$10,000	107	129	140	151	164	185	115	133	133	142	148	91	111	118	127	135	152	147	173	183	193	60	70	74	77	150	167	175	110	129	145	146	155	164
$10,001-$12,000	121	144	159	171	185	207	131	152	154	162	169	103	123	134	143	153	170	167	198	208	219	68	81	85	88	172	194	203	125	145	164	166	177	186
$12,001-$14,000	135	158	176	190	204	227	146	170	173	181	188	115	134	148	158	169	187	186	220	232	243	76	91	95	99	194	220	229	139	159	182	186	197	206
$14,001-$16,000	148	170	192	207	223	246	160	186	191	199	207	125	145	161	173	184	202	204	242	255	266	84	101	105	110	214	244	254	152	173	198	205	216	225
$16,001-$18,000	160	182	208	224	240	265	173	202	209	216	224	135	154	174	186	198	217	222	263	276	288	92	110	115	120	234	268	279	165	186	213	223	234	244
$18,001-$20,000	172	193	222	240	257	282	186	218	226	233	241	145	163	186	199	212	231	238	282	297	309	99	119	125	129	253	291	302	177	198	228	240	252	261
$20,001-$22,000	183	204	236	255	273	298	198	233	242	249	257	154	172	198	212	225	245	254	301	317	330	106	128	134	138	271	313	325	189	210	242	256	268	278
$22,001-$24,000	194	214	250	269	288	314	210	247	258	264	273	163	181	209	224	238	258	270	320	336	349	113	136	143	147	289	335	347	200	221	256	272	284	294
$24,001-$26,000	204	224	263	283	303	329	222	261	274	279	288	172	189	220	235	250	270	285	338	355	368	119	144	152	156	306	357	369	211	232	269	287	300	310
$26,001-$28,000	214	234	276	297	317	344	233	274	289	294	303	180	197	230	246	262	282	299	355	373	386	125	152	160	165	323	378	390	222	242	282	302	315	325
$28,001-$30,000	224	243	289	310	331	358	244	287	304	308	317	188	204	240	257	273	294	313	372	391	404	131	160	168	173	340	398	411	232	252	295	317	330	340
$30,001-$32,000	234	252	301	323	344	372	255	300	319	322	331	196	211	250	268	284	305	327	389	408	422	137	168	176	181	356	418	432	242	262	307	332	345	354
$32,001-$34,000	244	261	313	336	357	385	266	313	333	335	345	204	218	260	278	295	316	341	405	425	439	143	176	184	189	372	438	452	252	272	319	346	359	368
$34,001-$36,000	253	269	324	348	370	398	276	325	347	348	359	211	225	270	288	306	326	354	421	441	455	149	183	192	197	388	458	472	262	281	330	360	373	382
$36,001-$38,000	262	277	335	360	383	411	286	337	361	361	372	218	232	279	298	316	336	367	436	457	471	155	190	200	205	403	477	491	271	290	341	373	387	396
$38,001-$40,000	271	285	346	372	395	423	296	349	374	374	385	225	239	288	308	326	346	380	451	473	487	160	197	207	213	418	496	510	280	299	352	386	400	409
$40,001-$100,000 (See Step 3B)	14	14	17	19	20	21	15	17	19	19	19	11	12	14	15	16	17	19	23	24	24	8	10	10	11	21	25	26	14	15	18	19	20	20

[1] Local sales taxes are not included. Add an amount based on the ratio between the local and State sales tax rates considering the number of months the taxes have been in effect.

[2] Local sales taxes are not included. Add the amount paid.

[3] If the State sales tax rate becomes 5¾ percent, November 1, 1983, taxpayers can add three percent to the table amounts. The 1¼ percent local sales tax is included. If a ½ of 1 percent local sales tax for transportation is paid all year (Alameda, Contra Costa, Los Angeles, San Francisco, San Mateo, Santa Clara and Santa Cruz counties) taxpayers can add 8 percent to the table amount.

[4] Local 1 percent sales taxes are included. If public transportation sales taxes are paid, compute the allowable deduction by the method in footnote 1.

[5] If your local sales tax applies to food for home consumption check your local newspaper during mid-January for the correct deduction. Otherwise see footnote 1.

[6] Sales tax paid on purchase of electricity of 750 KWH or more per month, can be added to the table amounts.

[7] Sales tax paid on the purchase of any single item of clothing for $175 or more can be added to the table amounts.

[8] Sales tax paid on purchases of natural gas or electricity can be added to the table amounts. For local sales tax see footnote 1.

[9] Local sales taxes are not included. If paid all year add 26 percent of the table amount for each 1 percent of local sales tax rate. Otherwise use a proportionate amount. For N.Y. City add 107 percent of the table amount.

9 Logical Data Base Description

All but the simplest computer information systems involve many data stores. When these data stores must be shared by several application programs or when these data stores are intended to answer users' queries about their contents, it is important to specify the relationships among all the data stores as well as to define their information content.

When a collection of data stores is viewed as a whole, it is considered to be a *data base.* This chapter describes the data base environment and the tools used to specify users' requirements in that environment.

Systems analysts must ensure that each transformation in the set of applications programs has access to the data it needs, whether a transform requires an input from a data store or must save an output in a data store. Multiple users of a system are likely to need to access stores information in different ways.

Specifying requirements for access to a data base completely, consistently, and nonredundantly is the task of logical data base description.

This chapter is concerned with logical data base description within the data base environment. A logical data base description is composed of objects and their relationships. The principal graphical data base modeling tool, the data access diagram (DAD), depicts these objects and relationships. It may also be used to test whether or not a particular user need for access to information in the data base is supported. This chapter also discusses the portions of the data dictionary which support the logical data base description.

The examples used in this chapter are not taken from the FastFood Store case, because Chapter 12 shows how the data base description for the FastFood Store is derived. Instead, the examples for this chapter are taken from a university environment.

THE DATA BASE ENVIRONMENT

This section describes the principal components of a data base environment and how they interact. They include the users of the data base, their views of the data, the data base management system, the data base administrator, applications programs and applications programmers, and query languages.

An organization's data is a valuable resource. As such, it requires management, as does any valuable commodity. Like raw material, data needs to be inventoried. However, unlike raw material, data may be shared and reused. Thus, organizations that successfully manage their data extract value from it many times.

Users and User Views

Users of data within an organization may be classified according to their function, such as accounting, engineering, production, or distribution. They may also be classified by job title, such as clerk, technician, assembly-line worker, shop-floor supervisor, department head, vice president, or chief executive officer. The variety of these classifications reflects the diversity of the users and of the enterprise that employs them. The information requirements of each user's job lead him or her to see the data from one or more task-related perspectives known as user views.

A *user view* is a collection of information necessary to carry out one or more related information processing transformations for which a user is responsible. A user view is often associated with and identifiable by data flows involved in these transformations. Examples include a report, a query, or a transaction that captures the essentials of some important event. Description of a user view requires identifying the user, the tasks or transformations, and the necessary information. Making the data available to the task or transformation entails accessing the stored collection of information.

User views of data may overlap. Consider three users of telephone information. A real estate agent knows the street address and would like to find the name and telephone number of the occupant. A directory assistance operator responds to queries for telephone number, and possibly address, given a name. A telephone company billing clerk needs the name and address for a telephone number. These three users have quite different views of the same data.

Suppose there were three distinct telephone directories as in Figure 9-1. Each directory is ordered by the accessing requirements for each user view—one by name, one by telephone number, and one by street address. What difficulties are inherent in this physical arrangement of data for the telephone directories? Consider the problem from the perspective of the telephone company that produces the directories. If all three users' data access requirements could be satisfied from a single shared data source, the telephone company could cut by two-thirds the cost of producing directories. There is another benefit of the single-source approach and the concomitant reduction in redundant data—updating is easier. It is much easier to change one copy of a data element than it is to change many copies simultaneously.

Even if we could convince the users to share a common collection of data, we would still face the limitations imposed by the storage medium. In the telephone book example, in what order should the book be printed? If we choose to order it by telephone number, then access by patron name or street address must be made

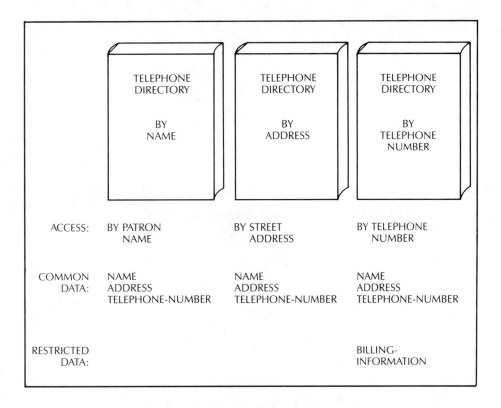

FIGURE 9-1. Three arrangements of a telephone directory.

through an intermediate cross-reference between name or address and phone number. Data sharing may also pose problems for security. Some users may wish to restrict access to some data, in this example, Billing-Information.

Data Bases and Data Base Management Systems

A *data base* is a systematically partitioned, reusable, integrated collection of data that can be shared by many diverse users as dictated by their individual needs.

The ability of many users with different data requirements to access the same data base is illustrated in Figure 9-2. Each user may gain access to a telephone directory data base. The directory assistance operator may act as an intermediary for a caller requesting a telephone number. The end user with a personal computer may access the telephone directory directly. The real estate agent can subscribe to a service that allows a computer terminal access to the telephone directory via microwave communications. The billing clerk uses a computer terminal connected directly to the computer system in which the data base resides. In this way all the different types of users are able to satisfy their specific requirements for information from the telephone directory.

In this environment an interface between the users and the data is needed. This interface should selectively access the data required by each individual use. The interface should support *logically independent* user views so that we can alter data required by one user's view without affecting the others. The interface should

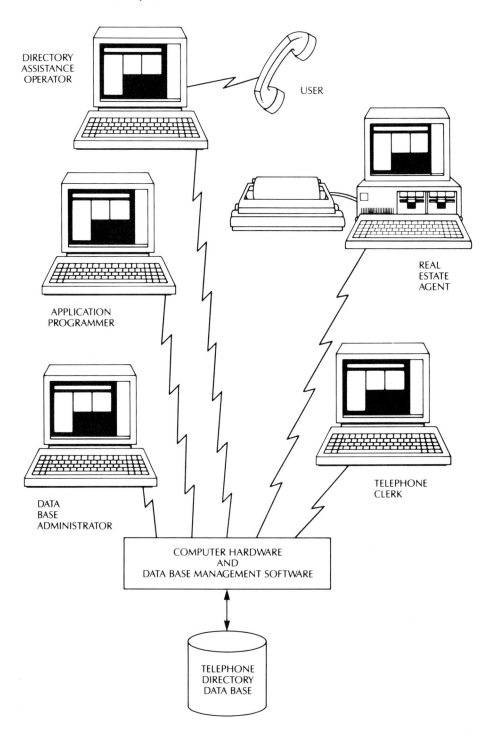

FIGURE 9-2. Multiple users of a telephone directory data base.

insulate users from the constraints and peculiarities of storage devices, providing physical independence. That is, we should be able to change storage devices or recording methods without affecting user access to data. Accomplishing this interface is a primary function of the data base management system. The computer system soft ware and hardware that manages a collection of information and provides for logical and physical data independence is called a **data base management system.** The data base management system provides logical and physical data independence while controlling redundancy of stored information. Other major roles of a data base management system are to enforce **integrity constraints**—restrictions that maintain consistency within a data base—and **security**—prevention of unauthorized access to the data.

The specialized personnel who supervise the data base management system (DBMS) fulfill the **data base administration** (DBA) role.

Roles and Responsibilities in the Data Base Environment

Before a scenario like that described by Figure 9-2 can take place, someone has to analyze the various users' requirements and describe these to the data base management system. Figure 9-3 presents an overview of this process.

The systems analyst interacts with the various users, using the tools of structured analysis, to cull from each the data requirements peculiar to each view. The analyst organizes the collection of the users' potentially overlapping data requirements into an abstract model. This **conceptual data model** presents a statement of user data requirements that is comprehensive, logical, and essential.* It is comprehensive because it covers all the data needed for each user view. It is logical because it is

*D. A. Jardine (editor), "The ANSI/SPARC DBMS Model," **Proceedings of the Second SHARE Conference, Montreal, 1976,** North Holland, New York, 1977.

FIGURE 9-3. People and data models in the data base environment.

Person or System	User	Analyst	Data Base Administrator	Data Base Administrator and Data Base Management System
Data models	Individual user view of data	Conceptual model of *all* user views	Logical model of *all* conceptual models	Physical and internal model for each logical model
Modeling tool		Data access diagram	Varies with specific DBMS Automated DD	Varies with specific DBMS Automated DD
Model components		Objects Relationships	Schema Subschemas	Storage schema (varies with specific DBMS)

free of any computer considerations. It is essential because it presents the minimum information requirements of each user view without any redundancy.

The analyst uses two tools to describe the data base requirements. A **data access diagram** (DAD) provides a graphical partitioning of the data base into **objects** and their **relationships**. An accompanying set of data dictionary entries describes the attributes of the objects, the relationships, and the operations used to manipulate the objects and relationships. A data access diagram is sometimes called a **data immediate access diagram** or a **data structure diagram**.

The systems analyst communicates the conceptual description of objects, relationships, and data base operations to the DBA. Some sophisticated users may also be able to deal directly with the DBA. The DBA reconciles this conceptual model with conceptual models from other parts of the organization and with descriptions of any existing data base systems. The DBA translates the conceptual model into a form acceptable to the DBMS. This includes a both a logical and a physical model of the data base.

A logical model comprises a schema and its subschemas. A **schema** is a version of the conceptual model coded to be understandable by the DBMS. A subset of the schema, called a **subschema**, is produced for each user view. The schema and subschema provide the logical data independence mentioned earlier.

The physical model of the data base is called a **storage schema**. It provides a detailed description of how the data should be allocated and managed on the mass storage devices. The purpose of the storage schema is to provide physical data independence. This ensures that the method of access to the data may be changed without adversely affecting the data base users.

The terminology introduced in this section evolved from researchers and developers of data base management systems. It is oriented toward data base design and administration. However, our goal here is data requirements analysis and specification. From this perspective we prefer to use the term **logical data base description** for what data base specialists call formulating a conceptual data model.

FIGURE 9-4. Two ways of accessing a data base.

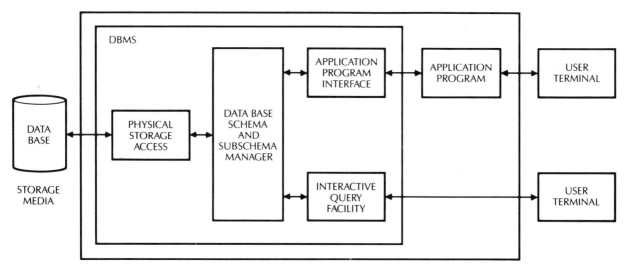

In the data base environment users may access data as indicated in Figure 9-4. There are two distinct ways to obtain the data. In one, the user interacts directly with a subsystem of the DBMS that allows requests to be made for the retrieval of data from the data base. In the other, the user must interact with the data base through an intermediary computer program. This program may be executed interactively or in batch mode, and it may be written in a higher-level language such as COBOL. Writing this computer program requires an *applications programmer* familiar with the appropriate higher level language, its interface to the DBMS, and a statement of the data requirements to support the user view for which the program is needed. Whether a user chooses to interact directly with the DBMS or to have an application program written, the models for data base description shown in Figure 9-3 must be built before the data base can be constructed and the information made accessible to users.

COMPONENTS OF A LOGICAL DATA BASE DESCRIPTION

Objects, Attributes, and Values

We are now ready to discuss the tools and conventions for logical data base description, using a university data base as an example. We will view a university from the perspective of a student and consider the important facts that a data base system should be able to record and report.

Initially, a student is admitted to the university. Subsequently, the student enrolls in classes, drops some classes before completing them, successfully completes some classes and earns grades in them, and ultimately graduates or leaves the university for other reasons. Some students may be placed on probation for not maintaining satisfactory grades.

In this system, what are the objects or entities about which we might wish to record facts? Certainly we would include students, professors, classes, and perhaps a few others. Some objects, such as a student or a professor, are tangible. Even though it is easier to identify these tangible objects, it is important to include more abstract objects, such as a class, in a complete logical data base description. Thus an *object* is some type of real-world entity about which we record information. In fact, the term *entity* is often used as a synonym for, or instead of, the term *object.*

How do we describe objects? An object can be characterized by its attributes.

An *attribute* is a named characteristic of an object that may take on a value. We can distinguish between different types of objects by object name and by each object's list of attributes. We can distinguish the object Student from Professor and Professor from Class as shown in Example (9.1).

Object:	Student	Professor	Class
Attributes:	ID	ID	ID
	Name	Name	Meeting-Time
	Address	Address	Meeting-Place
	Major	Title	
	Class-Level		

(9.1)

Note that some objects have similar attributes (at least in name), e.g., ID (as in IDentifier) for Student, Professor, and Class, or Name for Student and Professor. To uniquely identify an attribute of a particular type of object, we concatenate the object name with the attribute name, as shown in Example (9.2).

Object:	Student	Professor	Class
Attributes:	Student-ID	Professor-ID	Class-ID
	Student-Name	Professor-Name	Meeting-Time
	Student-Address	Professor-Address	Meeting-Place
	Student-Major	Professor-Title	
	Student-Class-Level		(9.2)

Our university data base will store information about many different students. In this collection of data about students it is necessary to be able to refer to and access data about each student individually. Data about an individual student is called an ***instance*** of the object Student. Thus an instance is a specific member of the collection of data associated with an object.

By assigning values to attributes, we can distinguish between two instances of the same object, as in Example (9.3).

Object: Student

Attributes	Values	Values
Student-ID:	41068	82704
Student-Name:	Louella Fernbee	Mortimer Snow
Student-Address:	123 Any St.	456 Some St.
Student-Major:	CIS	CS
Student-Class-Level:	Junior	Sophomore (9.3)

Thus the assignment of a value to each attribute of an object determines an instance of the object.

Notice that there are two perspectives of an object. One is abstract; it is characterized by naming the object and enumerating its attributes. The other is concrete; it is characterized by the association of a specific value with each attribute to distinguish an instance of the object.

We must be careful about the values assigned to an attribute. Does it make sense, for example, to assign the value blue to the attribute Height? No. Use of the term "height" implies a number, not a color. We must be careful that the values specified for each attribute are appropriate for that attribute. We must also be careful when identifying the range of values for an attribute. Suppose we specified integer values for the attribute Student-Height. We could then make a statement like Student-Height = 185, but not Student-Height = 185.5. You might also be puzzled about the values until you learned that the unit of measure was centimeters. Clearly, integer values are not a correct specification for Student-Height. A more appropriate specification might be: the value for Student-Height is real and in the range from 100.0 to 250.0 centimeters.

Usually there is one attribute that can uniquely identify an instance of an object. It is called the *key* of the object. Information system developers depend heavily upon the ability to uniquely identify each instance of an object for the purposes of retrieval and updating. In order to ensure proper recording of events of importance in the university system, keys (Student-ID, Professor-ID, and Class-ID) were adopted for the Student, Professor, and Class objects. Sometimes keys are a single number, as, for example,

Student-ID = 41068

At other times keys are described by the concatenation of several attributes. This is the case when a single attribute alone cannot uniquely identify an instance of an object. For example, Class-ID might be composed of Department-ID, Course-Number, and Section-Number, as in

Class-ID = CIS-4-01

By supplying a value for a key, we may gain access to an instance of an object and thus to the values of the other attributes for that instance.

Objects, Attributes, and Values in the Data Dictionary

How should we record objects, attributes, and values for logical data base description? In a data flow system model an object is a data store, and an attribute is a data element. The description of an object lists the attributes of the object, using data dictionary notation. For example,

Student = { <u>Student-ID</u> & Student-Name & Student-Address
& Student-Major & Student-Class-Level }

We indicate the key by underlining the attribute chosen to be the key. The braces indicate that there may be many instances of this object. Other authors call this an *attribute file description,* and each instance a *record.* This terminology is used because objects are represented on the data flow diagram as data stores. The permissible values for an attribute are specified as in a data element description. (See the discussion of data element definition in Chapter 7.)

Relationships

Information systems would not be very useful if they contained only data about objects. Most interesting systems also incorporate *relationships*, which represent associations or interactions between objects. A relationship usually exists between two different types of objects, because one object is logically connected to or associated with the other. Further, just as we have instances of objects, we also have instances of relationships.

Consider a relationship between the objects Student and Class, which we have called Enrolled-In. An instance of the relationship Enrolled-In exists when an instance of the object Student is paired with an instance of the object Class. Example (9.4) shows an instance of Student, Louella Fernbee, enrolled in an instance of Class, CIS-4-01.

Object: Student   Object: Class

Student-ID: 41068   Class-ID: CIS-4-01
Student-Name: Louella Fernbee   Meeting-Time: 2–3 PM MWF
Student-Address: 123 Any St.   Meeting-Place: Bus. Bldg. Rm. 203
Student-Major: CIS
Student-Class-Level: Junior

Relationship: Enrolled-In

Student-ID & Class-ID: 41068 & CIS-4-01   (9.4)

This instance of the relationship is described by pairing the key of the particular student in the relationship with the key of the particular associated class, as in

> 41068 & CIS-4-01

The abstract relationship Enrolled-In is described by listing the keys of the two related objects. In this case

> Enrolled-In = { Student-ID & Class-ID }

Again the braces indicate that there may be many instances of the relationship. Both attributes are underlined because they must be concatenated to identify each unique instance of the relationship.

Notice that the relationship has a direction. It is said to be rooted at one object, Student in this case, and terminated at another, Class. The relationship can be read in a stilted English as

> Student-ID Enrolled-In Class-ID

Usually object names will be nouns, and relationship names some form of verb.

We can depict objects and relationships as shown in the data access diagram of Figure 9-5. The partitioned boxes represent objects and their keys. The right-hand portion of each box is labeled with the name of the object, and the left-hand portion of each box is labeled with the name of the key. The directed arc represents a relationship. It is labeled with the name of the relationship. In this case, the double-headed arc indicates that this is a one-to-many relationship; that is, a student may be enrolled in many classes. If (however far-fetched this may seem) each student is

FIGURE 9-5. Data access diagram for the relationship Enrolled-In.

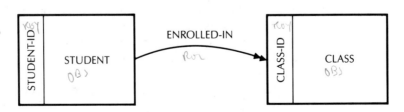

allowed to enroll in only one class, this restriction is shown with a single-headed arc, indicating a one-to-one relationship.

We stated earlier that relationships usually exist between two different types of objects. However, consider the relationship shown in Figure 9-6. The relationship Prerequisite-For means that a student must have passed a course before being permitted to enroll in a subsequent course which depends on knowledge of the previous course. In this case there is a relationship between two instances of the same type of object, Course. While some data base management systems have a difficult time dealing with this kind of relationship, as analysts we should not discourage or ignore these kinds of relationships just because their realization may be difficult.

Relationships in the Data Dictionary

A relationship among objects in the data base is a data store in the data flow system model. As the Enrolled-In relationship illustrates, it is a pairing of keys for the related objects. These keys are data elements (in some cases, combinations of data elements) in an ordered sequence. There may be many instances of a relationship in the data base. Thus the composition of a relationship is iterations of the paired keys, as in

Enrolled-In = { <u>Student-ID</u> & <u>Class-ID</u> }

Each relationship shown on a data access diagram appears in the data dictionary as an entry for a data store.

Because a relationship shows a link or correlation between two objects, some authors call the data store specifying a relationship a *correlative file.*

Logical Operations on Objects and Relationships

In a logical data base description it is not enough to describe the composition of objects and relationships. We must also specify the allowable operations on each object and relationship and the logical processing constraints on each operation.

The generic operations on an object are

1. Establish a new instance of the object.
2. Remove an existing instance of the object.
3. Modify the value of one or more attributes for an existing instance of an object.

FIGURE 9-6. Data access diagram for the relationship Prerequisite-For.

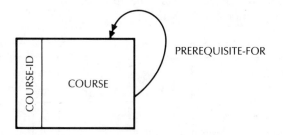

The generic operations on a relationship are

1. Establish a new instance of the relationship.
2. Remove an instance of the relationship.

Each operation makes some presumption about **existence**. The existence of an instance of an object or relationship is determined by the presence or absence of a record with the specified key. Operations that establish a new instance of an object or relationship should be allowed only if the instance does not already exist. The operation of removing or modifying an instance of an object makes sense only if that instance exists. That is, we should not be able to modify or remove something that does not exist. A similar restriction holds for removing instances of relationships.

Another related presumption is that of **uniqueness.** This means that each instance of an object or relationship appears once and only once in the logical data base description. This requires nonredundancy. There will not be more than one copy of an instance, ensuring that a key will always cause the same instance to be accessed. The restriction against creating a duplicate instance enforces uniqueness as well as presumes that the data base does not already contain the instance.

There is a further subtle presumption when removing an instance of an object. The operation of removing an object must not invalidate any relationship that the object participates in. For example, consider instances of the two objects Student and Class and an instance of their relationship as illustrated in Figure 9-7. Suppose we were to remove either the instance of the Student or the Class. Where would that leave the instance of the relationship, Enrolled-In? The removal of the instance of either object would invalidate the instance of the relationship. Preservation of the validity of relationships by restricting operations that remove objects is called *referential integrity.*

One responsibility of the data base management system is to ensure that the presumptions of existence, uniqueness, and referential integrity are enforced for every operation on each object and relationship. This is accomplished by checking conditions that must be satisfied both before and after the completion of the operation. Figures 9-8 and 9-9 give these preconditions and postconditions for operations on objects and relationships. Usually, the analyst does not need to explicitly specify these low-level preconditions and postconditions. The data base management system is expected to provide for their enforcement.

For this reason, the analyst's statements of preconditions and postconditions will not be like those of Figures 9-8 and 9-9. Rather, they will be more like those exemplified in Figure 9-10, which presents some operations on the object Student and the relationship Enrolled-In.

FIGURE 9-7. Diagram of an instance of a relationship between instances of the objects Student and Class.

FIGURE 9-8. Low-level preconditions and postconditions for operations on objects.

Operation:	Establish an instance of an object.
Precondition:	The instance of the object does not exist.
Postcondition:	The instance of the object exists.
Operation:	Modify attribute values for an instance of an object.
Precondition:	The instance of the object exists.
Postcondition:	The modified instance of the object exists.
Operation:	Remove an instance of an object.
Precondition:	The instance of the object exists.
Postcondition:	The instance of the object does not exist.
Postcondition:	The removal does not invalidate any established relationship.

FIGURE 9-9. Low-level preconditions and postconditions for operations on relationships.

Operation:	Establish an instance of a relationship.
Precondition:	The instance of each object participating in this instance of the relationship exists.
Postcondition:	The operation does not create a duplicate instance of the relationship.
Operation:	Remove an instance of a relationship.
Precondition:	The instance of the relationship exists.
Postcondition:	The instance of the relationship does not exist.

There are several points worth mentioning about the example in Figure 9-10. First, not every operation requires a precondition or postcondition. Also, there may be several preconditions or postconditions for an operation. Second, there may be several operations that have the same effect, such as the removal of an instance of an object or relationship. This is exemplified by the Graduate and Dismiss operations on Student. Third, to list all the operations that change an attribute for an object might prove quite cumbersome. Clearly, we need some guideline for determining which operations are important from a logical data base description standpoint. The guideline we suggest is that if a precondition or postcondition for an operation refers to other objects or relationships, then include the operation in the data base description; otherwise do not.

Operations and Conditions in the Structured Specification

In what part of the structured specification should we locate the operations on objects and relationships which are included in the logical data base description? Remember that we wish to minimize or eliminate redundancy. Note that information is stored in the data base or retrieved from it by the transforms in the system. They access the data base to get a necessary input or to store an output for subsequent use. To change their incoming data flows into their outgoing data flows, these transforms will use the data base operations to manipulate the objects and relationships in the data base. A transform description contains a statement of the

FIGURE 9-10.   Operations on Student and Enrolled-In.

Object:	Student.
Operation:	Accept Student to university.
Precondition:	Student has declared major.
Operation:	Graduate Student from university.
Precondition:	Student meets requirements for graduation.
Postcondition:	Student is an alumnus.
Operation:	Place Student on academic probation.
Precondition:	Student meets requirements for probation.
Operation:	Take Student off academic probation.
Precondition:	Student meets requirements for removal of probation.
Operation:	Dismiss Student from university.
Precondition:	Student must be on academic probation.
Operation:	Place Student on leave of absence.
Operation:	Take Student off leave of absence.
Precondition:	Student must be on leave of absence.
Relationship:	Enrolled-In.
Operation:	Enroll Student in Class.
Precondition:	Student is active (not graduated, on leave of absence, on probation)
Precondition:	Student has met the prerequisites for the Class.
Precondition:	Class is not full.
Operation:	Drop Student from Class.
Precondition:	Student enrolled in Class.

steps in the transformation and the business policies which govern it. Thus it is appropriate to locate the specification of an operation and its preconditions and postconditions together with the descriptions of a transform which uses the operation. If the guidelines in the previous section are followed, redundancy will be eliminated or minimized in most system descriptions. In a well-decomposed system, each of the important operations is likely to be associated with only one transform. For example, we would expect the Graduate operation to be used by only one transform in the student registration system, and the Dismiss and Enroll operations each by another transform.

LOGICAL DATA BASE ACCESS

In this section we demonstrate through an example how instances of objects and relationships may be accessed. Our example relies upon a physical file structure because information storage requires some medium. However, our concern is logical data base access, which is independent of this or any other particular physical file structure. Logical data base access requires only that the key for the desired record be supplied, and the record (the instance of an object or relationship) for that key will be returned. If the record containing the key supplied does not exist, that fact will also be indicated.

Consider a portion of the data base for Poor University. Each object and relationship in the data base is implemented physically by a separate 3 × 5 card file, as illustrated in Figure 9-11. The logical description of the same data base is shown in the data access diagram of Figure 9-12 along with the composition of each object.

There each object is represented as a rectangle, with its key shown at the left. Each relationship is shown as an arc. The direction of the arc depicts the direction of the

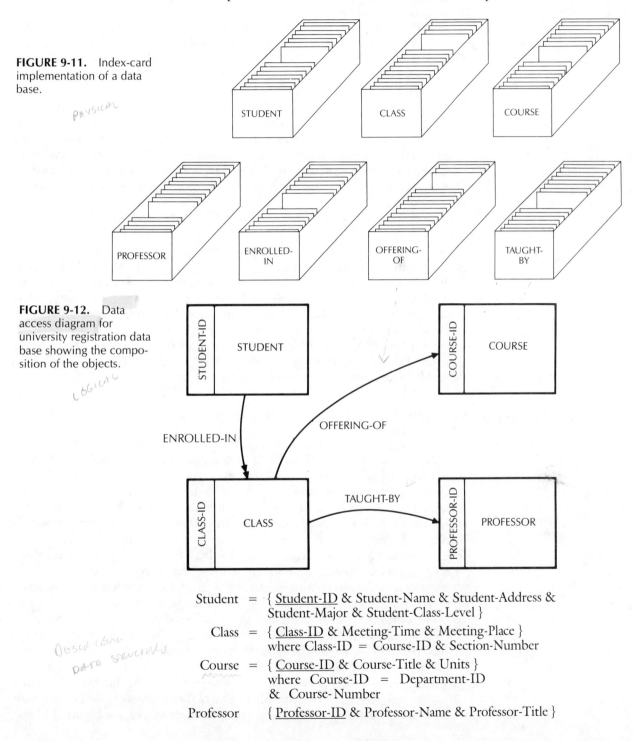

FIGURE 9-11. Index-card implementation of a data base.

PHYSICAL

FIGURE 9-12. Data access diagram for university registration data base showing the composition of the objects.

LOGICAL

DESCRIBING DATA STRUCTURE

Student = { <u>Student-ID</u> & Student-Name & Student-Address & Student-Major & Student-Class-Level }

Class = { <u>Class-ID</u> & Meeting-Time & Meeting-Place } where Class-ID = Course-ID & Section-Number

Course = { <u>Course-ID</u> & Course-Title & Units } where Course-ID = Department-ID & Course-Number

Professor { <u>Professor-ID</u> & Professor-Name & Professor-Title }

relationship. The double arrowhead shows that Enrolled-In is a one-to-many relationship, and the single arrowheads show that Offering-Of and Taught-By are one-to-one relationships.

Suppose we wish to change the value for some attribute, say Address, for student 41068. We open the Student file, leaf through until we find the card labeled **41068,** remove the card, change the Address, and return the card to its correct position. This access to the Student file is represented in Figure 9-12 as entering the rectangle for the Student object via its key, Student-ID.

To add a new student, we obtain a blank student card, fill it out (being careful to ensure that the Student-ID is unique), and place it in the Student file. To place a student on probation, we remove the student's card from the Student file and place it in the Probation file (not shown in Figure 9-11). A similar procedure exists for graduation of students. Thus the maintenance of instances of objects is straightforward. In all these examples we enter the data base directly at the Student object in Figure 9-12.

The maintenance of instances of relationships is also easy. Suppose we wish to enroll a student in a class. First, we ensure that both the student and the class exist by referring to the Student and Class files, respectively. Next, we fill out a card for the Enrolled-In relationship. Finally, we find the appropriate position in the Enrolled-In file, checking to see that an instance of this relationship does not already exist, and place the card there. Similarly, in Figure 9-12, the Enrolled-In relationship is represented by the arc from Student to Class. However, a valid instance of the relationship requires that both the student and the class exist. The data access diagram enforces this condition by requiring entry to the Student object via the Student-ID for the desired instance, then following the Enrolled-In arrow to the Class-ID for the desired instance of the Class object.

TESTING THE LOGICAL DATA BASE DESCRIPTION

A complete logical data base description must show all the accesses required to support all the user views. The data base description must be checked to be sure that it is complete.

A data access diagram shows all accesses to information in the data base. The data base may be entered at any object for which the value of a key is known. Other objects may then be accessed by following the relationships in the directions shown by the arcs in the diagram. As long as such an access path exists, objects (and thus their attributes) may be stored in or retrieved from the data base. Otherwise, there is no provision for access built into the data base requirement.

Of course, any attribute for any instance of an object can be found in a data base if we are willing to search long enough. The access paths in the data access diagram provide for reaching the data without an exhaustive search.

We continue with the Poor University example to discuss how we may test the logical data base description to see whether or not a particular user view is supported.

At the beginning of a school term students are supplied a list of the classes in which they are enrolled. An example of a Student Class List is illustrated in Figure 9-13. Each professor is provided with a roster of the students enrolled in each class that he or she teaches. We shall traverse the data access diagram of Figure 9-12 and

STUDENT CLASS LIST
Fall, 198_

Louella Fernbee
123 Any St.
Somewhere, Ca.

Student-ID: 41068
Major: CIS
Class Level: Junior

Course	Section	Units	Title	Time	Place	Instructor
CIS-4	01	4	Structured Analysis	MWF 2-3pm	Bus 203	I.B. Wise
ACC-2	02	4	Accounting II	TTH 6-8pm	Bus 110	B. Counter
PE-101	04	1	Swimming	MWF 3-4pm	Pool	A. Drowner

Total Units 9

FIGURE 9-13. Student class list.

follow the data base access paths which touch the objects and relationships necessary to produce the information for these two user views—one for the Student Class List and one for the Class Roster.

The accesses required to obtain the data for the Student Class List are shown in Figure 9-14. They should also be traced along the corresponding paths through Figure 9-12. Knowing the Student-ID of the student for whom we wish to produce a Class List, we enter the data base by accessing the Student object (Figure 9-14a). With this access we can obtain and supply the requisite student information: Student-Name, Student-Address, Student-Major, and Student-Class-Level. We next navigate from the Student object to the Class object via the Enrolled-In relationship (Figure 9-14b). This is accomplished by finding the first occurrence of this Student-ID in the Enrolled-In file and then retrieving from the Class file the instance of the Class indicated by the Class-ID on the Enrolled-In card. This is illustrated in Figure 9-14c.

By iteratively accessing each Class indicated by the Enrolled-In cards, we can supply the requisite Class information: Building, Room, and Meeting-Time. In a similar fashion we can provide the Course information (Course-Title and Units) through the Offering-Of relationship and the Professor information through the Taught-By relationship. This is illustrated in Figures 9-14d, e, f, and g.

What has this navigation exercise demonstrated? It has shown that the Student Class List user view is supported. The requisite objects and relationships exist to allow generation of the report.

Now consider a Class Roster, as in Figure 9-15. If we know the ID of the professor for whom we wish to produce a Class Roster, and wish to obtain the data for it, we enter the data base via the Professor object. How should we navigate from Professor to Class? We cannot use Taught-By; it goes in the wrong direction. (Recall that relationships are directed from one object to another and, in this sense, can be thought of as one-way streets.) Thus there is no path from Professor to Class. Clearly, an additional relationship is called for to support this user view. This relationship is

Teaches = { Professor-ID & Class-ID }

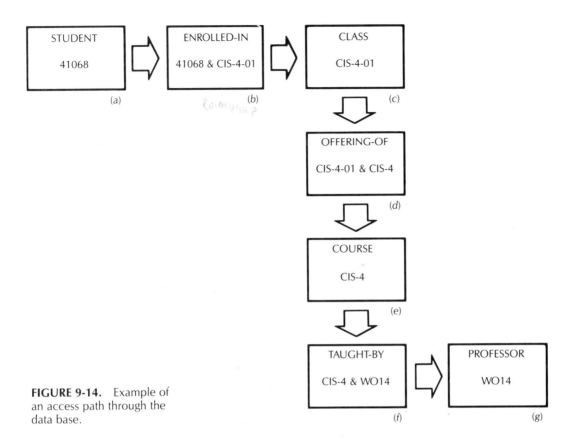

FIGURE 9-14. Example of an access path through the data base.

FIGURE 9-15. A class roster.

CLASS ROSTER
Fall, 198_

Class: CIS-4 Structured Analysis Methods   Instructor: I.B. Wise

Section: 01   Units: 4   Time: MWF 2-3pm   Place: Bus 203

Student ID	Student Name	Major	Class Level
41068	Fernbee, Louella	CIS	Junior
37608	Snow, Mortimer	CS	Sophomore
:	:	:	:

25 Students Total

Now we are able to provide all the classes that a professor teaches. However, there is still no path from Class to Student. The following relationship will provide the path:

Has-Enrolled = { <u>Class-ID</u> & <u>Student-ID</u> }

Thus, the updated logical data access diagram includes two new relationships, as depicted in Figure 9-16.

Note that relationships often come in complementary pairs. This is the case for Taught-By and Teaches, also for Enrolled-In and Has-Enrolled. To simplify the data access diagram, a complementary pair of relationships may be represented by a single arc with arrowheads at both ends, as in Figure 9-16. The arc is labeled by placing the name of each relationship near the arrowheads corresponding to the direction of that relationship.* While it is by no means a requirement that relationships be paired, the astute analyst anticipates the need for a user view requiring a complementary relationship although only one has been identified.

Before closing this section, we want to stress the importance of logical and physical data independence. If we review the relationships Enrolled-In and Has-Enrolled, we note that in the Poor University data base shown in Figure 9-11 these two relationships are implemented as separate files. This is due to the physical organization. The Enrolled-In file is maintained by Class-ID within Student-ID; the Has-Enrolled file is maintained by Student-ID within Class-ID. The redundancy in this scheme is obvious.

Consider the problems if these two files become inconsistent. For example, someone makes a mistake when enrolling a student in a class by placing a card in the

---

*The alternative is to show each relationship as a separate arc.

FIGURE 9-16. The university registration data base with added relationships.

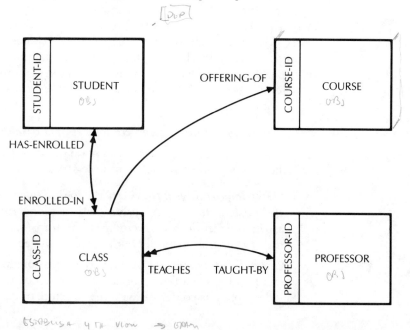

Enrolled-In file without placing a corresponding card in the Has-Enrolled file. The students shows up for the first class meeting. The professor takes roll. The student's name is not called, yet the student has a valid Class List indicating that he or she is properly enrolled. Clearly, some means of preventing this situation is required. Logically, there are two user views of the association between Student and Class, implying complementary relationships. But to prevent inconsistency and reduce redundancy, it would be preferable to have these two views supported physically by one file. This is one of the major roles of a data base management system: to provide for logical and physical data independence while controlling physical redundancy.

SOME TYPICAL QUESTIONS

We close this chapter with answers to some typical questions that arise here.

"Would you distinguish file from data store and data base from data base management system?"

We have tried to avoid use of the term "file," as it has such a strong physical connotation. For example, consider what file means to a secretary as opposed to a COBOL programmer. To a secretary a file is a set of cards revolving in a holder or the contents of a folder in a cabinet. It is accessed by rotating the cards or by opening the cabinet, pulling out the folder, and opening the folder. To the COBOL programmer a file is a collection of records whose structure is associated with an FD statement in the DATA DIVISION. It is accessed by OPEN, READ, WRITE, and CLOSE verbs. The term "data store" bypasses the preconceived notions that accompany the term "file" and emphasizes the composition of the information.

By "data base" we mean a systematic and logically organized collection of information. The data base is partitioned into objects and their relationships. Objects are described by attributes whose values are chosen from a defined range. The objects and relationships are manipulated through defined operations.

By "data base management system" we mean the combination of computer hardware and software used to manage the collection of instances of objects and relationships.

"What constitutes a complete logical data base description?"

A complete logical data base description specifies all the objects and relationships in the data base and the operations necessary for their maintenance.

"How important is it to specify the logical operations with preconditions and postconditions for objects and relationships? Isn't this redundant, as the operations are specified in the transform descriptions for those transforms that manipulate the objects and relationships?"

Our focus in logical data base description is somewhat different than when we produce DFDs. Data flow diagrams are concerned with the flow of data. A data base description is concerned with the structure of data objects and their relationships. The data base description must also include the allowable operations on each class of object or relationship. The preconditions and postconditions on these operations ensure the integrity of the data base so that it will continue to support the user views implied by the data access diagram. To avoid redundancy, the logical data base description is depicted in a data access diagram, but the definition of its components is incorporated in the data store and transform descriptions in the structured specification. The logical data base operations and their preconditions and postconditions are specified in connection with the transform descriptions. The description for each transform that manipulates an object or relationship is accompanied by a statement of the data base operations involved and the preconditions and postconditions for each.

> *"How are the data access diagram and its components related to the data modeling tools introduced in previous chapters?"*

The data access diagram (see Figure 9-16) depicts objects and their relationships. The DAD is used to test the data access requirements of the various user views that the data base is supposed to support. The composition of objects and relationships and the operations associated with them, and the preconditions and postconditions for these, are specified in the structured specification. The data stores for objects are shown in the DFDs. There is a data-store entry in the data dictionary for each object and relationship. Operations and their preconditions and postconditions are specified in association with the transforms which use them.

> *"How is the logical data base description derived?"*

This is one of the subjects of Chapter 12. Briefly, the logical data base description for a system is derived by examining the leveled set of DFDs along with the data dictionary and transform descriptions. We look for objects and relationships implied in each. Usually the structure of the objects is complex; they are composed of many embedded objects. One of the major efforts in the derivation of logical data base description is to simplify complex objects by decomposing embedded objects to form a larger number of simple objects and relationships.

SUMMARY

A data base management system comprises a combination of computer hardware and software. It provides logical and physical data independence; enforces uniqueness, existence constraints, referential integrity, security, and privacy; controls redundancy; and facilitates data sharing by many users.

The data base environment includes users and their views of data; the systems analyst, who is responsible for identifying user data requirements; the data base administrator, who supervises the data base management system; and the data base.

Each user's view of data arises from the information requirements of the user's job. The analyst captures the essential data requirements for each user view. A logical data base description is developed from these data requirements using two principal tools—the data access diagram and the data dictionary. Objects and relationships are represented on the data access diagram. The data store section of the dictionary is used to define the composition of the objects and relationships. Operations on the objects and relationships, along with the appropriate preconditions and postconditions for these operations, are specified in the transform dictionary in connection with the transform descriptions.

The logical data base description must be tested to ensure that the data access requirements for each user view are supported. This involves navigating through the objects and relationships in the data access diagram to see that the access paths required by each user view exist. It also means checking the data dictionary to see that the appropriate attributes are defined for the objects implied in the user view.

REVIEW

9-1. Define the following:
  a. user view
  b. data base
  c. data base management system (DBMS)
  d. data base administrator (DBA)
  e. object, attribute, and value
  f. relationship

9-2. State the purpose of the data access diagram (DAD). Identify the graphic symbols used to represent objects and relationships on the DAD. What is the significance of the arrowheads on a DAD?

9-3. State the reasons for braces and underlining when describing objects and relationships using DD notation.

9-4. With respect to logical data base operations, why should we be concerned with existence, uniqueness, and referential integrity?

9-5. Discuss the role of the DAD and its associated DD entries when testing whether or not a particular user view is supported.

9-6. In what part of the structured specification is each of the following components of a logical data base description defined or specified? Which of them are shown graphically in the structured specification, and where?
  a. Object
  b. Relationship
  c. Operation on an object or relationship
  d. Precondition or postcondition for an operation on an object or relationship

EXERCISES

9-1. Consider the DAD of Figure 9-16 and the following queries (user views):
  a. Which department is Prof. Wise a member of?
  b. List the name of each student majoring in the CIS Department. (Assume that Student-Major is the same as Department-ID).

c. Who are the faculty members in the department that student 41068 majors in?

d. Which students does Prof. Wise advise?

Identify any object(s) or relationship(s) that must be added to Figure 9-16 to satisfy these additional user views.

9-2. For each object or relationship identified in Exercise 9-1:

a. Add the object or relationship to the DAD.

b. Write the DD description for the object or relationship.

9-3. Test your updated data base description by identifying for each user view in Exercise 1:

a. The initial object accessed in the data access diagram.

b. Each subsequent relationship and object accessed.

9-4. For the user view implied by the schedule of classes in Figure 9-17:

a. Add the necessary object(s) and relationship(s) to the data access diagram.

b. Write DD descriptions for them.

c. Test the new data base description to verify that it will support this additional user view.

SCHEDULE OF CLASSES
Spring 198_

Accounting Department – Chair: E.D.P. Audit

Class	Course Title	Units	Time	Room	Instructor
ACC-1-01	Accounting I	4	MWF 9-10am	Bus 110	B. Counter
ACC-1-02	Accounting I	4	TTH 9-11am	Bus 110	B. Counter
⋮	⋮		⋮		

Computer Information Systems Department–Chair: B.I.T. Twidler

Class	Course Title	Units	Time	Room	Instructor
CIS-1-01	Intro. to Computer Sys.	4	MWF 9-10am	Bus 220	G.I. Neuit
⋮	⋮		⋮		
CIS-4-01	Structured Analysis Meth.	4	MWF 2-3pm	Bus 203	I.B. Wise

FIGURE 9-17. Portions of a class schedule.

10 Drawing Data Flow Diagrams

The discussion of data flow diagrams in Chapter 6 introduced the conventions for a leveled set of data flow diagrams, defined the terms used to describe this data flow model of an information processing system, and emphasized reading and understanding a set of data flow diagrams. This chapter is devoted to the more demanding activity of drawing a set of data flow diagrams. It presents the basic principles, process, and techniques for creating data flow diagrams. There is some unavoidable repetition of material from Chapter 6, usually with a fuller explanation related to how to draw data flow diagrams.

Part III explains how data flow diagrams are used throughout the analysis phase of the system life cycle. Data flow diagrams are used both to model existing information processing systems and to depict the requirements for new systems. The approach described in this chapter is primarily useful in modeling an existing system, when there are no data flow diagrams to use as a starting point. In most situations, diagrams for a new system are modifications of the current system description.

An organized presentation of the steps involved in drawing data flow diagrams, such as the one in this chapter, is important to the beginning student. In practice, however, there is great variability in the process of system modeling from project to project. Most projects will differ from the precise sequence prescribed here. As discussed in Chapter 5, specifying user requirements is an iterative process. Data flow diagrams, the system dictionary, transform descriptions, and data access diagrams are working tools to aid understanding and communication. The data flow diagrams and data dictionary are usually developed in parallel. Nevertheless, while the other parts of the system specification are essential to a complete system model, the data flow diagrams drive the process of system description.

Note, however, that in describing information processing systems, there is an important distinction between the process of structured analysis and its product. The process of structured analysis is important because it coordinates the efforts of users and analysts to produce a mutually understandable system description that is complete, precise, and free of errors. But a good process alone cannot guarantee success.

When the system model has been completed it exists objectively and is independent of the process by which it was created. The resulting data flow diagrams, as well as the other parts of the model, stand or fall on their own merits. Whether they were developed quickly or evolved through many cycles of laborious interaction between analysts and users cannot be determined from viewing the structured specification. The information processing system will be designed and constructed from the product of structured analysis—the structured specification. The process by which it was produced is then hidden and irrelevant; what counts is the result.

GETTING STARTED—THE INITIAL DATA FLOW DIAGRAM

Perhaps the maxim "Well begun is half done" is a bit optimistic as an estimate of progress; nevertheless, if you never start, you'll never finish. Even the most seasoned analyst begins a new project with some confusion and uncertainty. That is a natural human response when confronted by ill-structured complexity. The veteran performer has butterflies; the writer hesitates before the emptiness of a blank sheet of paper. Remember that an important function of a data flow diagram is to help understand the system by drawing a picture of it, and that you increase your understanding as you repeatedly criticize, revise, and improve the system description. So, lay aside your fears and begin.

The initial goal is to develop a high-level data flow diagram showing the essential features of the system. If the big picture is captured incorrectly, no amount of attention to detail will remedy the situation. Once the overall system description at the top few levels is reasonably correct, the framework for detailed decomposition has been established, and the remaining details will find a place in the system structure.

The following steps lead to an initial set of data flow diagrams:

1. Identify the documents and displays associated with the system.

The documents used in or produced by the system are the best starting point. They are the clues to the data flows—both internal and external. As more and more automated systems include cathode-ray tubes and other graphic devices, the information which they display electronically must not be overlooked and should be listed along with their paper counterparts. In the future, data flows transmitted by speech will also have to be noted.

Documents which accumulate or wait in the system may indicate the presence of a data store.

2. Determine the system boundary.

This is not always easy, as we learned through the "Warning of the Doorknob" in Chapter 1. A good guideline here, as in many other system modeling activities, is to do a little more than you think is necessary. It will be easier to shrink the system boundary later, if you make it too inclusive, than to get additional information about the surrounding data flows and transformations if the boundary must be extended.

3. Use the list of documents and displays to identify net system inputs and their associated sources as well as net system outputs and their associated destinations.

Every document entering the system (often called a "source document") is modeled by a data flow crossing the system boundary to enter the system. The person or organization supplying the document is its origin.

Every document leaving the system is modeled by a data flow crossing the system boundary toward its destination, the person or organization receiving it.

By their nature, display devices produce outputs from an automated system. The location of the overall system boundary determines whether they are also net system outputs. Displays often provide confirmation of interactive input to an automated system. Initial top-level data flow diagrams should not be concerned with details of interactive data entry. Rather the data flow diagram should depict the essential packets of information flowing into the system.

4. Draw a context diagram and check it for reasonableness.

Having listed as many net system inputs and outputs as you can, you are ready to draw a context diagram, including the sources and destinations. Then you should make a gross check for reasonableness. In your imagination trace the unseen paths relating inputs and outputs. Can each system output be derived from the contents of the incoming data flows? Does each input flow through the system to contribute to one or more of the outgoing data flows? In this way you may be able to discover system inputs or outputs that were initially overlooked.

5. Identify the principal data stores using the following clues:
   a. Look for files, manual or automated, in the existing system.
   b. Look for documents which must wait to be processed as a group. For example, in a payroll system time cards may be filled out, submitted, and checked daily or weekly but not processed until the end of the month.
   c. Look for collections of information (documents, tables, or reference information) required by the system. In many cases, these data stores are read regularly, but seldom modified, and therefore are easy to overlook. For example, a payroll system may require a table of withholding rates or amounts

in order to calculate the amount of income tax to be withheld from each employee's gross pay.

d. Identify objects (concrete or abstract) about which stored information is required. In a data base environment, these are likely to be the constituent data stores, as discussed in Chapter 9.

6. Draw a top-level diagram of the whole system.

At this stage, you should not worry if there are too many bubbles in Diagram 0 or if the partitioning is a bit uneven. Strive for completeness and consistency. You may:

a. Start with a system input and trace its flow forward through the system until it emerges transformed into a system output.

b. Start with a system output and trace its derivation backward through the system until you reach the system input(s) from which it was transformed.

c. Start with a data flow or transformation in the center of the system and work outward in both directions toward the system boundary.

In short, start with what you know and understand and work from there toward what is still unknown.

As you proceed, assign the best names you can to each component of the data flow diagram. Check your list of possible data stores. If one is shared by two or more transformations in the data flow diagram, draw the data store and its interfaces to the processes, showing the net flow over each interface. If a data store is not shared, can you associate it with one of the transforms? If so, you will know in what part of the hierarchy to put it when you decompose the system further by drawing the level-level diagrams.

When you have finished Diagram 0, you should check it to see if it makes sense, just as you did with the Context Diagram.

CONTINUING THE DATA FLOW DIAGRAM

Drawing the rest of the leveled set of data flow diagrams continues by a combination of improving what has just been done and decomposing the system further until the partitioning is complete. Here is one appropriate sequence for accomplishing this:

7. Improve the partitioning of the initial data flow diagram.

There is considerable practical experience, supported by psychological evidence, that a diagram with no more than about seven transforms is conceptually manageable. More complex diagrams should be repartitioned.*

*George Miller, "The Magical Number Seven, Plus or Minus Two: Some Limits on Our Capacity for Processing Information," *The Psychological Review,* vol. 63, no. 2, March 1956, pp. 81–97.

If there are too many bubbles, or if some of the bubbles represent very detailed transformations while others lump together several relatively significant transformations, you may wish to change the partitioning before proceeding further. You may choose to combine several bubbles from the initial diagram that depict related detailed transformations. Replace them with a single bubble in Diagram 0, and show the detailed transformations in a child diagram; that is, level down. Or, it may be appropriate to draw a new Diagram 0 with fewer bubbles than your first Diagram 0 and then split your initial diagram into children of the new Diagram 0; that is, level up.

USE 5 – 9 BUBBLES

8. Develop an initial data dictionary to support the data flow diagram you have produced.

   This should include at least the compositions of
   a. System inputs
   b. System outputs
   c. Data stores identified to this point, whether depicted in the data flow diagrams yet or not
   d. All internal data flows shown in the data flow diagrams thus far

9. Evaluate the data flow diagrams and make improvements and refinements based on that evaluation. Evaluation and refinement are discussed in Chapter 11.

10. Use walkthroughs to find mistakes.

   A walkthrough is a review of a system description by users or analysts. Its primary purpose is to identify errors in the system description so that they can be eliminated. Walkthroughs are discussed in detail in Chapter 11. The first walkthrough with the users should usually occur after the first few quick iterations of refinement as soon as the analyst is satisfied with the top few levels of the system description. Note that the first review of the FastFood system with Mr. and Mrs. O (Chapter 6) took place the week after the initial meeting between them and the analyst and involved only the Context Diagram, Diagram 0, and its children, even though lower-level diagrams had been drawn. If the top-level system description is incomplete, incorrect, or inconsistent, it must be corrected before further decomposition will be fruitful. Mistakes at the upper levels will affect the lower levels as well.

Completing the Data Flow Diagrams

11. Continue the decomposition to lower levels until the primitives are reached.

   How can you tell when to stop partitioning the system? There are two principal tests for knowing when the bottom level has been reached.
   a. Stop partitioning when it is impossible to continue.

   Partitioning transformations frequently involves simultaneous partitioning of data flows. As this occurs, eventually the data flows cannot be decomposed any further without violating the unity of the data flow. That is, further separation of the data elements composing the flow would render the fragments meaningless or useless. This is not a matter of counting data flows to see whether a low-level transform has one input and one output, but a question of whether an essential packet of related information will be fragmented if further decomposition is attempted.

    b. Stop partitioning when the scope of a transformation is small enough.

Clearly this is a subjective decision. Again, it is better to overdecompose than to underdecompose; it is always easier to level up than to level down.

A rule of thumb is:

Stop decomposing when the transform description will fit on a single page and therefore can be read at a glance.

Further evaluation, refinement, and walkthroughs will accompany the iterative development of the lower levels of the set of data flow diagrams.

Completing the System Description

As the lower levels of the data flow diagrams are drawn, the system dictionary must be expanded to include an entry for each added name. A transform description must be written for each primitive transform, as discussed in Chapter 8.

Completing the system description in terms of data flow diagrams, the data dictionary, transform descriptions, and data base descriptions does not necessarily mean that the entire structured specification is also complete. Chapter 14 describes what must be done to complete the specification.

The preceding discussion has outlined a procedure for producing data flow diagrams. It is summarized in Figure 10-1. For examples of applying this procedure, see Part III. The remainder of this chapter discusses principles for drawing data flow diagrams and presents some practical tips.

FIGURE 10-1. Summary of the steps in drawing a set of data flow diagrams.

1. Identify the documents and displays associated with the system.
2. Determine the system boundary.
3. Use the list of documents and displays to identify net system inputs and their associated sources as well as net system outputs and their associated destinations.
4. Draw a context diagram and check it for reasonableness.
5. Identify the principal data stores.
6. Draw a top-level diagram of the whole system.
7. Improve the partitioning of the initial data flow diagram.
8. Develop an initial data dictionary to support the data flow diagrams you have produced.
9. Evaluate the data flow diagram and make improvements and refinements based on that evaluation.
10. Use walkthroughs to find mistakes.
11. Continue the decomposition to lower levels until the primitives are reached.

PRINCIPLES FOR DRAWING DATA FLOW DIAGRAMS

The following principles will aid in drawing data flow diagrams:

Connections Among Components

The movement of information interconnects transforms and data stores. It also connects the system to the origins and destinations of information in the environment. Thus data flows are required as the interfaces between other types of system components. The conventions for permissible connections among the various types of components are shown in Figure 10-2.

Data flows may connect a transform to any other type of system components. Data stores, origins, and destinations may not be connected to each other directly, but require an intervening transform. Direct flow between external entities is, by definition, outside the scope of the system and is not shown.

Balancing from Level to Level

Net data flows on a child diagram must be equivalent to the data flows to and from the parent bubble. If the flows on both diagrams are identical, this is easy to check. If decomposition of both data flows and transformations occurs on the lower-level diagram, the data dictionary must be consulted to find out whether the lower-level data flows are equivalent to those on the level above.

When Diagram 0 of the FastFood Store (Figure 10-4) is compared with the Context Diagram (Figure 10-3), the data flows match except for Financial-Statements in the Context Diagram and Balance-Sheet and P-&-L in Diagram 0. The

FIGURE 10-2. Permissible connections among components of a data flow diagram.

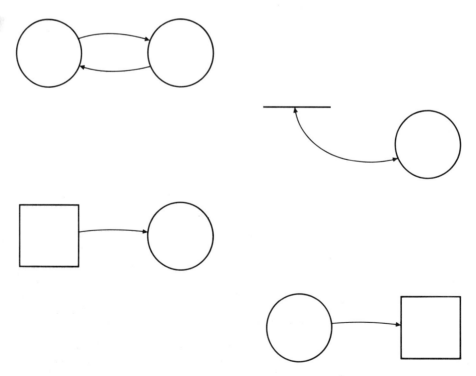

CONTEXT DIAGRAM: FASTFOOD STORE

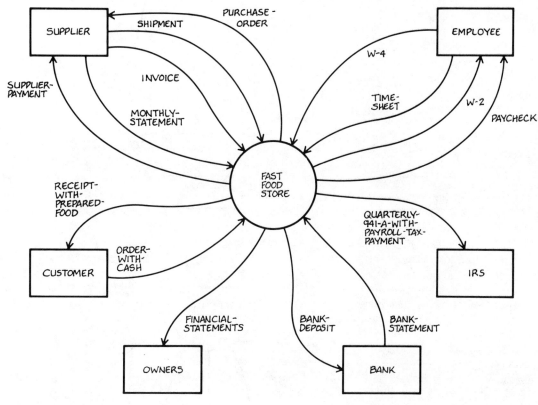

FIGURE 10-3. Context Diagram: Mr. and Mrs. O's FastFood Store.

data dictionary shows that

$$\text{Financial-Statements} = \text{Balance-Sheet \& P-\&-L}$$

Reference to the data dictionary is also necessary when more than one name is given to a single data flow. For Diagram 1 (Figure 10-5) to balance with the Context Diagram, Purchase-Order and P.O. must be aliases or synonyms for the same data flow. Packing-Slip and (Checked)-Packing-Slip must also be synonyms. When aliases are used in a data flow diagram, the equivalence of names must be stated in the data dictionary entries for both names. Clearly, aliases are a source of redundancy in the data dictionary, and their use should be minimized. + DECOMPOSITION

Hierarchy and Decomposition

Decompose evenly

It is desirable to partition evenly. This means that all the transforms on the same diagram should be of about the same complexity. This is difficult to define precisely, but some examples of uneven partitioning should illustrate the concept.

Figure 10-6 might represent Mr. O's view of the FastFood Store. There is one very low-level transform: Check-In Shipment, which corresponds to Bubble 1.1.

DIAGRAM 0: FASTFOOD STORE

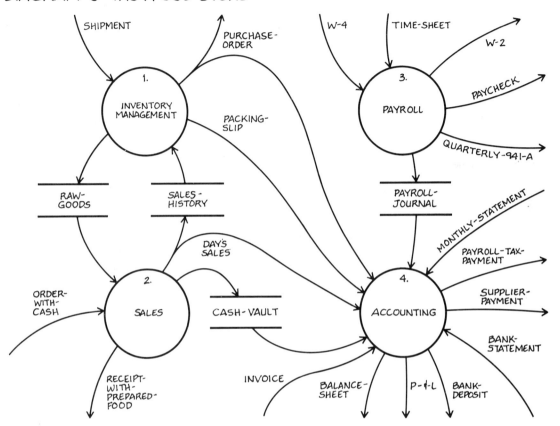

FIGURE 10-4.   Diagram 0: FastFood Store.

There is also Sales (Bubble 2 in the original set of diagrams). The remaining bubble includes Everything Else. Diagrams such as this show an analyst's understanding of bits and pieces of a system with no clear idea of what happens in the rest of the system. The inability to provide a better name than Everything Else also indicates the analyst's ignorance.

A more likely example of uneven partitioning is shown in Figure 10-7. Sales has been decomposed into only two parts. However, the name of Bubble 2.1' suggests that there are really three transforms inside. The fact that the analyst gave such a precise name to Bubble 2.1' has also provided a clue to the appropriate decomposition.

Even partitioning does not mean there that must be the same number of transforms on every diagram or the same number of levels of decomposition throughout the system. The partitioning should depict the structure of the system; it should not impose an arbitrary and unnatural structure upon the system.

The guidelines for stopping the decomposition were stated in step 11 above. As we have seen, the transform description for 4.5.2: Produce P & L (Figure 10-8) follows our rule of thumb by fitting on one page. At this point we anticipate that 4.5.1: Produce Balance Sheet is also a primitive transform, as illustrated in Figure

DIAGRAM 1: INVENTORY MANAGEMENT

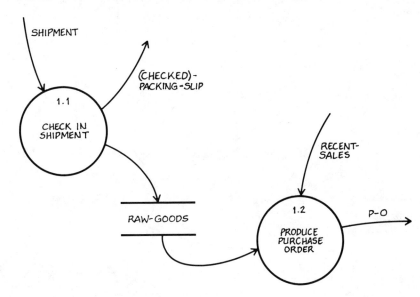

SHIPMENT

(CHECKED)-
PACKING-SLIP

1.1

CHECK IN
SHIPMENT

RECENT-
SALES

RAW-GOODS

1.2

PRODUCE
PURCHASE
ORDER

P-O

FIGURE 10-5. Diagram 1:
Inventory Management.

6-14. If its transform description turns out to be more than a page, we are free to decompose it further.

**Data Store
Conventions**

As discussed in Chapter 6, interfaces to data stores show the net information flow. In Diagram 0 (Figure 10-4) Day's-Sales, a daily dollar total of food sold, flows from Sales to the Sales-History data store, where it waits to be used by Inventory Management. Thus, Sales writes information to the data store, and Inventory Management reads information from the data store.

If an arrow to or from a data store is not labeled, the data flow is presumed to be a complete record. If the flow comprises only a portion of the record, then the arrow is labeled so that the composition of the flow can be defined in the data dictionary.

At the system boundary, direct data flow between a source or destination and a data store is not normally shown. We expect to find a transform between the external entity and the data store. In most systems, information in a data flow is at least checked or edited before it is placed in a data store. Information in a data store is usually extracted selectively or formatted before it goes to its receiver in the form of a report or display. In any case, the data flow diagramming conventions prohibit a data flow between an external entity and a data store without an intervening transformation. If the transform seems trivial, or introducing it, artificial, then an alternative is to shift the system boundary slightly and show the data store as an external entity. This solution is not appropriate when more than a single data flow crosses the boundary.

Data stores shared by transforms and their interfaces to these transforms are always shown. As a consequence, they appear at the highest level where the

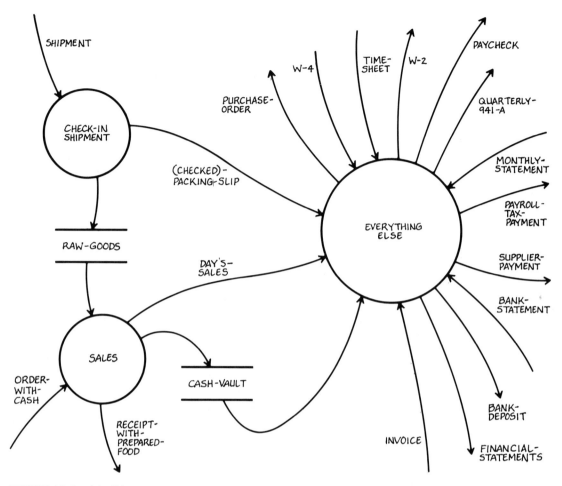

FIGURE 10-6. Mr. O's view of the FastFood Store.

transforms appear, and all other accesses to these data stores are made by subordinates of these transforms.

For example, Sales-History is shown in Diagram 0 because it is shared by Sales and Inventory Management. Any other accesses to Sales-History will be made by subordinates of bubbles 1 and 2, but not by subordinates of bubble 3 or 4, because Diagram 0 shows no data flow between Sales-History and bubble 3 or 4.

Day's-Orders, a data store containing a copy of each individual order for the day (Figure 10-9), appears in Diagram 2. It is **local** to Transform 2, being shared by Transforms 2.3 and 2.4. An alternative way of stating this is that the data flows to and from Day's-Orders are **internal** to Transform 2. In terms of our magnification analogy, Day's-Orders, along with the other internal details, disappears when Transform 2 is reduced to become Bubble 2 in Diagram 0.

Every data store is shown at least once. As a consequence, a data store accessed by a single transform only will be accessed by a primitive transform. When you have reached the primitive transforms, be sure to check that all data stores are shown.

DIAGRAM 2: SALES

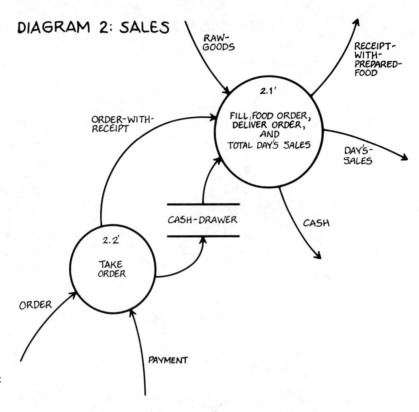

FIGURE 10-7. Uneven partitioning of Transform 2: Sales.

FIGURE 10-8. Diagram 4.5: Produce Financial Statement.

DIAGRAM 4.5 : PRODUCE FINANCIAL STATEMENTS

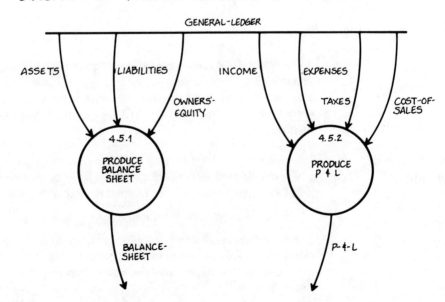

DIAGRAM 2: SALES

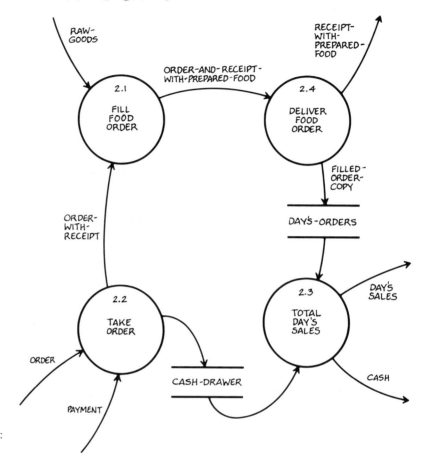

FIGURE 10-9. Diagram 2:
Sales.

For the sake of completeness, it is a good time to show every data store some-where in the system description, no matter how few levels a set of data flow diagrams may currently have. As the decomposition proceeds, the data stores that are not shared will migrate downward. (Of course, data stores never appear in a context diagram, as they are internal to the system.)

Conservation of Data Flow

Conservation of flow is a basic principle of consistency in any system modeled by a network. It requires a balance between what flows into a system and what flows out. In a physical system with internal storage, this means that

Flow out = flow in − amount stored

Each car entering a freeway must continue to travel along the freeway or must exit. Otherwise, cars would magically appear and disappear. Similarly, water entering a distribution main leaves the main to serve the utility's customers (or is lost through leaks).

In the case of information processing systems, in which the flows are logical rather than physical, the principle of conservation of flow may be stated as follows:

Data flow out = known transformation of data flows in

The data elements in the data flows entering any transform in the system must be sufficient to permit their transformation into every data element in the data flows leaving the transform.

Because the principle of conservation of data flow holds for every transformation in the system wherever it is located in the hierarchy, it is true for the system as a whole.

At a gross level, every transform must have at least one input and one output.

Checking conservation of flow at this level requires consulting the data dictionary. To check conservation of flow for Bubble 4.5.2 (Figure 10-8), we would need to look at the data dictionary entries for Income, Expenses, Taxes, and Cost-of-Sales to see whether together they contain all the data elements specified in the transform description shown in Figure 10-10.

Conservation of flow also applies to data stores. If information keeps flowing into a data store but never flows out, it may mean (in order of probability):

a. The analyst has failed to find out what use is made of the information in the data store.

b. The information is being kept for posterity (or because a permanent record is required) but no regular specific uses can be identified. (For example, see Reconciled-Bank-Statements in Figure 10-11.)

FIGURE 10-10. Transform description for Transform 4.5.2: Produce P & L.

TRANSFORM 4.5.2: Produce P & L

DESCRIPTION:

Determine Period-Ending-Date.

Calculate Cost-of-Goods-Sold = Beginning-Inventory + Purchase − Ending-Inventory.

Calculate Gross-Profit-on-Sales = Income-from-Sales − Cost-of-Goods-Sold.

For each Operating-Expense:

Add the Operating-Expense to Total-Operating-Expenses.

Calculate Net-Operating-Income = Gross-Profit-on-Sales − Total-Operating-Expenses.

Calculate Gross-Income = Net-Operating-Income + Other-Interest-Income.

For each Interest-Expense:

Add the Interest-Expense to Total-Interest-Expenses.

Calculate Net-Income-Before-Tax = Gross-Income − Total-Interest-Expenses.

Calculate Net-Profit-or-Loss = Net-Income-Before-Tax − State-Income-Tax − Federal-Income-Tax.

DIAGRAM 4: ACCOUNTING

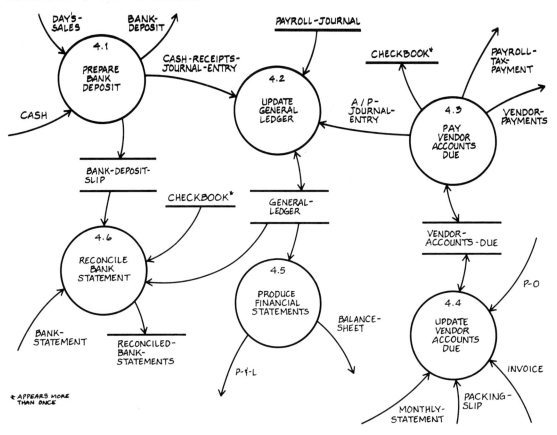

FIGURE 10-11. Diagram 4: Accounting.

c. No one needs or uses the stored information; the data store is unnecessary and should be eliminated.

If information leaves a data store but never enters, the implications are often less serious, but the situation should still be investigated. Some possibilities are:

a. The analyst has omitted incoming flows which store or modify information in the data store.

b. In the steady state, the data store is used for reference only. Updating the data store never occurs or happens very infrequently. In this situation, the analyst may defer depicting the initial loading or occasional updating of the data store until the steady-state data flow diagram has been completed and checked.

c. Updating of the data store is the responsibility of someone outside the system.

Emphasis on Logical Rather than Physical Description

This is done in two ways:

a. Avoid showing material flow; instead show the related information flow. Flow of material is usually accompanied by flow of data that includes everything of interest in the information processing system.

Consider the flow of cash from Sales to Accounting via the Cash-Vault (Figure 10-4). The material flow consists of bills and coins. The corresponding information flow is a count of the number of coins or bills of each denomination. We expect the dollar total of cash added to the vault each evening to equal the Day's-Sales.

b. Eliminate transforms which are entirely physical and represent only changes from one information-carrying medium to another.

Figure 10-12 depicts a purely physical transformation. The information content of the source document has not changed. Only the information-carrying medium is different—cards instead of paper.

Naming Components

Assigning useful, meaningful, communicative names to the system components is one of the most critical parts of the modeling process. Names carry the burden of communication in a data flow diagram. The more easily a data flow diagram can be read without constant reference to the data dictionary, the better. Obscure, arbitrary, or idiosyncratic names inhibit understanding.

Component names should be concrete and system-specific.

For example, **Transaction** is a very general name; it could refer to a time card in a payroll system or a customer payment in an accounts receivable system or even a food order in the FastFood Store. **Banking-Transaction** is a more specific name; at least it tells us the nature of the system to which it belongs. But in a data flow diagram for a bank, the name Banking-Transaction is still not very helpful. **Deposit** or **Withdrawal** is a considerable improvement. It denotes a specific data flow within a specific system.

Similarly, the name **Master-File** tells us nothing more than **More-or-Less-Permanent-Data-Store.** Instead, its name should tell us what is in it, such as **Professor, Student,** or **Class.**

Avoid vague verbs such as **Process, Handle,** and **Do.** Instead, choose precise transform names such as **Compare Invoice and Purchase Order, Produce Customer Statement,** and **Record Customer Payment.**

FIGURE 10-12. A purely physical transformation.

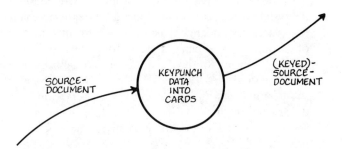

When possible, choose names the users use, as long as they are meaningful. In this way the users and the analysts will be speaking the same language. If the users' names do not convey the contents of the data flows and data stores or the essential actions of the transformations, then assign a better name and retain the users' name as an alias (or vice versa).

Remember the principles for assigning good names summarized in Chapter 6:

- Data flows are moving packets of data; they take on the name of these packets.
- Data stores are resting places for data; they take on the name of the data that reside there.
- Transforms change incoming data flows into outgoing data flows; their names state the activity involved in the transformation and incorporate the names of the inputs and outputs.
- Origins and destinations are producers and consumers of data flows; they take on the name of the outside entity they represent.

Qualify names for uniqueness. That is, modify a basic data flow or data store name to show what distinguishes it from other system components with similar names. Names used in a data flow diagram may be as long as necessary; they are not limited to an arbitrary number of characters as in many programming languages. General-Ledger-Current-Asset-Account-Name and General-Ledger-Current-Asset-Account-Number in the FastFood Store system have 41 and 43 characters, respectively.

It is common for different data flows to have the same composition. Some people like to take advantage of this fact to reduce the size of the data dictionary as well as indicate on the data flow diagram that the compositions are identical. They enclose the qualifiers that make the names unique in parentheses and have a single entry in the data dictionary for the unqualified name.

For example, (Checked)-Packing-Slip would not be found in the data dictionary for the FastFood Store. Its composition is the same as that of Packing-Slip, which has an entry in the data dictionary.

When the data elements are reached, it is desirable for their names to be self-defining. Then the data dictionary entry for that data element need not include further description or explanation.

Minimum Redundancy

It is important to minimize redundancy in the set of data flow diagrams as well as in the rest of the structured specification. Each redundant element makes modifying the system description more difficult by requiring changes to be made in more than one place. Redundancy makes it easier to introduce inconsistency if one or more of the duplicated elements is left unchanged when the system description is modified.

The following kinds of redundancy are permissible within data flow diagrams:

When drawing a lower-level diagram, it is necessary to duplicate data flows from the parent diagram. This preserves conservation of flow at the lower level and depicts the required balancing from level to level. If incoming or outgoing data flows access data stores, the data stores must be shown unless the data flows are named.

It may be necessary to depict a data store more than once within a single data flow diagram to eliminate crossing of data flows. In this case, a special symbol such as * or + should call the reader's attention to the duplicated data store.

All other redundancy, such as additional cross-references between diagrams, should be avoided.

Redundancy among the parts of the structured specification is also undesirable. The following guidelines many help in deciding where various features of a system model belong:

The **movement** of data is depicted in a **data flow diagram.** Here the emphasis is on the data flow as an interface between other system components, showing the connectivity and the paths of movement. The information content is implied by a well-chosen name.

The **composition** or information content of data flows and data stores is defined precisely in the **data dictionary.**

The **algorithm** for deriving outputs from inputs—what transformation is performed on the incoming data flows, not how the transformation is implemented—is rigorously detailed in a **transform description,** but only for a primitive transform.

SOME HELPFUL HINTS

Here are a few more tips for producing clear, correct data flow diagrams. Follow them where they are helpful, but use them flexibly to promote understanding and communication. To make them a rigid set of rules can become counterproductive, if not self-defeating.

1. **Where to show origins and destinations.** Sources and sinks are usually shown on only one high-level diagram—on the context diagram if there is one, otherwise on Diagram 0. Anything else is redundant.

2. **Show the big picture first: suppress details.** Concentrate on presenting a good model of the system in its steady state. Ignore trivial inputs, such as a date supplied by the computer operating system. Logically, the date is probably part of a more comprehensive packet of information, anyway. Ignore trivial rejects and detailed error processing. A representation such as that shown in Figure 10-13 will suffice initially. If the context makes it obvious, the rejected data flow need not even be named.

What is a trivial reject? Here is DeMarco's rule for identifying which error conditions are important enough to be shown on the early iterations of a set of data flow diagrams:

> If the error requires no undoing of past processing, ignore it for the moment; if it requires you to back out previous updates or revert a file or files to a previous state, then do *not* ignore it.*

*Tom DeMarco, **Structured Analysis and System Specification,** Yourdon Press, New York, © 1978, p. 68. Reprinted by permission of Yourdon Press.

DIAGRAM 2: SALES

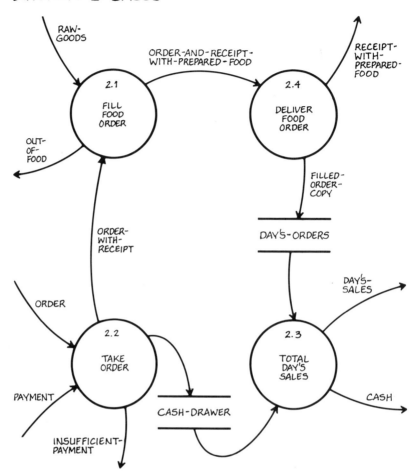

FIGURE 10-13. Revised DIAGRAM 2: Sales, showing rejected orders.

Also ignore the updating of a data store whose contents seldom change as well as the initial loading of a data store which for all practical purposes is read-only.

To complete the system model, these details must be filled in, as described in Chapter 14. Do not forget them; just defer them.

3. ***It is all right to have more than 9 bubbles in Diagram 0.*** For lower-level diagrams, the magic number 7 plus or minus 2 is a useful guide to the number of transforms shown. This guideline is often relaxed for Diagram 0, because it presents the top-level decomposition of the entire system in a single picture. Here as many as a dozen or so bubbles may be permitted if the additional decomposition clarifies the system model. There is an interaction between the complexity of the connections and the number of transforms. If some bubbles have many connections, fewer bubbles are desirable; if the number of connections is small, more bubbles can be tolerated visually. If increasing the number

of bubbles will decrease the number of interfaces to some of the bubbles, the resulting diagram will have greater clarity.

4. *Avoid aggregation that conceals or confuses.* Combining related data flows at the upper levels of a data flow diagram can reduce the number of interfaces and thus help communicate the big picture. However, lumping together unrelated data flows is usually detrimental. A good test for an appropriate degree of aggregation is whether you can think of clear, specific names for the combined data flows. Overly general names such as Incoming-Transactions and Printed-Reports suggest too much aggregation.

5. *Minimizing the number of interfaces.* Simplifying the interfaces in a set of data flow diagrams is another way of reducing the complexity of a system description. It is desirable to minimize the number of data flows between a transformation and its neighbors. It is also important to eliminate superfluous data elements, restricting the content of each data flow to what is necessary to the data store or transform to which it moves. One technique for reducing the number of interfaces is *topological repartitioning.* It is discussed in Chapter 11.

EVALUATING DATA FLOW DIAGRAMS

Throughout the process of system modeling, it is necessary to evaluate the evolving system description during and after each iteration. Mistakes and problems discovered during one iteration will drive the search for improvements to be made in the next iteration.

The criteria for evaluating a system model are explained in Chapter 11. Informal review and evaluation is continuous; formal reviews (walkthroughs) tend to occur when an iteration is complete and are also discussed in Chapter 11.

SUMMARY

This chapter suggests a process for drawing a set of leveled data flow diagrams, consisting of 11 steps:

1. Identify the documents and displays associated with the system.
2. Determine the system boundary.
3. Use the list of documents and displays to identify the system inputs, outputs, origins, and destinations.
4. Draw a context diagram and check it for reasonableness.
5. Identify the principal data stores.
6. Draw a top-level diagram of the whole system.
7. Improve the partitioning of the initial data flow diagram.
8. Develop an initial data dictionary to support the data flow diagram.
9. Evaluate the diagram, and make improvements and refinements based on that evaluation.

10. Use walkthroughs to find mistakes.

11. Continue the decomposition to lower levels until the primitives are reached.

It also summarizes principles for drawing data flow diagrams, illustrated by examples from the FastFood Store system. These principles are:

Data flows are required as the interfaces between other types of system components.

Data flows on a child diagram must be in balance with those for its parent bubble.

Partition as evenly as the system will permit.

Every data store must be shown at least once in the set of diagrams; a data store shared by transforms always appears on a diagram where those transforms are shown.

Conservation of data flow holds for every transform in the system. It also holds for every data store not used as an archive.

Emphasize logical rather than physical system characteristics.

Component names must be unique and should also be concrete and system-specific.

Eliminate or minimize redundancy throughout the structured specification.

REVIEW

10-1. How are documents and displays in an information processing system represented in a data flow diagram? How are these documents and displays related to the other components of a data flow diagram? Give examples from a system with which you are familiar.

10-2. List 4 clues to identifying the data stores in an information processing system.

10-3. Give 2 criteria for deciding when primitive transforms have been reached in the process of decomposition.

10-4. Pairs of components of a system model are listed below. For each pair, state whether it is legitimate to connect the components with a data flow. If not, why not?
   a. Origin to data store
   b. Transform to data store
   c. Data store to data store
   d. Transform to transform
   e. Transform to destination
   f. Origin to destination

10-5. List 4 conventions related to data stores in a leveled set of data flow diagrams.

10-6. State the principle of conservation of data flow.

10-7. State in what part of the structured specification you would expect to find the following:
   a. Movement of data
   b. The information content of a data flow
   c. An algorithm for deriving a data flow

11 Evaluation and Refinement of a System Description

This chapter discusses the criteria for evaluating a system description and presents a formal evaluation technique—the structured walkthrough. A system description is reviewed and evaluated with respect to completeness, consistency, correctness, and communicability. Problems identified through formal or informal evaluations can then be corrected. The system description may be further improved and refined by keeping the transformations independent of each other and ensuring that each transformation and its subordinates carry out a unified task. This chapter addresses each group of criteria and gives specific recommendations for the components of a system model—data flow diagrams, the system dictionary, transform descriptions, and the data base description.

CRITERIA FOR EVALUATING A SYSTEM DESCRIPTION

The criteria used to evaluate a system description may be grouped into four major categories: completeness, consistency, correctness, and communicability—the four C's of evaluation.

Completeness

The concern for completeness addresses the question "Is everything in the system description that needs to be there?" The answer to this question is determined by the specific system and the requirements of its users. A complete system description requires the system boundary to be sufficiently inclusive. In particular, those parts of a system that may be changed or automated should be included in the system description.

Diagram 0 and the Context Diagram are used to judge whether the system boundary is appropriately located. If the system boundary has been drawn too tightly, the analyst may wish to know what happens in detail inside an origin or destination. Important flows of information may occur between external entities. In such situations, the boundary should be extended to incorporate those details and data flows. Otherwise, by definition, they are outside the scope of study and interest.

On the other hand, if the boundary is too inclusive, the origins and destinations will be remote from the focus of study and interest. In that case, analysts and users may realize that some of the transformations shown in Diagram 0 will have no interaction with the proposed new system and therefore can be eliminated from the system description.

For example, the financial statements for the FastFood Store are currently prepared by an accountant rather than by Mrs. O. Yet Figures 6-1 and 6-2 show that the production of financial statements is part of the FastFood Store system being studied. Diagram 4 (Figure 6-6) and its subordinate diagrams provide further detail. Because the information requirements for producing the financial statements for the FastFood Store are included in the system description, Mr. and Mrs. O and the analyst can consider automation or other alternatives as part of the system study.

As applied to a leveled set of DFDs, completeness means that there are no missing data flows, data stores, transforms, origins, or destinations.

Data stores that are ready-only or write-only have to be questioned as to whether all the data flows to and from them have been included. If a read-only data store is not modified as part of routine system operation, it should be considered incorrect or a net system input. A similar argument holds for write-only data stores. The key phrase is "as part of routine system operation." Consider a vendor catalog used in the FastFood Store system to order food or supplies. It arrives irregularly and is not modified as part of routine system operation. Thus, it may be considered a net system input. The retention of payroll records for seven years as mandated by the government constitutes an example for write-only data stores.

Completeness of the system dictionary may be tested by ensuring that each data flow, data store, transform, origin, and destination in the set of DFDs has a corresponding entry in the relevant portion of the system dictionary. Each lower-level named data structure must also have a data dictionary entry, as must every data element in the system. There must be a data store entry in the data dictionary for each object and relationship in the data base description. Each primitive transform must have a transform description. This is a tedious task and a promising candidate for automation.

The data base description is evaluated for completeness by checking that all objects and relationships required in the system are included. Each object's attributes must be defined, and the necessary operations (as implied in the transform description) for each object and relationship must be specified with their appropriate preconditions and postconditions.

While achieving completeness, the analyst is concerned with eliminating excessiveness and redundancy. Otherwise, the statement of users' requirements will incorporate duplication or unnecessary detail. Thus, the analyst strives to strike a delicate balance between including in the structured specification everything necessary to describe the system and being parsimonious in the description. Parsimony in the structured specification is achieved through eliminating extraneous or redundant components from the system description. Similarly, extraneous or superfluous data in a data flow or data store should be eliminated. The use of graphics and concise notation instead of narrative also contribute to parsimony. *FRUGAL*

Consistency

Consistency within a system description measures compatibility of its components. Within a set of DFDs, consistency is maintained through conservation of data flow, balancing between parent and child diagrams, consistent naming of components and appropriate leveling, partitioning, and connectivity.

In the data dictionary, consistency is achieved through the use of names that occur in the real world, through the elimination of aliases, and through a uniform notation for describing data structure.

Consistency for transform descriptions implies a uniform style for the logic description techniques—Structured English, decision tables, and decision trees.

For the data base description, consistency may be judged by checking that each object and relationship on the DAD matches what is in the rest of the system description. Each object in the data base description should appear as a data store on the data flow diagrams. The composition of the relationships in the data dictionary should match the arrows shown in the data access diagram. The access patterns stated in the data flow diagrams and transform descriptions should also be consistent with the data access diagram.

Correctness

In practice, it is easier to identify errors than to try to prove that a system description is correct. Thus, the principal test for correctness lies in the question "Is it demonstrably wrong?", where "it" is the component or components being evaluated.

Correctness of a system description may be judged from two perspectives: syntax—the concern for proper form and structure of the system model—and semantics—the concern for proper meaning and interpretation of the symbols used in the system model. Of the two, syntax is by far the easier to evaluate. Correctness of form can be determined by referring to the system description alone. An analyst, even one unfamiliar with the system, can tell whether the modeling rules have been violated or the notation has been misused. Semantic correctness, on the other hand, requires comparing the model to reality and involves the users. That is, the analyst depends upon the users to determine how accurately the model corresponds to the reality it purports to represent. That is why, with the analyst's help, users must be able to understand data flow diagrams. Sometimes a simple counterexample is sufficient to show that a data flow diagram is inadequate. At other times, a detailed investigation is needed.

A DFD is obviously wrong if any data flow makes an illegal connection between DFD components. A flow is wrong if it is a control flow masquerading as a data flow. Material flows ought not to be shown on data flow diagrams; they can usually be replaced by genuine data flows containing the essential information about the material flow.

The principal syntactic concern in the data dictionary is with correct use of DD notation for the description of data structure. A similar concern holds for the control structures of a transform description, as discussed in Chapter 8. Semantic correctness of a data dictionary description or transform description is judged by comparing it to the real-world entity it describes.

The correctness of the data base description requires an accurate description of objects and relationships. Objects in the DBD must be simple and minimal. They must have a key. A relationship must connect two defined objects. It concatenates the key of the object at which the relationship is rooted and the key of the object at which the relationship is terminated. The necessary and sufficient operations on objects and relationships should be made explicit in the descriptions of the transforms that directly manipulate those objects and relationships. Security and integrity requirements may dictate additional conditions for operations on objects and relationships.

Communicability

The principal concerns here are the conceptual clarity of the system description and the clarity with which the description is presented. All the effort put into completeness, correctness, and consistency can be compromised if effective communication is not also achieved. The effectiveness of communication is measured by the quality of the names used and their correspondence to identifiable real-world entities. Effective communication also depends upon the neatness, legibility, reproducibility, and graphic organization of the system description as well as on the presentation quality of the complete structured specification.

Good names make an extremely important contribution to efficient and effective communication. By choosing names that capture the meaning or contents of a component, we reduce the need to refer to the system dictionary. Wishy-washy names—handle, process, do, information, data—are to be avoided, as are jargon, cryptic names, or form identifiers such as W-2 and 941-A.

Effective graphic organization of DFDs means the elimination of crossing lines and a proper balance of components (more components imply fewer connections). Neat lettering and sharp, clear, easily reproduced drawings increase the graphic quality of the diagrams. Similar standards of graphic clarity apply to the data access diagram.

Effective communication in the data dictionary means the use of symbols for informative communicating with systems professionals and expanded notation for communicating with users. Decomposition of complex data structures into meaningfully named constituents also enhances communication.

In transform description the most important decision is the choice of representation—Structured English, decision table, decision tree, or hybrid.

For data base description the data access diagram should present only the minimum information necessary to support the user views. Each object and relationship on the DAD must be named. Objects must have their keys identified. Only the required access paths through the data base should be shown. The details of the objects and relationships are relegated to the system dictionary.

One major challenge is to minimize the complexity of presentation. This complexity of communication is measured in terms of the number and complexity of components and the number and complexity of interfaces. Structured analysis stresses the use of partitioning, hierarchy, aggregation, and generalization to minimize complexity.

REFINEMENT OF DATA FLOW DIAGRAMS

In addition to the criteria of completeness, correctness, consistency, and communicability, there are some further important considerations as the data flow diagrams are iteratively revised. These additional considerations help improve the data flow diagrams by examining the partitioning of the transforms and the data flow between transforms.

Independence of the Transformations

The first group of considerations is intended to maximize the independence of the transformations. This is motivated by the desire to have a system (and a system description) that is easy to modify. Minimizing the information dependencies among transforms will localize and limit the effects of changes to the system.

In a system in which the transforms are completely independent of each other, the transforms operate in parallel, each receiving its own input and producing its own output. There are no data flows connecting them and creating information dependencies among them (Figure 11-1). Such a system is fully decomposed; each of its transforms can be modified without affecting any of the others. This situation is ideal from the perspective of independence of the transformations.

In most systems it is necessary for the transforms to work together to derive the system's outputs. Thus they will be connected by data flows, and some will be dependent upon previous transforms for their inputs. In such cases, the goal is to accomplish all the required transformations while minimizing the number of connections and the amount of data moving across each connection.

Topological Repartitioning. One technique for reducing the number of interfaces is ***topological repartitioning.*** It is based on the topology, or connectedness, of the network and looks at the interfaces abstractly. Because of the abstraction involved, the use of this technique should be tempered by an understanding of the system being modeled.

Here is how it works: Figures 11-2 and 11-3 show data flow diagrams that are good candidates for topological repartitioning. Transform 3 is highly connected to the rest of the diagram and has many inputs and outputs. Diagram 3, however, shows that the transform consists of three disconnected or independent parts. This suggests that Diagram 0 can be rearranged into parts with fewer interfaces.

The transforms and data flows are shown abstractly because topological repartitioning looks only at the network and not at the meaning of its parts.

The first step is to expand Diagram 0, replacing Bubble 3 by its subordinates. The result is shown in Figure 11-4. Then we look to see what is the smallest number of data flows we can cut to break the network into separate pieces. As the figure

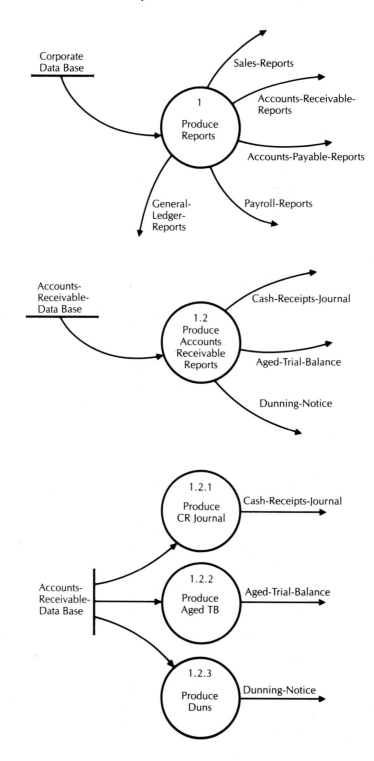

FIGURE 11-1.
Decomposition into independent transforms.

DIAGRAM 0

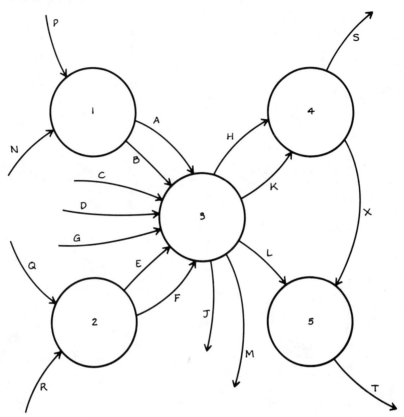

FIGURE 11-2. Data flow diagram before topological repartitioning, Diagram 0.

shows, cutting arc K separates Bubble 3.3; cutting arc H creates another piece; arc X now separates Bubble 4; and arc L separates Bubble 5.

The final step is to collect each of these clusters of bubbles into a single bubble and draw a new Diagram 0, renumbering the transforms. (See Figure 11-5.) The lower-level diagrams 1′ and 2′ must also be redrawn.

Limiting Shared Data.   Another way of minimizing the interdependence of the transforms is to reduce the amount of data shared via data stores. The techniques for logical data base description presented in Chapter 9 and illustrated in Chapter 12 remove nonessential data from the data stores, limiting the composition of each object to its own essential attributes.

Directness of Data Flow.   It is also important for data to move as directly as possible from the transform where it is produced to the transform where it is needed (or to a data store) without passing through other transforms which do not use it.

For example, Figure 11-6 shows Z flowing through Transform 1 even though

DIAGRAM 3

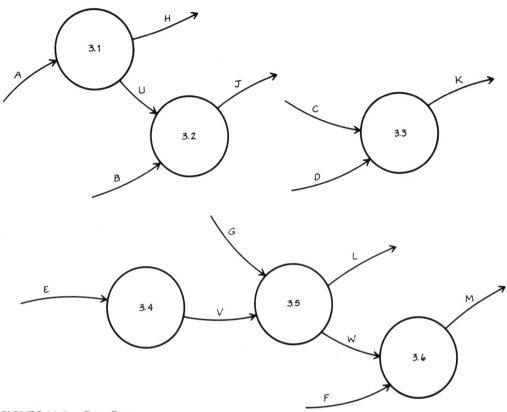

FIGURE 11-3.   Data flow diagram before topological repartitioning, Diagram 3.

Z is not needed to produce B, the output of Transform 1. However, Z is the essential input to Transform 2, which produces Y. Z should flow directly into Transform 2, as shown in Figure 11-7.

Aggregation of Related Transforms

Another important aspect of refining the system description deals with the way transforms are aggregated in a set of data flow diagrams. Because the diagrams and the transforms have a hierarchical structure, they may be viewed from the top down as a successive decomposition of the system or from the bottom up as a successive aggregation of the primitive transforms. When considered from the bottom up, transforms should be combined to form higher-level transforms only if they are closely related. Each transform in an aggregation should work closely with its neighbors to contribute to the joint performance of a single more inclusive transformation.

In many businesses the organizational structure is a result of such factors as history, politics, and geography. Information processing tasks may be allocated to

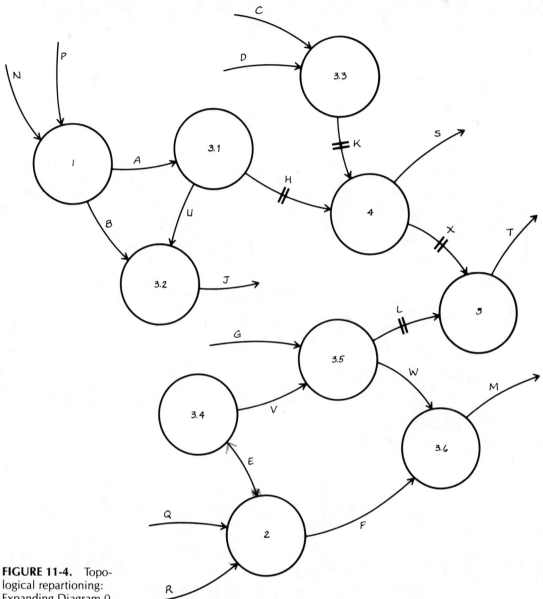

FIGURE 11-4. Topological repartioning: Expanding Diagram 0.

individuals or groups based on their ability or availability to level work loads or for other reasons that have little to do with the logical or inherent relationship of these tasks to each other.

However, in specifying users' information processing requirements, the underlying unity of a related group of transforms should be emphasized in the process of aggregating the diverse components at each level. This is a matter of putting

DIAGRAM 0

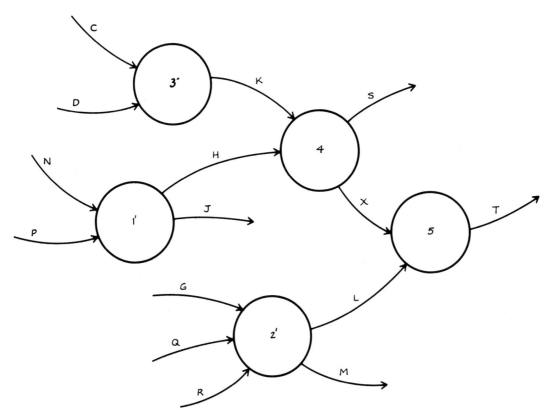

FIGURE 11-5. Topological repartitioning: Revised Diagram 0.

FIGURE 11-6. Unnecessary indirect data flow.

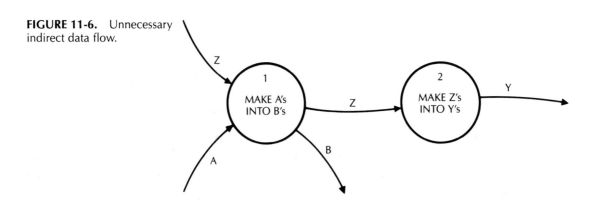

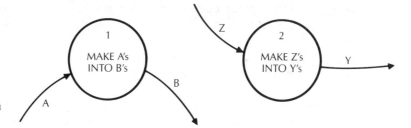

FIGURE 11-7. Direct data flow.

together things that belong together because they must work together rather than because they are similar in nonessential ways. For example, transforms which update different and unrelated data stores should not be combined just because they both update data stores. The fact that unrelated transforms may occur at about the same time is not a good reason to aggregate them. Aggregation of transforms which do not contribute to a single problem-related task is likely to result in increased system complexity because of increased connectivity and unnecessary data flow.

Here are some useful tests for checking how good a job of aggregation you have done:

One test is to look at the names of the transforms. If a transform name is an imperative sentence with a single specific verb, that implies a unified aggregation. On the other hand, if the sentence requires several different verbs, that suggests that several unrelated transforms have been improperly combined. (Sometimes at the top levels, where there are only a few bubbles, it is difficult to find a single appropriate verb. In that case, it is better to use several specific verbs which convey what is happening below than to choose a single, vaguer verb which camouflages the realities at the lower levels.)

Another test is to look for generalized transform names such as **Update Files** or **Print Reports.** It is not likely that all the data stores in the system are so closely related that a single transform is required to update them. A transform that updates data stores should be limited to the accesses necessary to support a single user view. Similarly, a transform that generates a report should be combined with transforms that provide the data for the report rather than with other report-producing transforms.

A final test is to look at lower-level diagrams. If a single transform on an upper level breaks apart into entirely disconnected pieces below, perhaps these independent pieces should not have been aggregated. Admittedly, on Diagram 0 and its children it is necessary to limit the number of bubbles for the sake of clear communication. Sometimes the consequence is lumping together transforms we would otherwise prefer to keep separate.

WALKTHROUGHS

A walkthrough is a method for quality assurance and control during the development of the structured specification. This portion of the chapter discusses what a walkthrough is, why it is held, who should participate, roles and responsibilities of walkthrough participants, and how walkthroughs are conducted, and presents a comprehensive checklist for quality assurance of a system description.

What Is a Walkthrough?

There are several terms for walkthrough, such as "structured walkthrough" and "review." No matter what they are called, walkthroughs are directed toward the same result—identification of deficiencies in quality and of areas that should be improved. The important word here is "identification." We will see later that resolution of the defects identified is not an objective of a walkthrough. Correction of defects is left to the author of the work product.

A walkthrough is a group review of a work product for the purpose of judging its completeness, correctness, and adherence to standards. This definition suggests that a constructive critique of the work product will be an output of a walkthrough.

There are many differences in walkthroughs depending upon who the participants are, the type of product being reviewed, and the phase of the system life cycle. Walkthroughs may be conducted with users or analysts or managers or designers or programmers or auditors. Products reviewed in walkthroughs may include: components of the structured specification, a project plan, a system design, computer program code, and test plans. Thus walkthroughs may be conducted during various phases of the system development life cycle.

By any name, a walkthrough is criticism in the sense of evaluation and appraisal. It attempts to detect defects as early as possible so that they can be remedied quickly, thereby improving the quality of the product being evaluated. There is a strong need to make the criticism constructive. This implies a need for a healthy, supportive environment so that critical comments will help the author improve the product. One final admonishment: it is only the work product that is judged in a walkthrough—the author is always innocent because he or she is never on trial.

Why Have Walkthroughs?

The economic benefits of walkthroughs provide an important justification for them. It is less expensive to catch defects early in the system development life cycle rather than later. There are two reasons for this. One, errors early in system development have a much broader effect. An error in the structured specification may affect many computer programs, while an error in a computer program is more limited. Two, the cost of fixing an error rises dramatically as a project proceeds further in the system development life cycle.

Other less tangible benefits of walkthroughs include: the development of team identity, group responsibility for the quality of the work products, and sharing of creativity. Enforcement of standards and monitoring of project progress are other benefits of walkthroughs.

Walkthroughs are held to address the quality issues of: identifying of errors, omissions, and improvements; judging the correctness, readability, and understanding of the product; and adhering to standards.

User walkthroughs during analysis focus on the validation of DFDs as an accurate representation of the users' current or future business operations. Internal walkthroughs with other analysis team members focus upon the completeness, correctness, and consistency of the structured specification. Management reviews focus on project progress.

There are several important distinctions between management reviews and the two kinds of walkthroughs identified above. A management review is held less often, usually at the end of a major phase. The focus of a management review is on project progress with less emphasis on the quality issues for the product produced. It is concerned with comparison of actual versus projected progress as measured by resource utilization and products produced. A management review may involve a decision to continue or terminate the project. However, the decision made as a result of walkthroughs by users or analysts is whether to accept or reject the work product. We will say very little more about management reviews in this chapter, but will discuss in some detail relevant issues for walkthroughs.

Who Should Participate in a Walkthrough and What Are their Roles and Responsibilities?

We mentioned earlier that users, analysts, managers, designers, programmers, auditors, and consultants may participate in a walkthrough depending on its purpose. In this section we identify the generic roles for walkthrough attendees and their associated responsibilities before, during, and after the walkthrough.*

Figure 11-8 summarizes the roles and responsibilities of walkthrough attendees.

The presenter is usually the producer or author of the product to be reviewed. The presenter must be careful to let the product stand or fall on its own merits. Therefore, introductory remarks should be kept to a minimum.

The coordinator's role is critical to the success of the walkthrough. The coordinator must supervise the preparation for the walkthrough, ensure proper decorum and adherence to the agenda during the walkthrough, and follow up by distributing the management summary and detailed action list.

The secretary must be able to take intelligent notes without interrupting the walkthrough. This suggests that a stenographer is not appropriate because a detailed transcription of the walkthrough is not called for; rather, an understanding of the context, purpose, and terminology is required.

Participants may be assigned other roles: user representative—responsible for protecting the users' interests; standards enforcer—responsible for checking that the work product adheres to predefined organizational standards; maintenance evaluator—responsible for judging the maintainability of the work product; auditor—responsible for assessing the adequacy of controls.

Information about a walkthrough is recorded on the **walkthrough report.** The walkthrough report is divided into two parts to serve two different audiences. These are: a summary of the walkthrough for project management; a detailed list of action items for the producer(s) of the work product.

Figure 11-9 presents the multipurpose **walkthrough summary.** It presents the before and after checklists for the coordinator. It acts as a cover memo for work products delivered to participants to notify them of walkthrough particulars. It provides the walkthrough agenda. It may also serve as a management summary.

*This material is based upon Edward Yourdon, *Structured Walkthroughs,* Yourdon Press, New York, 1978.

FIGURE 11-8. Roles and responsibilities before, during, and after a walkthrough.

	BEFORE	DURING	AFTER
PRESENTER	1. Produce product. 2. Choose product that can be reviewed in 30 to 60 minutes. 3. Choose coordinator. 4. Suggest participants. 5. Suggest agenda.	1. Seek out problems. 2. Thank participants.	1. Fix problems. 2. Start cycle again.
COORDINATOR	1. Confirm product is ready for walkthrough. 2. Choose participants. 3. Assign responsibilities. 4. Select meeting place and time. 5. Notify participants and send work product for preview. 6. Check response and receipt of work product for each participant.	1. Act as moderator. 2. Keep to the subject matter. 3. Maintain decorum.	1. Distribute management summary. 2. Distribute detailed action list. 3. File management summary and detailed action list.
SECRETARY		1. Take down detailed action items.	
PARTICIPANTS	1. Agree to participate. 2. Review work product.	1. Offer constructive criticism. 2. Decide disposition of work product. 3. Sign Walkthrough Report.	

WALKTHROUGH SUMMARY

To: _____   Date: _____

From: _____
                              Walkthrough Coordinator

Re: Please review the attached material and bring your marked-up copy for a walkthrough
   to be held at _____
                                              Location
   on _____ at _____ for _____
        Date   Time   Duration

        This is ☐ the first walkthrough of   _____
                                                     Work Product Name
               ☐ a subsequent walkthrough of   _____
                                                     Work Product Name
        for: _____
                                              Project
        produced by: _____
                                        Responsible Analyst(s)

A. Coordinator's prewalkthrough checklist.

  1. Confirm with producer(s) that material is ready.
  2. Choose attendees.
  3. Assign responsibilities.
  4. Select meeting time and place.
  5. Notify participants and send work product.
  6. Receive responses and check receipt of work product.

B.	Participants	Responsibilities	Reply Received	Work product
1.	_____	_____	___	_____
2.	_____	_____	___	_____
3.	_____	_____	___	_____
4.	_____	_____	___	_____
5.	_____	_____	___	_____
6.	_____	_____	___	_____

C.   Walkthrough agenda

  1. ☐ All participants agree to abide by the same set of rules.

  2. ☐ First walkthrough for work product led by presenter.

      ☐ Subsequent walkthrough of work product: item-by-item check-off of previous
         action list.

  3. ☐ Creation of new action list from contributions by each participant.

  4. ☐ Disposition of work product:
              ☐ accept work product as is.
              ☐ conditionally accept [revision by producer(s); requires no further
                walkthroughs.]
              ☐ revise work product and reschedule subsequent walkthrough.

  5. ☐ Participants' signatures:
              _____   _____   _____
              _____   _____   _____

D. Coordinator's postwalkthrough checklist:

  1. Forward copy of Walkthrough Summary to project management.
  2. Forward copy of Detailed Action List to each participant.
  3. File copy of Walkthrough Summary and Detailed Action List in project file.

FIGURE 11-9. Walk-
through report:
walkthrough summary.

Figure 11-10 shows the **detailed action list** portion of the walkthrough report. This is the document on which the walkthrough secretary records action items. It is used by the author as a checklist of items to be resolved. For subsequent walkthroughs, it serves as an agenda item.

Conduct of the Walkthrough

There are two major concerns for the conduct of a walkthrough—work product quality control and cost-effective use of each participant's time and effort.

Quality control checklists for the various components of the structured specification are provided in the following section.

The following measures can enhance the cost-effectiveness of walkthroughs conducted during the analysis phase:

- Emphasize detection not correction. Problems identified in the walkthrough should be resolved afterward.
- Participants must agree to abide by the same set of rules.
- Participants should know what is to be accomplished and what is expected of them.
- Participants should be charitable critics. The benefit of the doubt should go to the author. Some organizations insist that participants offer at least one positive and one negative comment.
- Participants should be objective. Do not assume or insist that your method is best. Rather ask: Will the proposed method work? Is there a better solution? Is that solution vastly superior?
- Participants should notify the author privately prior to the walkthrough of any flaw so serious that it should be corrected before the walkthrough or of anything that might otherwise embarrass the author. In this case, the problem should be remedied before the walkthrough is held, even if a postponement is necessary.
- It is better to reschedule a walkthrough if participants are not prepared. It is better to terminate and reschedule a walkthrough when a substantial deficiency is identified.
- Keep walkthroughs as short as possible.
- Keep the number of participants to a manageable number—three to six.
- Make the walkthrough only as formal as necessary.

FIGURE 11-10. Walkthrough report: detailed action list.

WALKTHROUGH DETAILED ACTION LIST			
Fixed	Action Item #	Issue raised in walkthrough.	(Optional short explanation of correction.)
___	___	_____	_____
___	___	_____	_____
___	___	_____	_____

User Walkthroughs

For user walkthroughs, avoid the use of computer jargon. Especially avoid acronyms—a DFD is a picture of the system, the DD is a list of data used or stored in the system, and transform descriptions are statements of user policies or procedures. Stress the need for the user to verify that the DFD is an accurate representation of the operation of the system. Remember, the user is the expert in his area.

Conduct an initial user walkthrough on the user's own turf. Keep it short (less than a half-hour). Be careful not to overwhelm the user. Use physical references (people, departments, document names, etc.) where necessary to help orient the user. Do not distribute materials prior to this initial user walkthrough—users need gentle and controlled introduction to the tools of structured analysis. When talking to a user who is not familiar with data flow diagrams, begin by referring to them as pictures of the system and gradually introduce the term *data flow diagram.*

For subsequent user walkthroughs, ease the user into a more formal approach. Predistribute materials for review. Suggest an agenda for each meeting—list of questions and/or points for discussion. Try to wean the user from dependence on physical features in the system description. Encourage the user to mark up his or her copy of the DFD and to write questions directly on the diagrams.

Quality Control Checklists for System Descriptions

We conclude this chapter with checklists which summarize the criteria for the various components of the structured specification. These are intended to be used by analysis team members in preparation for internal walkthroughs. Figure 11-11 presents the checklist for a leveled set of data flow diagrams.

FIGURE 11-11. Data flow diagram checklist.

For each diagram in a leveled set of DFDs:
    Is the diagram number and name correct?
    Is the diagram understandable?
      Are there any crossing lines?
      Is the number of components reasonable?
      Has the asterisk been used to indicate duplication?
    Is the diagram complete and correct?
    For each component of a diagram:
      Select the appropriate case:

        Origin/destination
          Is there a system dictionary entry for the origin/destination?

        System input/output
          Is there a DD entry for each system input/output?
          Is the composition complete and correct?
          Does the system input originate at the proper origin?
          Does the system output terminate at the proper destination?

(continued on next page)

FIGURE 11-11   *(continued)*

Transform
    Is the transform number and name correct?
        Is the transform number unique?
        Is the name a simple sentence using an imperative verb and object?
        Does the name capture the meaning of subordinates or contents?
    Is the incoming and outgoing data correct?
        Are there unnecessary or superfluous data flows?
        Are there missing data flows?
        Do data flows balance between parent and child?
    Is the transformation performed essential?
    If the transform is primitive
            Does the transform description exist?
        Otherwise (the transform is not primitive)
            Does the child diagram exist?

Data flow
    Is the data flow properly named?
        Is the name unique?
        Does the name reflect the data flow's contents?
    Does DD entry exist?
    Is composition correct?
    Does the data flow make a legal connection between components?

Data store
    Is the data store properly named?
        Is the name unique?
        Does the name reflect the data store's contents?
    Does DD entry exist?
    Is its composition correct?
        Has the key been identified?
    Is the data store essential? Could it be replaced by a data flow?
    Is the data store shown at proper level?
        Is it accessed by two or more transforms?

Figure 11-12 presents the checklist for the data dictionary.

FIGURE 11-12.   Data dictionary checklist.

For the data dictionary:
    Are the entries for data elements, data flows, and data stores alphabetized?
    Does each data store appear at least once in the leveled set of data flow diagrams?
    For each data element:
        Is the specification of values and meanings or range of values complete and correct?

FIGURE 11-13.   Transform dictionary checklist.

For the transform dictionary:
  Are the entries in transform number sequence?
  For each primitive transform:
    Is the transform description correct?
      Are the constructs used properly?
      Are the constructs explicitly scoped?
      Are the data elements, data flows, and data stores named in the transform descrip-
      tion also defined in the data dictionary?
      Is the choice of representation (structured English, decision tree, or decision table)
      appropriate?

Figure 11-13 presents the checklist for the transform dictionary.
Figure 11-14 presents the checklist for the data base description.
These four checklists together cover the entire system description.

FIGURE 11-14.   Data base description checklist.

For the data base description:
  For each object:
    Is the object name unique?
    Is there a DD description?
    Is the composition complete and correct?
      Has the key been identified?
      Are there any embedded objects?
    Have the appropriate logical operations been identified?
    Have the preconditions and postconditions been specified for each operation?
    Does the object appear somewhere on a data flow diagram?
  For each relationship:
    Is the relationship name unique?
    Is there a DD description?
    Is the composition correct?
    Is the one-to-many or many-to-many indication on DAD correct?
    Is there a relationship in the opposite direction?
      Has it been specified?
      Should it be specified?
  For each user view:
    Does the requisite access path exist?
      Are the attributes of the objects on the access path sufficient to fulfill the
      information requirements?
      Are there any unnecessary objects and/or relationships on the access path?

SUMMARY

Ensuring a high-quality system description requires a determined effort to detect and eliminate errors in the system model—the earlier, the better. The relevant criteria for this effort fall into four major categories—completeness, consistency, correctness, and communicability—the four C's of evaluation.

Checking completeness in a data flow diagram requires determining whether any data flows, data stores, and transformations have been omitted. It also examines whether the system boundary is sufficiently inclusive.

Checking consistency looks at conservation of flow, balancing among levels, and consistent labeling of the components of the data flow diagram.

Checking correctness compares the system model with the present or future reality it is intended to represent. Users are the arbiters of whether the data flow diagram corresponds to the reality of their information processing system.

Checking communicability involves evaluating the extent to which the data flow diagrams and data base description hinder or help all the participants in the analysis process to understand the system. It is partly a matter of neatness, graphic organization, and clarity in drawing and presenting the diagrams. It is also a matter of the conceptual clarity of the system description.

As the system description is iteratively evaluated and revised, further refinement can be achieved by making the transformations as independent as possible and by aggregating only those transformations which belong together.

Evaluation of system descriptions during analysis involves not only informal reviews but formal quality-control reviews as well.

A group review of a work product in order to judge its completeness, correctness, consistency, and adherence to standards is called a walkthrough. Walkthroughs during systems analysis focus on the various system models to identify problems for subsequent resolution. Among the participants in a walkthrough are the author of the work product being reviewed, a coordinator, and a secretary.

REVIEW

11-1. Name the four criteria for evaluating a description of an information processing system.

11-2. Why is it important to maximize the independence of the transform descriptions in the system? What limits the degree to which the transforms can be made independent of each other?

11-3. Identify some pitfalls in aggregating transformations.

11-4. Define a walkthrough.

11-5. How does a walkthrough differ from a management review?

11-6. Discuss the differences among user walkthroughs, internal walkthroughs, and management reviews.

11-7. What are some of the concerns the analyst should keep in mind when conducting a user walkthrough?

EXERCISES AND DISCUSSION

11-1. Identify as many errors as you can find in the abstract data flow diagram in Figure 11-15. Also identify portions of the diagram which are not necessarily errors but which may indicate problems in the system description.

FIGURE 11-15.

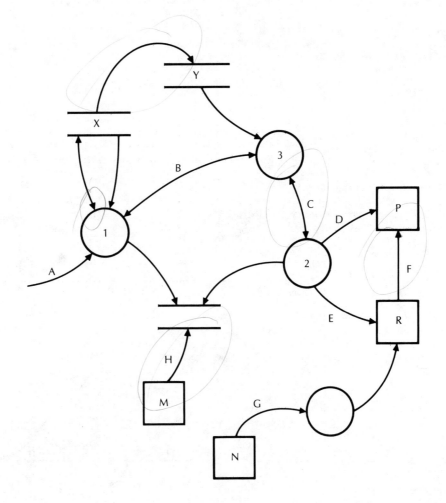

11-2. Improve the diagram shown in Figure 11-16 based on the following assumptions:
  a. Data flow Z is a necessary input for Transform 4 only.
  b. Data flow Z is a necessary input to Transform 1 as well as Transform 4.

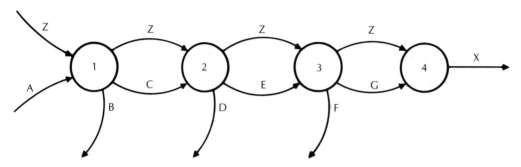

FIGURE 11-16.

11-3. Comment on the decomposition of Transform 1 shown in Figure 11-17. Also comment on any problems in the composition of data flows A and X.

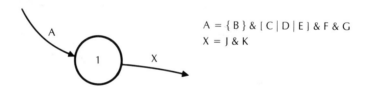

$$A = \{ B \} \& [C | D | E] \& F \& G$$
$$X = J \& K$$

FIGURE 11-17.

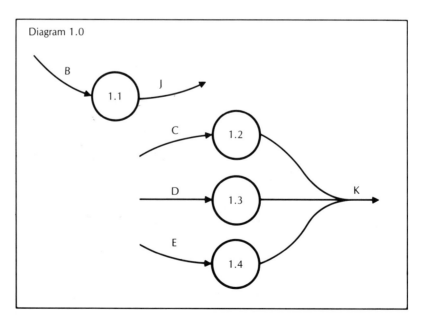

Diagram 1.0

11-4. Figures 11-18 through 11-20 show part of a set of data flow diagrams and the composition of selected data flows.
   a. Evaluate these diagrams and the data dictionary.
   b. Revise these diagrams to correct the mistakes and problems you found.
   c. Also check for violations of the data dictionary conventions.

Diagram 0: Payroll

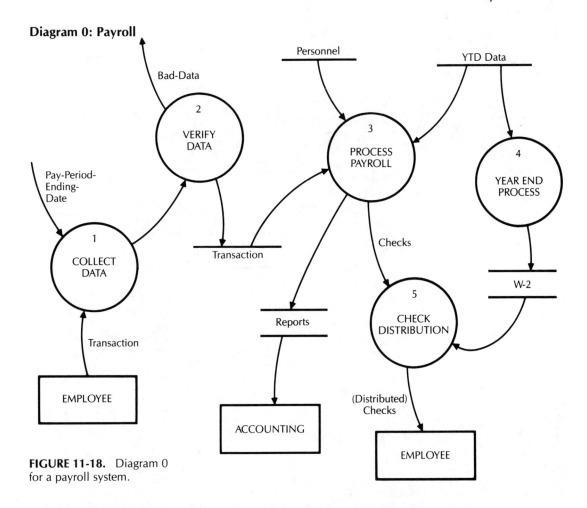

FIGURE 11-18. Diagram 0 for a payroll system.

Hints for Diagram 0:
   Check for legal connections.
   Check for specific transform names of the form verb and object.
   Check for specific data flow names.
   Check for bubbles which make no meaningful transformation.
   Check for data stores not shared by transforms.
   Check for unnamed data flows.
   Check for extraneous or unnecessary data flows.

Diagram 3: Process Payroll

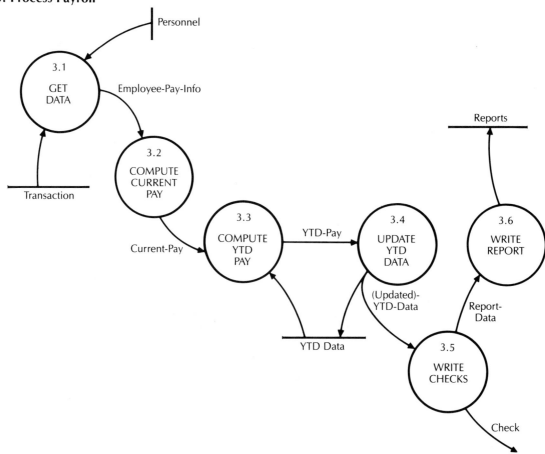

FIGURE 11-19. Diagram 3 for a payroll system.

Hints for Diagram 3:
Check for conservation of flow by comparing the data flow diagram and the data dictionary.
Check that data stores are shown at the proper level.
Check for bubbles which make no meaningful transformation.
Check for specific data flow names.

Hints for Diagram 3.2:
Check for conservation of flow by comparing the data flow diagram and the data dictionary.
Check for violations of the data flow diagramming conventions.

Diagram 3-2: Compute Current Pay

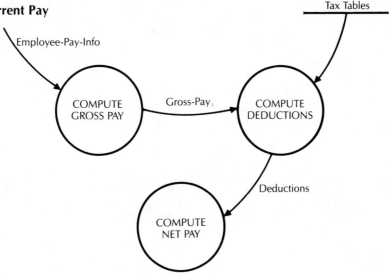

Data Dictionary Excerpts:

$$\begin{aligned}
\text{Transaction} &= \text{Pay-Period-Ending-Date \& Employee-Number \& } \{ \text{ Work-Date \&} \\
&\quad\ \text{Hours-Worked } \} \\
\text{Employee-Pay-Info} &= \text{Employee-Number \& } \{ \text{ Work-Date \& Hours-Worked } \} \text{ \& Pay-rate} \\
\text{Gross-Pay} &= \text{Hours-Worked} \times \text{Pay-Rate} \\
\text{Deductions} &= \{ \text{ Deduction-Type \& Deduction-Amount } \} \\[6pt]
\text{YTD-Data} &= \{ \text{ Employee-Number \& } \{ \text{ Pay-Period-Ending-Date \&} \\
&\quad\ \text{Total-Hours-Worked \& Pay-Rate \& Gross-Pay \& Deduction \&} \\
&\quad\ \text{Net-Pay } \} \} \\
\text{Check} &= \text{Pay-Period-Ending-Date \& Employee-Number \&} \\
&\quad\ \text{Employee-Name \& Hours-Worked \& Pay-Rate \& Gross-Pay \&} \\
&\quad\ \text{Deduction \& Net-Pay} \\
\text{Report} &= [\text{ Earnings-Register } | \text{ Deduction Register } | \text{ Check Register }]
\end{aligned}$$

FIGURE 11-20. Diagram 3.2 and data dictionary excerpts for a payroll system

11-5. Review the data flow diagrams for the FastFood Store found in Chapter 6, and identify any problems you find. (Resolving these problems will be dealt with in Part III.)

11-6. Discuss the purpose(s) of walkthroughs.

11-7. Discuss the importance of the roles and responsibilities of walkthrough participants.

11-8. Discuss why management is presented with a walkthrough summary rather than a list of detailed findings. Do you agree? Justify your position.

11-9. Walkthroughs have been accused of being expensive. Explain the basis for this accusation. Identify and discuss some measures to limit walkthrough expense.

PART III

APPLYING THE TOOLS: THE ANALYSIS PROCESS

Part III explains the process of structured analysis. Figure III-1 shows the activities involved in establishing the requirements for an information processing system, as presented in Chapter 5. In this process four kinds of system descriptions are developed—two for the existing system and two for the proposed system. These are the Physical and Logical Current System Descriptions and the Logical and Physical New System Descriptions. Each is produced from its predecessor in the sequence of transformations shown in Figure III-1.

Part III discusses all four of these system descriptions—Physical Current, Logical Current, Logical New, and Physical New. Chapter 12 focuses on the development of the physical current system model and its logical equivalent. This corresponds to Transform 2.1: Describe the Current System and Transform 2.2: Pare the Current System Description to Essentials in Figure III-1. Chapter 13 concentrates on modeling the new system. This corresponds to Transforms 2.3: Establish New System Requirements, 2.4: Define Alternative Scopes of Automation, and 2.5: Select the Best Scope of Automation in Figure III-1. Chapter 14 addresses the completion of the major work product of structured analysis—the structured specification. This corresponds to Transform 2.6: Complete and Package the New System Requirements. Chapter 15 discusses how the completed specification is modified to accommodate subsequent changes in the users' requirements. Chapter 16 concludes Part III with a discussion of alternatives to the full systems analysis process described in the preceding chapters.

Why are four kinds of system model necessary? Figure III-2, which extracts them from Figure III-1, provides a framework for answering this question.

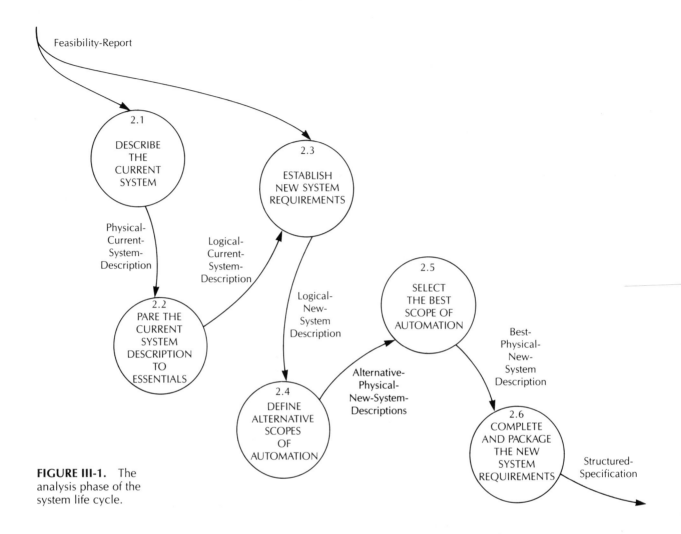

Feasibility-Report

2.1
DESCRIBE
THE
CURRENT
SYSTEM

2.3
ESTABLISH
NEW SYSTEM
REQUIREMENTS

Physical-
Current-
System-
Description

Logical-
Current-
System-
Description

2.2
PARE THE
CURRENT
SYSTEM
DESCRIPTION
TO
ESSENTIALS

Logical-
New-
System
Description

2.5
SELECT
THE BEST
SCOPE OF
AUTOMATION

Best-
Physical-
New-
System
Description

2.4
DEFINE
ALTERNATIVE
SCOPES
OF
AUTOMATION

Alternative-
Physical-
New-System-
Descriptions

2.6
COMPLETE
AND PACKAGE
THE NEW
SYSTEM
REQUIREMENTS

Structured-
Specification

FIGURE III-1.   The analysis phase of the system life cycle.

Users initiate information system development because they wish to change the existing system. At the outset there is a great deal of enthusiasm and pressure to move to the new system as quickly as possible. The enthusiasm and pressure may be due to anticipated economic benefits or increased employee morale, legal mandates, or a variety of other reasons. As Figure III-2 indicates, users would like to proceed directly from the physical current system to the physical new system—that is, from the system as it is currently implemented to an implementation of the new system. Unfortunately, at this stage users have only vague and unstructured perceptions of the new system, which are inadequate as a basis for the implementation. Users are often unsure of what their true requirements are. Occasionally they are too sure of what they want, jumping to conclusions without having carefully considered all their implications.

The analyst, on the other hand, would prefer to start with an abstraction of the new system that captures the essential inputs, transformations, and outputs. From

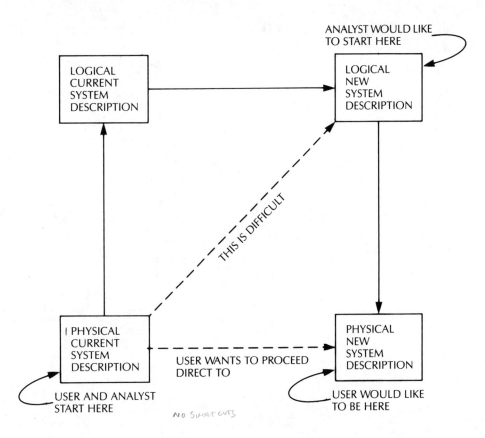

FIGURE III-2. Four system descriptions used in structured analysis.

this abstract model—the Logical New System Description—he can prepare alternative physical descriptions of the new system for consideration by the users. Management can then weigh the social, political, economic, and environmental factors affecting each alternative and select the best alternative for implementation. This is illustrated in Figure III-3.

In some cases, it is desirable to produce alternative Logical New System Descriptions. These may show differences in which transforms are removed from the current system, differences in which transforms are added to the new system, or differences in the scope of automation. Each of these logical alternatives may then lead to several physical descriptions as shown in Figure III-4.

The problem for the analyst is that preparation of the Logical New System Description requires attention to many details—identification of good and bad features of the existing system to know what to change; identification of transformations to be added, deleted, or modified; identification of the essential inputs, outputs and transforms for the new system; and determination of the scope of automation.

The development of a logical description of a new system directly from a physical description of the current system, if there is one, is extremely difficult. The analyst

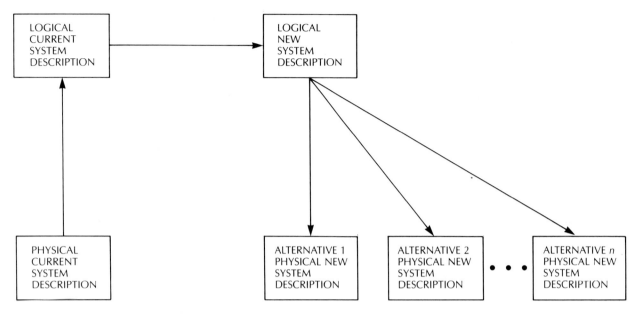

FIGURE III-3. The generation of Alternative Physical New System Descriptions.

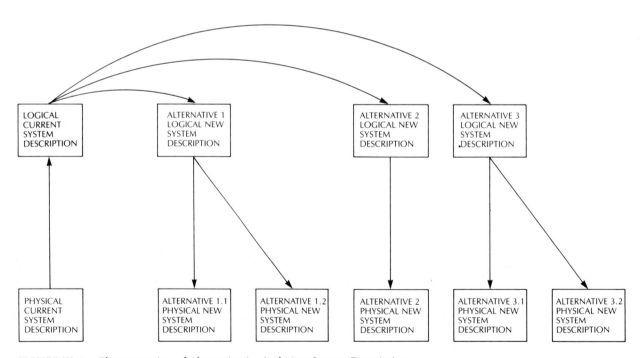

FIGURE III-4. The generation of Alternative Logical New System Descriptions.

is deluged with details. Some are essential to the new system; some are not. How is the analyst to decide which is which?

To overcome this difficulty the analyst is forced to attack the problem incrementally. Before producing the Logical New System Description, the analyst develops a logical description of the current system—an abstract model of its essential inputs, transformations, and outputs. However, even this task is difficult. There is usually no one person with a complete, coherent, and consistent understanding of the existing system to the degree of detail required for the logical system model. Managers do not know the details of current system operation; clerks do not see the big picture. Moreover, each person views the operation of the system in terms of the artifacts and procedures peculiar to his or her activities. Distinguishing the essence of a system from its physical manifestation is something that users are not regularly called upon to do. Thus the analyst must start with a physical model of the current system, a description of the implementation of the existing system, as indicated in Figure III-2.

12 Modeling the Current System

WHAT IS NEEDED?

Establishing the requirements for a new information processing system, as outlined in the introduction to Part III, begins by producing a complete description of the current system. *Completeness* here implies not so much a high degree of detail as it connotes inclusiveness of the essential transforms, data flows, and data stores. Beginning with a physical or implementation-dependent system description, the analyst tries to move as rapidly as possible toward a logical or implementation-independent and essential system description.

Thus the primary task of modeling the current system is to differentiate between that which is essential and that which is unnecessary. A system component may be considered unnecessary for many reasons. It may exist only for historical reasons: "That's the way we've always done it." It may depend on information storage media: hard-baked clay tablets. It may incorporate inconsequential or trivial detail: "I *always* start with a sharp pencil." It may refer to time cycles: "We do our filing during the middle two weeks of the accounting cycle because that's when we're least busy." A system component is considered necessary, however, because it is fundamental or essential. **The omission of an essential transform, data flow, or data store would render the system incapable of performing its functions and fulfilling its purposes.**

There are additional benefits from studying the current environment. During this period of analysis users and analysts can develop a rapport. The user can gain confidence in the analyst. Mutual education can take place—the analyst comes to

229

understand the users' system; users learn the analyst's tools and approaches to analysis.

Moreover, as Edward Yourdon indicates:

> Quite often, the modeling process itself is an agent of change. The act of modeling may highlight errors, redundancies, or inefficiencies in the users' organization. This is particularly likely during the development of the **physical** model. The user may be perfectly content with his fundamental business policy, but he may find ways to improve the **implementation** of that policy by studying an accurate physical model—long before the analysts have proposed a new automated system!*

One of the analyst's prime concerns at this stage of analysis is how much of the current system to look at and at what level of detail it should be studied. Some guidance is given in the Feasibility Report. Detailed discussion of the Feasibility Report is deferred until Chapter 17. Briefly, however, there is a section of the Feasibility Report that gives the charter for the system study. This charter determines the domain of the study. Enough of the environment surrounding the system must be studied to ensure that all interfaces relevant to the proposed areas of change can be identified. This is abstractly illustrated in Figure 12-1, where the prospective area of change, identified in the Feasibility Report, is totally contained within the domain of the system study.

If the analyst is fortunate, the existing system is documented with modern system description tools, such as those described in this book. The data flow diagrams, data dictionary, transform descriptions, and data base description for the last major system change, if accurate and up to date, can provide the requisite description of the current system. To judge the accuracy and currency of this description, the analyst still needs to check it against the way things are actually done. This task is much easier when the analyst is using familiar tools that provide a reasonably accurate description of a system with which the analyst has some experience.

All too often, this is not the case. The different users' views of the system are hidden in documents, reports, procedure manuals, job descriptions, organization charts, and the minds of individuals participating in the system. These difficulties are compounded when the information is incomplete, inconsistent, out of date, and unstructured. Descriptions of the automated parts of the system are not much better. These may consist of system flowcharts, computer programs, input and output formats, file descriptions and layouts, job control language, and procedures for data preparation, dissemination, and computer operation. The principal drawbacks of these forms of documentation for an analyst who is trying to understand a system are that they focus at too low a level of detail, show flow of control, and emphasize physical formatting requirements.

Although the task of structuring all this into a complete, coherent, and correct system description is filled with adversity and frustration and requires hard work, the tools of structured analysis introduced in Part II can help the analyst cope effectively with ill-structured complexity. Further, there is a tremendous sense of accomplishment and fulfillment when this task is completed. So the analyst should

---

*Edward Yourdon, **Managing the System Life Cycle**, Yourdon Press, New York, N.Y., © 1982, p. 74. Reprinted by permission of Yourdon Press.

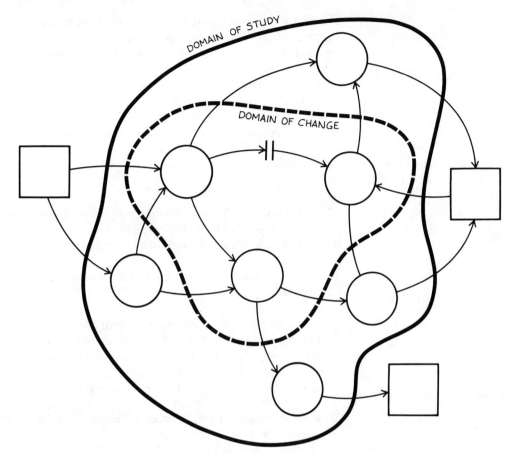

FIGURE 12-1. The domain of change contained within the domain of study.

put the existing documentation to work, recasting it into a system description comprising a leveled set of data flow diagrams, a data dictionary, a set of transform descriptions, and a data base description.

DEVELOPING THE PHYSICAL CURRENT SYSTEM DESCRIPTION — strategies

Chapter 10 gave some guidelines for drawing data flow diagrams. In this section we point out characteristics of DFDs that describe the current system in physical detail. We illustrate the iterative refinement that must be applied to the Physical Current System Description to make it complete, correct, and consistent.

There are some side effects and unavoidable consequences of trying to describe a system for the first time. These influence an analyst, whether veteran or novice, to take a defensive posture. Even though the analyst is interested in capturing the essence of *what* is being done in the current system, it is difficult to avoid being influenced by *how* things are currently done. A natural tendency is to include everything because it is difficult to distinguish the essential from the nonessential.

CHAPTER 10
INCORPORATE OUTPUT

SAMPLES 10 STORES
WITH DFDS

The analyst sets out trying to avoid too much physical detail, but in the end the analyst's defensive posture, coupled with the unstructured complexity, usually yields an initial system description that is incomplete, inconsistent, and incorrect. Perhaps a difference between the veteran and novice analyst is the veteran's ability to recognize this and make a conscious effort to iteratively refine the system description.

There are two principal strategies for developing an initial Physical Current System Description and then refining it. The first strategy works backward from the system outputs; the second works forward from the system inputs.

Working from the Outputs Backward to the Inputs

Some authors claim that this is the only effective strategy for developing a system description.*

We are less insistent. Nevertheless, starting with the outputs is an effective approach that enjoys widespread use. It is presented here as a series of questions addressed by the analyst in the order given below.

1. ***What essential outputs must the system produce to satisfy its users' requirements.***[†]

   In answering this question, the system outputs are modeled as data flows and their compositions defined in the data dictionary. The destination for each system output is identified. Implicit in this question is the supposition that the system's boundary has already been defined. If that is not the case, you must answer the question, **What is the system boundary?**

2. ***What transformations are necessary to produce the essential system outputs?***

   These transformations are shown on the data flow diagrams. The policy governing each primitive transform is recorded in a transform description.

3. ***What inputs are necessary for the transforms to produce their outputs?***

   These inputs are modeled as data flows. Some are net system inputs, originating in the environment and necessitating the identification of their origin. Other data flows are provided from previously stored data internal to the system. This gives rise to the following question.

4. ***What essential memory must the system maintain in order to carry out the essential transformations?***

   The system's essential memory is modeled with data stores, which are partitioned into objects and relationships as discussed in Chapter 9. This viewpoint leads to the data base definition—What are the objects, relationships, and operations with their preconditions and postconditions? With the requirement for

---

*Kenneth T. Orr, *Structured Systems Development*, Yourdon Press, New York, N.Y., 1977, p. 12.

[†]The notions of essential inputs, outputs and memory are introduced by John F. Palmer and Steve McMenamin in Chapter 3 of ***Essential Systems Analysis,*** Yourdon Press, New York, N.Y., 1984.

essential data stores comes the necessity for other transforms that maintain these data stores. These custodial or maintenance transforms carry out the logical operations on the objects and relationships in the data base.

Note that the system transforms consist of two types—those that contribute to the production of some essential system output and those that maintain essential data stores.

Even in the most modest system, it is difficult to list *all* of the systems outputs, develop *all* the essential transformations, and model the inputs and memory, *all* in parallel. Thus these tasks are accomplished iteratively, working with a small number of system outputs backward until each system output has been accounted for.

Stimulus-Response: Working from the Inputs Forward to the Outputs

In this second strategy, the system is viewed as reacting to external stimuli (inputs) with preplanned responses (outputs).*

Thus the arrival of a data flow triggers a predetermined set of transformations in order to generate the appropriate output.

For example, a withdrawal from a bank account causes the amount of the withdrawal to be subtracted from the account balance. At the end of the month, this transaction will be itemized as part of the account statement. A request for an airline reservation leads the reservations clerk or travel agent to search for an available flight, quote the applicable fare, record the reservation, and perhaps even issue a ticket and boarding passes. Reading a drop form causes a university registration system to remove a student from a class and generate a request for a refund of fees, if one is due.

The sequence of questions for this strategy is as follows:

1. *What are the stimuli and the response(s) to each stimulus?*

    These are modeled as data flows—a stimulus as a system input and a response as a system output. Obviously it is necessary to identify the origin for each stimulus and the destination of each response.

2. *For each stimulus-response pair, what are the transformations that take place along the path from the stimulus to the response?*

    These are modeled as a sequence of transforms on the data flow diagrams.

3. *What are the essential data stores?*

    These stem from the necessity to retain the details of some previous stimulus or response. They give rise to the same questions regarding essential memory found in the previous strategy.

The application of these two strategies for initial system modeling should yield similar system descriptions. The choice of strategy depends upon the system. If there are obvious connections between the inputs and outputs, the stimulus-response approach may be more appropriate. When the connection between inputs and outputs is unclear or complex, or when one input generates many outputs (or vice versa), working backward from the outputs may be easier. Certainly there are sit-

*The stimulus-response strategy is introduced by John F. Palmer and Steve McMenamin in Chapter 3 of **Essential Systems Analysis**, Yourdon Press, New York, N.Y., 1984.

uations where both strategies may be applied concurrently to form an initial system description.

No matter which strategy is used, lay out an initial partitioning of the system, showing the connections between inputs, transforms, outputs, and data stores. Then these data flow diagrams are ready to be reviewed and refined.

A word of encouragement: the DFDs illustrated here have already undergone some initial refinement. Expect your first DFDs to be messy—too busy, with too many components on a page, crossed lines, and questionable names. Do not throw away one of these messy diagrams until you have identified what is wrong with the diagram and what is worth keeping. Then recopy the diagram, and in the process clean it up. This is exactly the procedure followed to develop the diagrams in this chapter for the FastFood Store example.

Refining the Initial Physical Current System Description

There is no single correct approach to examining a system description. Find the problem spots by using completeness, consistency, correctness, and communication—the criteria for judging a system description discussed in Chapter 11. When evaluating your own diagrams, do not invent problems that do not exist just because there is an example here. In fact, if you strive to record the essential logical basis for the system under study, less cleanup should be required. Note too that it is generally easier to identify nonessential parts of a physical system description than it is to specify what is essential. This is true because so much of what is essential is system-dependent.

PARING THE CURRENT SYSTEM DESCRIPTION TO ESSENTIALS

We now shift our viewpoint from comparing the Physical Current System Description with reality and refining it to abstracting its essence—the Logical Current System Description. In fact, this transition moves along a spectrum from physical to logical. An experienced analyst tries to make the physical description of the current system as logical as possible. Physical details are limited to those that help analysts and users thoroughly understand the current system. The two strategies outlined above help create an initial system model already directed toward the essentials. Each successive refinement can contribute to a less implementation-dependent model of the current system.

The Logical Current System Description is a complete, coherent, and consistent system model stripped to its essential data flows, essential data stores, and essential transforms. How to determine what is essential is discussed below for each of these system components. In deriving the Logical Current System Description, the system components may be examined in any order. Several interleaved iterations are usually necessary. Whatever the sequence, the result is an integrated, logical model.

Essential Transformations

A transformation is essential for two reasons:

1. *It is required to carry out the system's fundamental information-processing functions and to fulfill the users' business policies.*

2. **It is required to maintain an essential data store.**

This is very much system-dependent. A clue to the essentiality of a transform is given by the answer to the following question: Is this transform independent of technology? That is, must the transformation still be done no matter how it is implemented or who does it? An affirmative answer indicates that the transform is essential.

Transforms that accomplish edits, audits, approvals, and system performance measurements are likely candidates for removal as nonessential. In a perfect world with error-free data and totally honest people, they would not be needed. In implementing a real-world system, users, managers, and EDP auditors are all rightly concerned with minimizing errors and preventing undetected fraud. For the time being, however, eliminating these important but unnecessary transforms from the system model can reduce its complexity and permit a clear view of what is essential. They will be reintroduced when the physical description of the new system is developed.

Essential Data Stores

A logical data base description, discussed in Chapter 9, is a conceptual model of stored data in terms of objects, attributes, values, relationships, operations, and conditions.

DeMarco provides one approach to derivation of the logical data base description in Chapter 19 of **Structured Analysis and System Specification.** In this section we provide an alternative approach based upon logical object analysis.* †

The objective of logical object analysis is to provide a minimal nonredundant description of the essential memory for the system under study. Just as the DFD drives the modeling of the system, objects drive the modeling of data. This is so because objects are the primary bearers of information in a data base. The following steps outline an approach to developing the logical data base description.

1. **Identify the objects and relationships.**

This is probably best accomplished by starting with the different user views found in the Physical Current System Description. Parts of speech used to label components of the Physical Current System Description provide clues to the role of each component in the data base description and the category to which it belongs, as illustrated in Figure 12-2.

Once an initial determination of the appropriate category has been made, subsequent questions follow for each category as indicated in Figure 12-3.

2. **Draw the data access diagram.**

3. **Simplify the objects.**

This is accomplished in several ways. One is to eliminate data elements whose values can be calculated from other data in the data base. Another is to identify

*David Kroenke, **Database Processing: Fundamentals, Design, Implementation**, 2nd edition, Science Research Associates, 1983, Chapter 5.

†Matt Flavin, **Fundamental Concepts of Information Modeling**, Yourdon Press, New York, N.Y., 1981.

FIGURE 12-2:   Correspondence of parts of speech and data base components.

Part of Speech	Data Base Component
noun	object, attribute
preposition, noun followed by preposition, certain verb forms	relationship
verb, noun derived from verb	operation
adjective, quality	value

and separate objects contained within other objects. A clue to the presence of an embedded object is a data store whose structure includes nested iterations. Another clue to the presence of an object is a subsidiary data structure common to several of the data stores in the data base. Or, if the key for one object appears as a data element in the composition of another object, a relationship between the two objects is often implied. Examples of simplifying objects in these ways appear later in this chapter for the FastFood Store data base description.

4. *Check the attributes of each object.*

    This is accomplished by answering the following questions for each attribute:

    • Is there only one value of this attribute for each distinct instance of the object?

    • Is the attribute dependent solely on the entire key for the object?

5. *Identify the operations on the objects and relationships and the essential preconditions and postconditions for each operation.*

FIGURE 12-3:   Questions about each type of data base component.

Data Base Component	Question
attribute	Of which object? What set of values?
relationship	What is the root object? What is the destination object? Is there a relationship in the opposite direction?
operation	What is the object or relationship operated upon? What are the precondition and postconditions?
object	What are its attributes? Does it have any embedded objects?

6. *Specify the complete data dictionary entry for each object, relationship, and attribute.* Be sure that the operations are described in the transform descriptions for the primitive transforms that maintain the objects and relationships.

7. *Test the logical data base description.*

This is accomplished as described in Chapter 9. Quite often, the result of this step is the identification of some previously undiscovered object or relationship. When this occurs, we have to go back to one of the previous steps.

Some analysts prefer to develop the data access diagram and data dictionary in parallel by successively refining them as new objects, relationships, and attributes are discovered.

The Essential Data Dictionary

Most of the names used in the data dictionary for the current system are evaluated as a natural consequence of examining the transforms and developing the logical data base description. However, the remaining definitions must also be examined to remove what is nonessential. Derivable and extraneous data elements are eliminated. Subsidiary data flows are factored out for reuse.

Essential Transform Descriptions

Reducing the transform descriptions to their essentials is best left until after the essential data flows and transforms have been identified, the logical data base description has been derived, and the data dictionary has been evaluated. Some transforms will have been eliminated; only the essential primitive transforms remain. The data dictionary entries will have been pared to their essential composition and assigned logical names. Thus what remains is to separate policy from its implementation. When describing transforms, we are interested in what the underlying policy is and not in how it is carried out. Though this goal is clear, in practice it is sometimes impossible to divorce a policy from the technology that implements it.

Constructing the Logical Current System Description

The final task is to revise the current system description, merging the essential data flows, transforms, data stores, data dictionary, and data access diagram to form a complete Logical Current System Description. This involves drawing a leveled set of DFDs that depicts the essence of the current system and incorporates a data store for each object identified in the logical data base description. These data flow diagrams are supported by the system dictionary (especially the data dictionary and transform dictionary), which completes the Logical Current System Description. This Logical Current System Description is then evaluated for overall completeness, consistency, correctness, and communication.

THE FASTFOOD STORE—CURRENT SYSTEM DESCRIPTION

Developing a Refined Physical Current System Description

The following discussion of refining a current system description is based on the FastFood Store system introduced in Part II. It is necessarily dependent upon this system and its description. In general, there is no single correct procedure for developing a refined current physical system description.

Before the discussion begins, you need to be aware of a conversation between the analyst and Mrs. O. that took place after the dialogue of Chapter 6. This dialogue shows how the user and the analyst work together to produce a system description that is complete, correct, consistent, and communicative. Before reading the dialogue, you may want to lay out copies of Figures 6-2, 6-6, and 6-7. Then, as you read, you may annotate the copies of the figures as suggested in the dialogue.

1	Mrs. O:	I had a real problem with your Diagram 4. But I'm not sure where to start our discussion.
2	Analyst:	Perhaps I can help. I have a list of questions. Let's lay out Diagram 0, Diagram 4, and Diagram 4.5 here on the table. I see you have written some questions on your copy. In our initial meeting, Mr. O said that you are in charge of all the accounting.
3	Mrs. O:	That's correct. He thinks accounting is nothing more than bookkeeping. He even questions why Payroll is separate here on Diagram 0. But when it comes to accounting and finance, I just don't pay much attention to him. I can't remember the last time he wrote a check, much less balanced the checkbook for either the store or our personal checking account.
4	Analyst:	How do you view your accounting responsibilities?
5	Mrs. O:	Well, you have sort of captured my views here on Diagram 4. But I question some of your names. Our CPA (certified public accountant) maintains the General Ledger and produces financial statements as you have shown here. However, I send him the whole Accounts Payable Journal and the whole Cash Receipts Journal, not just individual entries. I prefer the way you have indicated that Bubble 4.2 references the Payroll Journal. If we cross out Entry on each of these two flows, that pretty well fixes the problem.
6	Analyst:	Okay. I'll also make these name changes in the Data Dictionary. What about this Payroll-Tax-Payment coming out of Bubble 4.3?
7	Mrs. O:	Oh, I remember our conversation on paying vendor accounts. I probably misled you by mentioning paying vendors and payroll taxes in the same breath.
8	Analyst:	What if we strike Vendor from the name for Bubble 4.3?
9	Mrs. O:	That seems okay. But I refer to the General Ledger to find out what the amounts are for the various payroll taxes. I refer to invoices and statements when paying vendors.
10	Analyst:	That suggests two separate functions—one for paying payroll taxes and one for paying vendors. Here, let me sketch it for you [draws Figure 12-4].
11	Mrs. O:	Yes, that's much better. Each has its own information source—the General Ledger for Payroll and the statements for vendors. That reminds me—every so often we get a credit memo from a vendor. I received one just yesterday. You had better show that somewhere, don't you think?
12	Analyst:	Oh yes. Probably coming in here to Bubble 4.4. Will you give me a copy of a credit memo to go along with the other system documents I have collected?

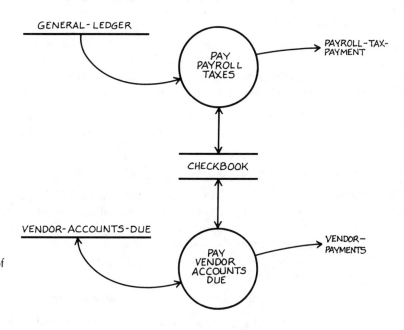

SKETCH OF VENDOR & PAYROLL PAYMENTS

FIGURE 12-4. A sketch of vendor and payroll payments.

13	Mrs. O:	Sure. I think there is another area where I may have misled you. When I reconcile the bank statement, I compare the checkbook balance after reconciliation with the cash balance in the General Ledger. This is only a reasonableness check, and I've never been out of balance. In fact, I'm not even sure why I bother to do it, except for that. Certainly, the bank and I have made mistakes in the past, but these have always been offset by adjusting entries in the General Journal kept by the CPA.
14	Analyst:	Are you suggesting that the reference to the General Ledger by Bubble 4.6 is unnecessary?
15	Mrs. O:	Yes.
16	Analyst:	Speaking of the General Ledger, did you check with your CPA to see if I could investigate that aspect of your accounting?
17	Mrs. O:	Yes I did. He is reluctant to talk with you about it. He wondered if we were considering taking our business elsewhere. After I convinced him that for the time being we were not, he stated that his firm uses a proprietary computerized system. He indicated that the proprietary nature of the computer programs precludes him from allowing you to see them. Frankly, I think he views you as a potential competitor.
18	Analyst:	Then, perhaps it would be better if I did not talk with him. Did you ask him about the processing steps in preparing the financial statements?

19	Mrs. O:	Yes I did. I wrote them down for you. He said they are the same steps as in a manual system. First comes journalizing, which is what I do.
20	Analyst:	Journalizing?
21	Mrs. O:	Writing down an entry in a journal. Next, the journals are posted to the General Ledger. Then a trial balance is prepared. After checking the trial balance, adjusting entries are prepared, journalized, and posted. The adjusted trial balance is prepared. Financial statements are prepared. Finally the books are closed.
22	Analyst:	Thank you. This should clear up a number of details. So, the major parts of your accounting system are Banking, Payroll, Accounts Payable, and the General Ledger.
23	Mrs. O:	Yes, that's correct.
24	Analyst:	What if I draw a circle around Bubbles 4.1 and 4.6, a circle around Bubbles 4.2 and 4.5, a circle around 4.3 and 4.4, and merge in Bubble 3 from Diagram 0 like this [points to Figure 12-5]. Then, the connections between the various parts of accounting are the Checkbook, the Cash Receipts Journal, the Accounts Payable Journal, and the Payroll Journal. Do you think this is a more accurate picture of how you view accounting?
25	Mrs. O:	Well, it's kind of messy . . . but I think it is better than what we had before.
26	Analyst:	Good. It is important that I get a clear understanding of how the system can best serve your needs. Thank you.
27	Mrs. O:	You're welcome.

Dialogues such as this provide the basis for evaluating the current system description. After incorporating the modifications indicated in the dialogue above, we can refine the current system description further.

We start refining the current system description by looking at all the diagrams at once. This can be accomplished by pinning them to a wall or by laying them out on the floor in hierarchical fashion (see Figure 6-13). Now we can judge the **overall consistency of decomposition.** A branch with noticeably fewer levels may indicate incomplete decomposition, suggesting the need for more investigation. For example, the Payroll branch looks suspicious. It merits more study. (It turns out that we did decompose Diagram 3: Payroll but saw no need to include it in Part II.) A branch with noticeably more levels may indicate overdecomposition, suggesting that some aggregation or generalization may be in order. We do not mean to imply that every system must be decomposed to exactly the same level of detail for every branch in the hierarchy of diagrams. To the contrary, each portion of the system description should be decomposed to the degree of detail appropriate to it, following the guidelines for when a primitive transform has been reached.

Next we focus on the Context Diagram (Figure 12-6). Checking completeness, we search for missing or superfluous inputs, outputs, origins, or destinations. We also judge the appropriateness of the system boundary. Satisfied for the moment with completeness, we check consistency by examining the **balancing** between the

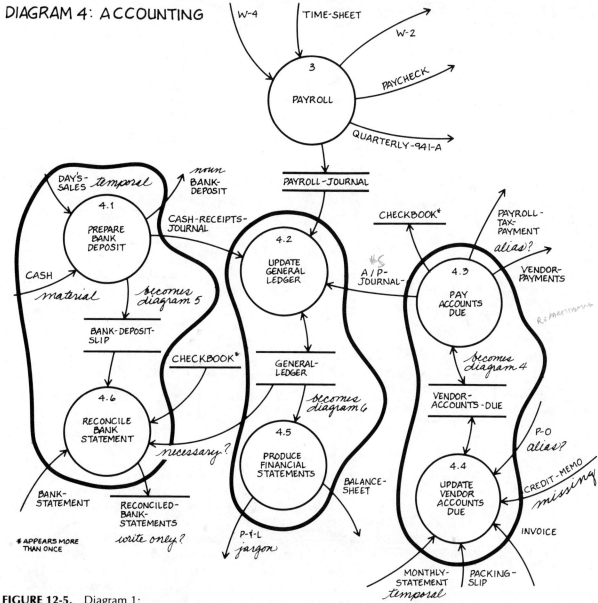

FIGURE 12-5. Diagram 1: Accounting (annotated and repartitioned).

Context Diagram and its child, Diagram 0 (Figure 12-7). We also examine the consistency of data flow names. We evaluate the correctness of the Context Diagram by comparing it with the real-world system it purports to represent. We also check correctness by searching for questionable flows—flow of material or flow of control instead of data flow. Receipt-*With-Prepared-Food* and Order-*With-Cash* certainly sound like flow of material. We mark these as questionable, possibly in color,

CONTEXT DIAGRAM: FAST FOOD STORE

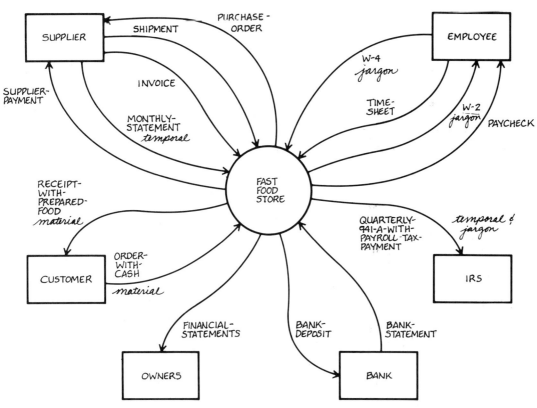

FIGURE 12-6. Annotated Context Diagram.

and label what is wrong so that when we redraw the diagram they may be corrected. Checking the communication effectiveness of the Context Diagram, we find that the graphic organization seems all right. However, the names of some of the system input and output data flows could be better. There are two problems here—the temporal references in **Monthly**-Statement and **Quarterly**-941-A and the coded form names **W-2, W-4,** and **Quarterly**-941-A. Again, we flag these for correction.

We now focus on Diagram 0. Of course, Diagram 0 exhibits some of the problems identified in the Context Diagram. These are annotated in Figure 12-7. Additional data flows have been identified as having temporal and arbitrary names, as in **Day's**-Sales, and **P-&-L.**

Just as data flow names can be questionable, there can also be problems with data store names. Raw-Goods and Cash-Vault are examples of stores for material. The data store name, Sales-**History,** contains a temporal reference in the word History.

Communication problems can also arise with transform names. The transform names in Diagram 0 are nouns indicating departments where certain functions take place. Although not totally incorrect, these names would be much better if they

physical

DIAGRAM 0: FAST FOOD STORE

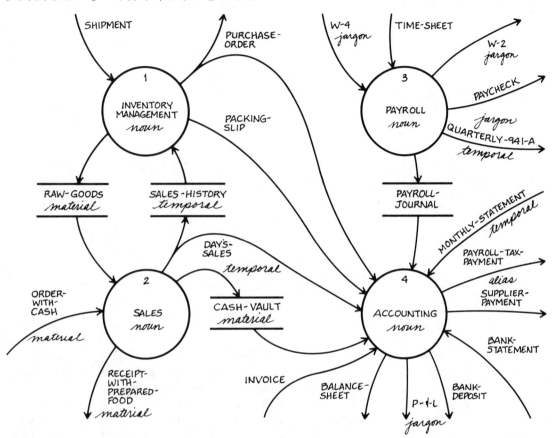

FIGURE 12-7. Annotated Diagram 0: FastFood Store.

were imperative sentences—Manage Inventory, Produce Payroll, and Record Sales.

Bubble 4 looks awfully busy, as Mrs. O. noted in Chapter 6. This is symptomatic of partitioning functions by department. Bubble 4 is definitely a candidate for further partitioning. It is not clear at this level how all the data flows for this transform are related. We defer detailed discussion of this problem for the moment.

We also need to consider the data dictionary; look at the entry for Supplier-Payment below.

Supplier-Payment = Vendor-Name & Vendor-Address & . . .

The apparent inconsistency stems from the two different names used by Mr. and Mrs. O. Mr. O calls them suppliers; Mrs. O calls them vendors. Each name is consistent with its user's perspective—Mr. O is concerned with supplies; Mrs. O is concerned with paying for those supplies. Clearly, there is a communication problem when synonyms arise.

Aliases also arise unintentionally through careless naming. Consider the balanc-

DIAGRAM 1: INVENTORY MANAGEMENT

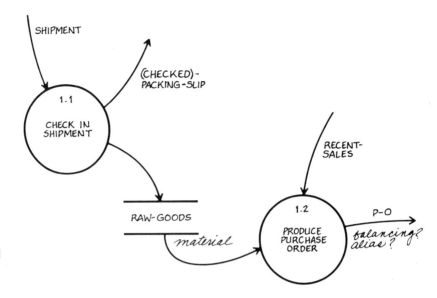

FIGURE 12-8. Annotated Diagram 1: Inventory Management.

ing between Bubble 1: Inventory Management in Diagram 0 (Figure 12-7) and Diagram 1: Inventory Management (Figure 12-8). Are Purchase-Order and P.O. synonyms? Which name should be used in the data dictionary? Obviously, one name must be chosen as the principal name, with the other indicated as an alias.

At this point we have probably identified enough problems to justify redrawing at least the Context Diagram and portions of Diagram 0. Use your own judgement as to when to suspend problem identification and correct the problems. The figures in this chapter illustrate the result of the process, omitting several iterations of intermediate diagrams. To give you a feel for this process, we describe the cleaning-up process for Diagram 4: Accounting. Study the differences between this version of the system description and the initial system description carefully.

Figure 12-9 presents a revised Context Diagram. Noteworthy modifications include replacing Supplier by Vendor, changing data flow names to be more indicative of their contents, and introducing a previously omitted data flow—Credit-Memo.

Figure 12-10 presents a thoroughly reworked Diagram 0. Changes include numerous modifications to names of data flows, data stores, and transforms; replacement of material flows with data flows; and repartitioned transforms that better illustrate the user's view of the system. Some physical references remain—Checkbook, Time-Sheet, Paycheck-and-Stub. Some questions also remain—Is it important to distinguish vendors of goods from vendors of services? On balance, however, this is a much improved representation of the system.

Figure 12-11 shows Diagram 1: Manage Inventory, with revised names for data flows and data stores, based on the annotations to Figure 12-8.

CONTEXT DIAGRAM

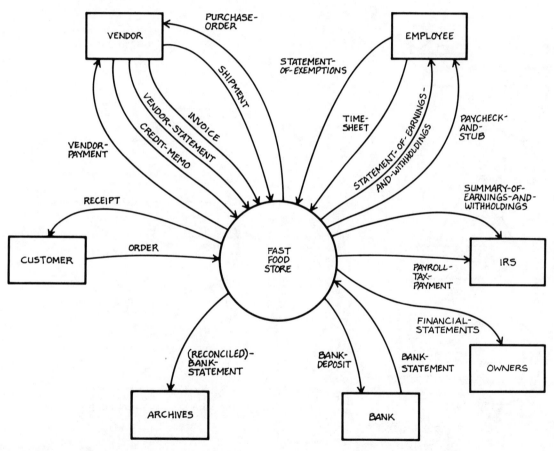

FIGURE 12-9. Revised Context Diagram.

DIAGRAM 0: FAST FOOD STORE

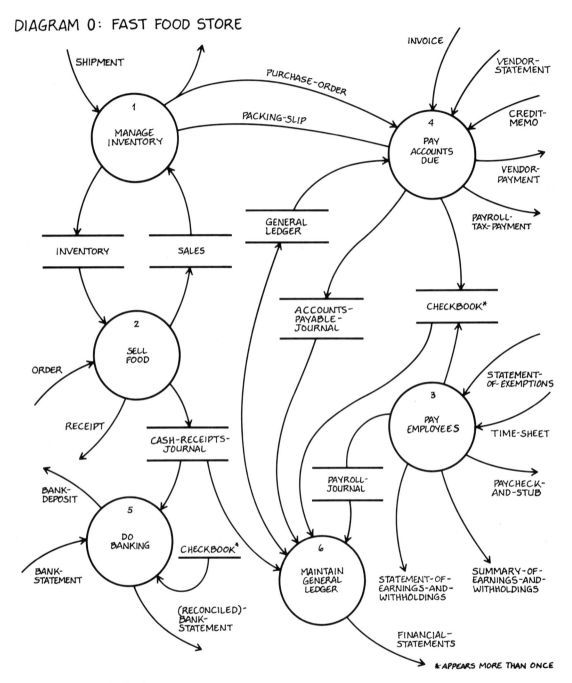

FIGURE 12-10. Revised Diagram O: FastFood Store.

DIAGRAM 1: MANAGE INVENTORY

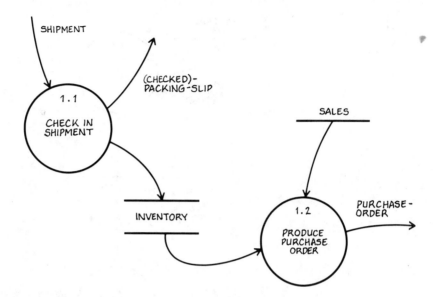

FIGURE 12-11. Revised Diagram 1: Manage Inventory.

The partitioning of Figure 12-5 (the early version of Diagram 4) results in three new diagrams—Diagram 4: Pay Accounts Due, Diagram 5: Prepare Bank Deposits and Reconcile Statements, and Diagram 6: Maintain General Ledger. These diagrams are shown in Figures 12-12, 12-13, and 12-14.

Note also that many changes to data dictionary names were necessitated by the modifications to the DFDs. Since only the names were changed, with their composition remaining the same, the affected parts of the data dictionary are not shown here. Later, when compositions must be modified, these changes to the data dictionary will be shown.

Deriving a Logical Data Base Description for the Accounts Payable Subsystem

The remainder of this chapter describes the steps involved in deriving an equivalent, pared-down, logical model of the current system. Much of the effort goes into producing a logical description of the data base in terms of objects and relationships.

To illustrate this process, this discussion concentrates on the Vendor-Product-Info and Vendor-Accounts-Due data stores and the data flows entering Bubble 4.3 on Diagram 4: Pay Accounts Due (Figure 12-12).

We start with the following data dictionary excerpts from the Physical Current System Description:

Vendor-Product-Info = { Vendor-Name & Vendor-Address
                        & Vendor-Telephone-Number & Vendor-Rep
                        & { Vendor-Catalog } }

DIAGRAM 4: PAY ACCOUNTS DUE

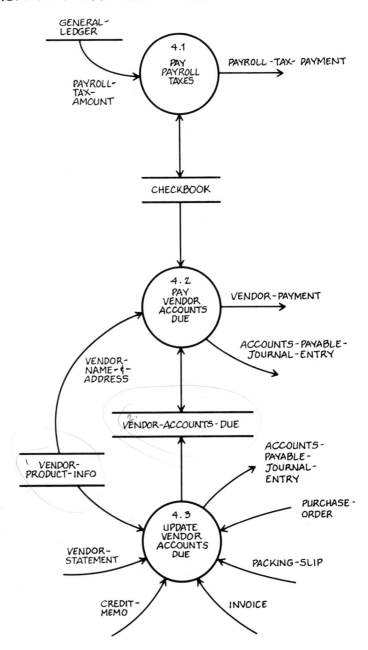

FIGURE 12-12. New Diagram 4: Pay Accounts Due.

DIAGRAM 5 : PREPARE BANK DEPOSITS AND RECONCILE STATEMENTS

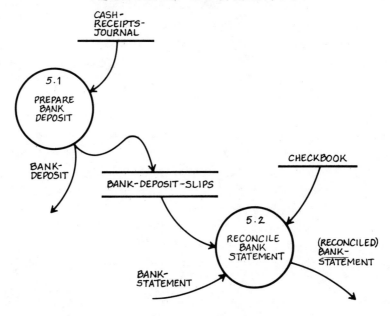

FIGURE 12-13. New Diagram 5: Prepare Bank Deposit and Reconcile Bank Statement.

$$Vendor\text{-}Catalog = \{ Item\text{-}No \ \& \ Taxable\text{-}Indicator$$
$$\& \ Product\text{-}Description$$
$$\& \ Unit\text{-}Of\text{-}Measure$$
$$\& \ Minimum\text{-}Order\text{-}Quantity$$
$$\& \ Unit\text{-}Price \}$$

We tentatively identify two objects, Vendor and Catalog. Note that Vendor-Catalog appears as an iterated data structure in the definition of Vendor-Product-Info. Such an iterated structure is often called an ***embedded repeating group.*** When Vendor-Product-Info is separated into a Vendor object and a Catalog object, the repeating group is removed:

$$Vendor = \{ \underline{Vendor\text{-}Name} \ \& \ Vendor\text{-}Address$$
$$\& \ Vendor\text{-}Telephone\text{-}Number \ \& \ Vendor\text{-}Rep \}$$

$$Catalog = \{ \underline{Catalog\text{-}Vendor\text{-}Name} \ \& \ Vendor\text{-}Catalog \}$$

that is:

$$Catalog = \{ \underline{Catalog\text{-}Vendor\text{-}Name}$$
$$\& \{ Item\text{-}No \ \& \ Taxable\text{-}Indicator$$
$$\& \ Product\text{-}Description$$
$$\& \ Unit\text{-}Of\text{-}Measure$$
$$\& \ Minimum\text{-}Order\text{-}Quantity$$
$$\& \ Unit\text{-}Price \ \} \}$$

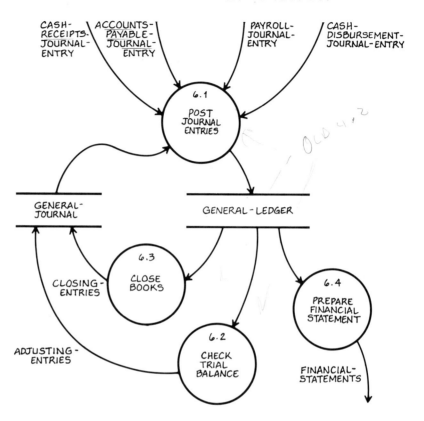

DIAGRAM 6: MAINTAIN GENERAL LEDGER

FIGURE 12-14. New Diagram 6: Maintain General Ledger.

Note that Vendor-Name, the key for the outer portion of the data structure, is retained when the repeating group is removed, as it may be required as part of the unique identifier for a catalog.

Figure 12-15 shows Vendor and Catalog along with their tentative attributes. The Vendor object looks fairly simple, and all its attributes appear to contribute solely to the description of a vendor.

That is not the case with Catalog. It still contains an embedded repeating group.

FIGURE 12-15. Initial description of Vendor and Catalog objects.

VENDOR

Vendor-Name	Vendor-Name Vendor-Address Vendor-Telephone-Number Vendor-Representative

CATALOG

Catalog-Vendor-Name	Catalog-Vendor-Name { Item-No Taxable Indicator Product-Description Unit-of-Measure Minimum-Order-Quantity Unit-Price }

Knowledge of the system helps to decide what to do next. While Vendor-Name will distinguish between catalogs published by different vendors, it is incapable of uniquely identifying an instance of the object Catalog. When information contained in a catalog changes, a vendor often publishes a new catalog, identified by a new date. We have discovered a previously omitted data element, Catalog-Date, which must be part of the key for Catalog:

Catalog = { Catalog-Vendor-Name & Catalog-Date
                   & { Item-No & Taxable-Indicator
                     & Product-Description
                     & Unit-Of-Measure
                     & Minimum-Order-Quantity
                     & Unit-Price } }

We also establish a basis for two relationships—Publishes and Published-By:

TWO WAY

Publishes = { Vendor-Name
                 & Catalog-Vendor-Name & Catalog-Date }

Published-By = { Catalog-Vendor-Name & Catalog-Date
                   & Vendor-Name }

(See Figure 12-16.)

The embedded repeating group within Catalog is treated similarly, as shown in Figure 12-17, producing the new object Catalog-Item, the revised object Catalog, and the complementary relationships Lists and Listed-In:

Catalog-Item = { Catalog-Vendor-Name & Catalog-Date
                   & Catalog-Item-Number
                   & Taxable-Indicator
                   & Product-Description
                   & Unit-Of-Measure
                   & Minimum-Order-Quantity
                   & Unit-Price }

Catalog = { Catalog-Vendor-Name & Catalog-Date }

Lists = { Catalog-Vendor-Name & Catalog-Date
         & Catalog-Vendor-Name & Catalog-Date & Catalog-Item-Number }

Listed-In = { Catalog-Vendor-Name & Catalog-Date
                 & Catalog-Item-Number
                 & Catalog-Vendor-Name & Catalog-Date }

You might question whether the Listed-In relationship is one-to-one instead of one-to-many. However, if you consider the whole three-attribute key for Catalog-Item, you see that a Catalog-Item is Listed-In only one Catalog.

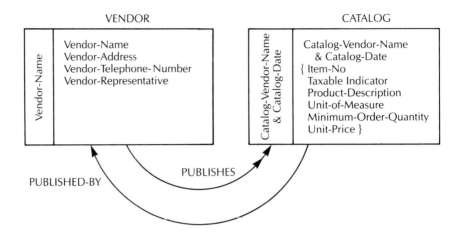

FIGURE 12-16. Partial decomposition of Vendor and Catalog.

FIGURE 12-17. Further partitioning of Catalog.

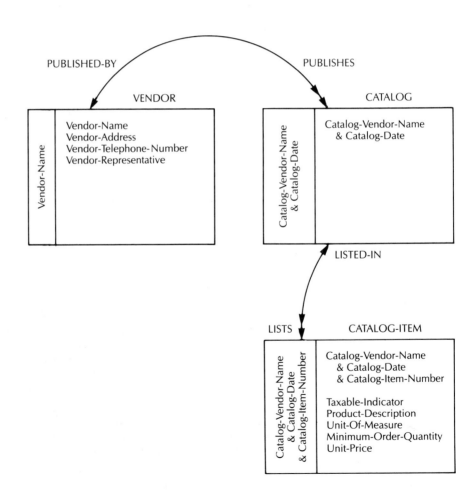

It is not coincidental that there are three levels of repetition in the original data dictionary definitions for Vendor-Product-Info and Vendor-Catalog and three objects in the resulting decomposition. Note that the key of each outer repetition is concatenated with the key of the inner repetition when nested structures are decomposed. In general, there is one object for each level in the nested description and at least one relationship that connects the objects derived from adjacent levels.

We might expect the Vendor-Accounts-Due data store to be a veritable rat's nest of complex objects, cross-references, and redundant data. Consider the highly stylized renditions of the documents stored there (Figure 12-18).

FIGURE 12-18. A Purchase-Order, Packing Slip, invoice, statement, Credit-Memo, and payment.

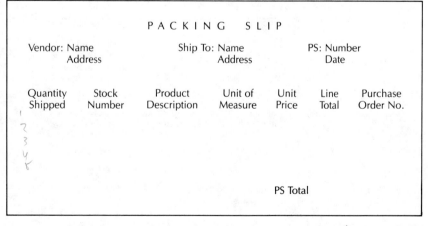

continued on next page

FIGURE 12-18. *(cont.)*

CREDIT MEMO

Vendor: Name   Customer: Name   Credit Memo: Number
        Address   Address   Date

        Invoice   Credit   Credit
        Number   Explanation   Amount
        Credited

                                Credit Memo Total

CHECK

Vendor: Name   Customer: Name   Check: Number
        Address   Address   Date

        Invoice   Invoice   Invoice
        Number   Date   Total

                                Check Amount

INVOICE

Vendor: Name   Customer: Name   Invoice: Number
        Address   Address   Date

Quantity   Stock   Product   Unit of   Unit   Tax   Line   Packing   Purchase
Invoiced   Number   Description   Measure   Price   Total   Slip No. Order No.

                                Invoice Total

FIGURE 12-18. *(cont.)*

```
                    S T A T E M E N T   O F   A C C O U N T

        Vendor: Name              Customer: Name          Statement Date
                Address                   Address


            Statement                                          Previous Balance
            Date                                               Forward


            Invoice                    Invoice                    Invoice
            Number                     Date                       Total


            Credit Memo                Credit Memo                Credit Memo
            Number                     Date                       Total


            Payment                    Customer                    Check
            Date                       Check Number                Amount


            Statement                                             New Balance
            Date                                                  Forward
```

At first glance, the high-level data dictionary description for this data store appears deceptively simple.

Vendor-Accounts-Due = { <u>Vendor-Name</u>
                                  & { [Purchase-Order |
                                      Packing-Slip |
                                      Invoice |
                                      Statement |
                                      Credit-Memo |
                                      Payment] } }

We decompose this data store following the same procedure as before and add to the objects and relationships already identified. However, two new problems challenge us here—how to cope with the redundancy and how to manage the cross-references. We initially partition by establishing one object for each kind of document. We note that each document references attributes of Vendor, an object already defined. Thus our first attempt at minimizing redundancy might be to establish a relationship between each new object and Vendor. But, is there a direct relationship between these, or will an indirect relationship through simpler objects reduce the complexity of the data base description?

We put off answering this question for the moment. First, we simplify each of the newly identified objects.

The Purchase-Order, Packing-Slip, and Invoice all appear to have similar contents. Therefore we give a detailed treatment of Packing-Slip and leave the simplification of Purchase-Order and Invoice as end-of-chapter exercises.

Packing-Slip = Packing-Slip-Vendor-Name
                 & Packing-Slip-Vendor-Address
                 & Ship-To-Name & Ship-To-Address
                 & Packing-Slip-Number
                 & Packing-Slip-Date
                 & { Quantity-Shipped
                     & Packing-Slip-Item-Number
                     & Product-Description
                     & Unit-Of-Measure
                     & Unit-Price
                     & Packing-Slip-Line-Total
                     & Purchase-Order-Number }
                 & Packing-Slip-Total

Packing-Slip-Line-Total and Packing-Slip-Total may be eliminated because they can be calculated from other data. In this case, because Mr. and Mrs. O have only one store, Ship-To is an alias for Customer-Address and should be noted as such. (In other situations, Ship-To could refer to a different location from the billing address, and the two would not be synonyms. However, Sold-To or Bill-To might appear as aliases for Customer.) The embedded repeating group has data in common with Catalog-Item. This observation allows us to eliminate Product-Description, Unit-Of-Measure, and Unit-Price as attributes of Packing-Slip-Line if we establish a relationship between Packing-Slip-Line and Catalog-Item. Packing-Slip-Line is still too physical a name, so we choose Packing-Slip-Item. After factoring out the embedded repeating group, Packing-Slip-Item, and incorporating the other changes noted, the result is

Packing-Slip = { <u>Packing-Slip-Vendor-Name & Packing-Slip-Number</u>
               & Packing-Slip-Date }

Packing-Slip-Item = { <u>Packing-Slip-Vendor-Name</u>
                      <u>& Packing-Slip-Number</u>
                      <u>& Packing-Slip-Item-Number</u>
                      & Quantity-Shipped
                      & Purchase-Order-Number }

Packing-Slip-Contains = { <u>Packing-Slip-Vendor-Name</u>
                         <u>& Packing-Slip-Number</u>
                         <u>& Packing-Slip-Vendor-Name</u>
                         <u>& Packing-Slip-Number</u>
                         <u>& Packing-Slip-Item-Number</u> }

Packing-Slip-Refers-To = { <u>Packing-Slip-Vendor-Name</u>
<u>& Packing-Slip-Number</u>
<u>& Packing-Slip-Item-Number</u>
<u>& Catalog-Vendor-Name</u>
<u>& Catalog-Date</u>
<u>& Catalog-Item-Number</u> }

These objects and relationships are shown in Figure 12-19.

FIGURE 12-19. De-
composition of Packing
Slip.

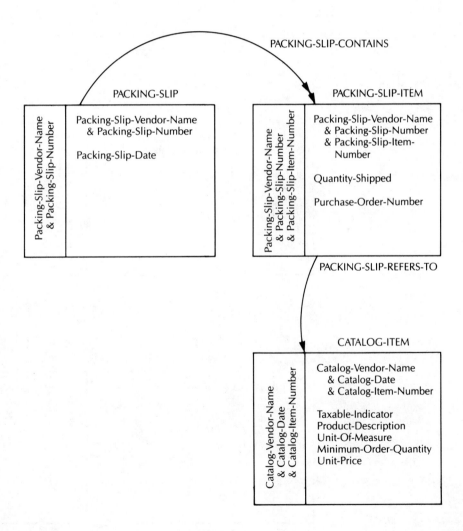

In Packing-Slip-Item we can eliminate the physical cross-reference to Purchase-Order-Number by establishing a relationship between Packing-Slip-Item and Purchase-Order and still provide the logical access path illustrated in Figure 12-20.

Packing-Slip-Item = { <u>Packing-Slip-Vendor-Name</u>
<u>& Packing-Slip-Number</u>
<u>& Packing-Slip-Item-Number</u>
& Quantity-Shipped }

Ordered-On = { <u>Packing-Slip-Vendor-Name</u>
<u>& Packing-Slip-Number</u>
<u>& Packing-Slip-Item-Number</u>
<u>& Purchase-Order-Number</u> }

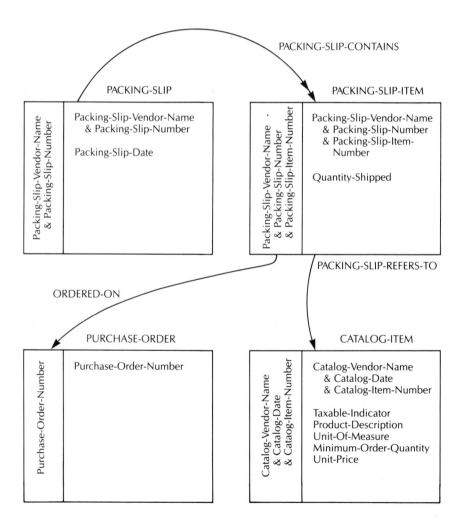

FIGURE 12-20. The substitution of a relationship for the pointer Purchase-Order-Number.

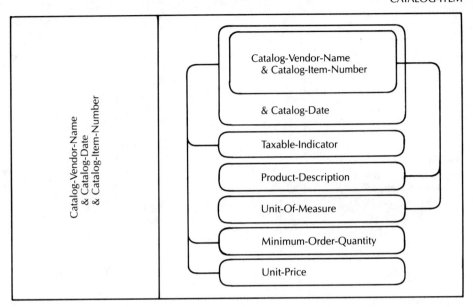

CATALOG-ITEM

FIGURE 12-21. Dependencies within Catalog-Item.

There is still a subtle problem with Catalog-Item. Step 4, Check the attributes of each object, uncovers the difficulty. Figure 12-21 illustrates that the attributes Taxable-Indicator, Minimum-Order-Quantity, and Unit-Price are dependent upon the whole key for Catalog-Item. That is, they are fully determined by the three attributes that compose the key. This is not the case for Product-Description and Unit-Of-Measure, which are independent of the Catalog-Date.

Thus we have uncovered an embedded object, Stock-Item, where:

Catalog-Item = { Catalog-Vendor-Name & Catalog-Date
                 & Stock-Item-Number
                 & Taxable-Indicator
                 & Minimum-Order-Quantity
                 & Unit-Price }

and

Stock-Item = { Catalog-Vendor-Name & Stock-Item-Number
               & Product-Description
               & Unit-Of-Measure }

The result of factoring it is illustrated in Figure 12-22. As before, when the embedded object is removed, a relationship is added to the data access diagram:

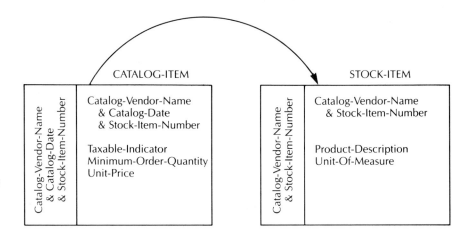

FIGURE 12-22. The data access diagram resulting from the dependencies shown in Figure 12-21.

Catalog-Item-Refers-To = { <u>Catalog-Vendor-Name</u>
<u>& Catalog-Date & Stock-Item-Number</u>
<u>& Vendor-Name & Stock-Item-Number</u> }

We are close to completing the data base description for Accounts Payable. What remains to be done is to complete the data dictionary descriptions for the objects and relationships, draw a complete data access diagram, and test the data base description. Figure 12-23 presents the DAD developed thus far for the Accounts Payable subsystem. Figure 12-24 indicates the objects and relationships accessed by several user views in this system.

We leave the Vendor-Statement, Credit-Memo, and Vendor-Payment object and relationship definitions and DAD testing as exercises.

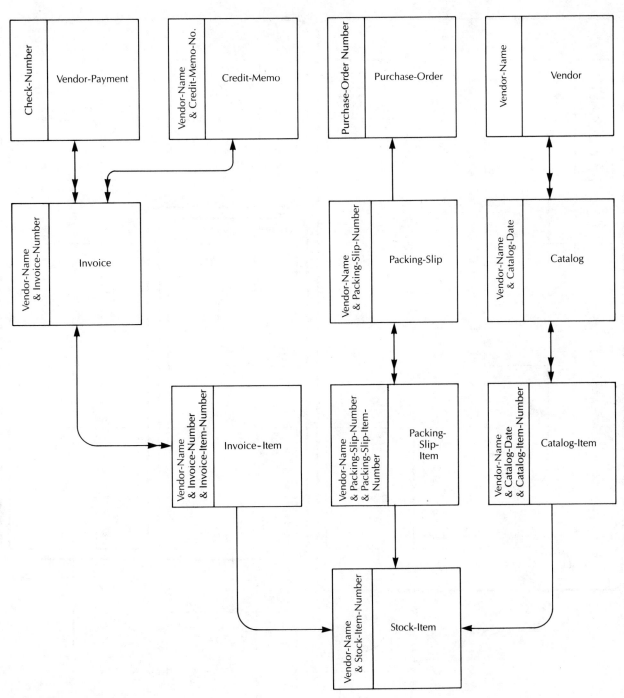

FIGURE 12-23. Data access diagram for the objects and relationships discussed thus far.

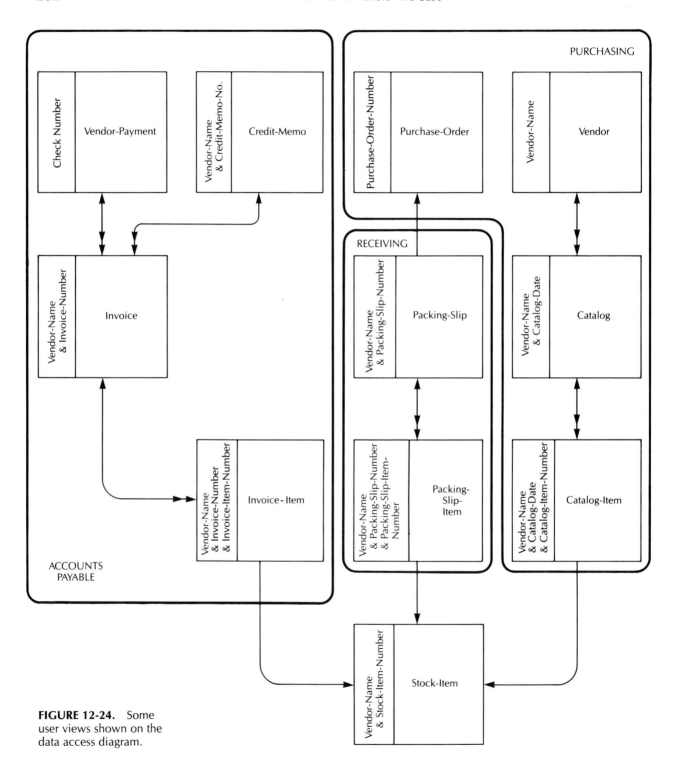

FIGURE 12-24. Some user views shown on the data access diagram.

Parts of the Logical Current System Description

We now present the results of examining the transforms and integrating the data stores for these logical objects into the Current System Description.

Diagram 4.1: Maintain Accounts Payable Data Base, Figure 12-25, resulted from the introduction of data stores for essential objects, elimination of data stores not essential to Transform 4.1 (Purchase-Order, Packing-Slip, Vendor-Statement), and removal of nonessential transforms (cross-checking between Invoices, Purchase-Orders, and Packing-Slips). Figure 12-25 also illustrates the parallel decomposition of data and transformation that often indicates a good job of identifying system essentials.

DIAGRAM 4.1: MAINTAIN ACCOUNTS PAYABLE DATA BASE

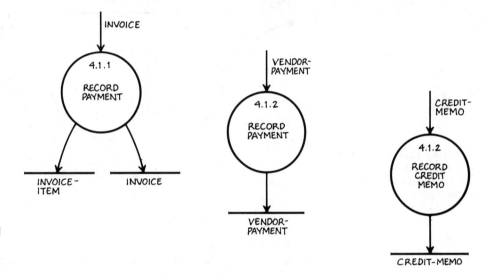

FIGURE 12-25. Diagram 4.1: Maintain Accounts Payable Data Base (Logical Current System Description).

Diagram 4 (Figure 12-26) has been simplified by aggregating the objects Invoice, Invoice-Item, Vendor-Payment, and Credit-Memo into Accounts-Payable-DB. The details of the Accounts-Payable-DB are of interest only at the level of Diagram 4.2. Bubble 4.2: Pay Vendor accesses this data base to determine the amount due a vendor. What happened to Vendor-Statement? We leave this as an end-of-chapter discussion question.

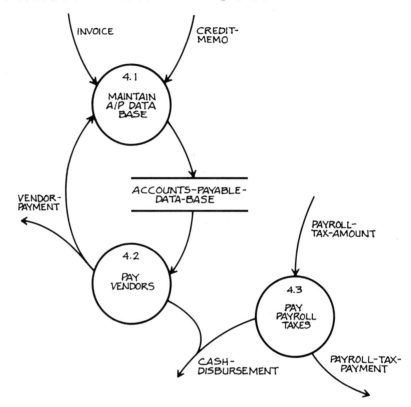

DIAGRAM 4: PAY ACCOUNTS DUE

FIGURE 12.26. Diagram 4: Pay Accounts Due (Logical Current System Description).

Diagram 0 (Figure 12-27) shows the effects of repartitioning and re-leveling Diagram 4. Compare Figures 12-10 and 12-27 as well as Figures 12-12 and 12-27. Everything related to bank statement reconciliation has been removed as non-essential checks. The General-Journal has replaced the individual journals for Payroll, Cash-Receipts, Accounts Payable, and the register portion of the checkbook.

DIAGRAM 0: THE FAST FOOD STORE

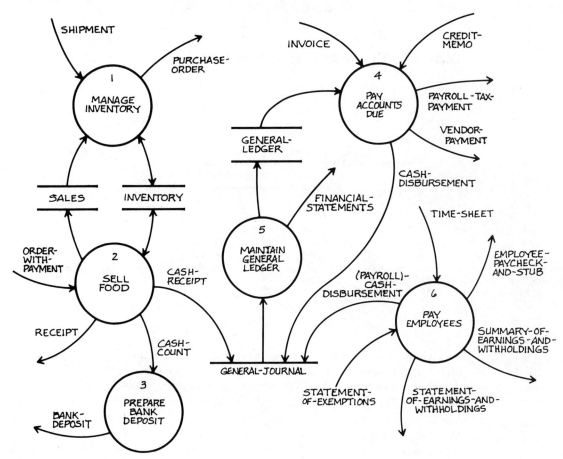

FIGURE 12-27. Diagram 0: The FastFood Store (Logical Current System Description).

Diagram 1: Manage Inventory (Figure 12-28) depicts the essential transformations for ordering and receiving goods.

DIAGRAM 1: MANAGE INVENTORY

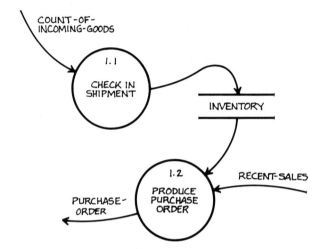

FIGURE 12-28. Diagram 1: Manage Inventory (Logical Current System Description).

Diagram 1.2: Produce Purchase Order (Figure 12-29) shows access by primitive transforms to data stores for logical objects. Note that the transforms are the same as those of Diagram 1.2 in the Physical Current System Description (see Figure 8-14). This suggests we did a good job of identifying essential transformations originally. The changes in this diagram result only from derivation of the logical data base description.

Diagram 2: Sales in the Physical Current System Description (Figure 6-3) is reminiscent of the daily scenario of events at the FastFood Store. The essential data and transformations for Sales in the Logical Current System Description are illustrated in Figure 12-30.

DIAGRAM 1.2: PRODUCE PURCHASE ORDER

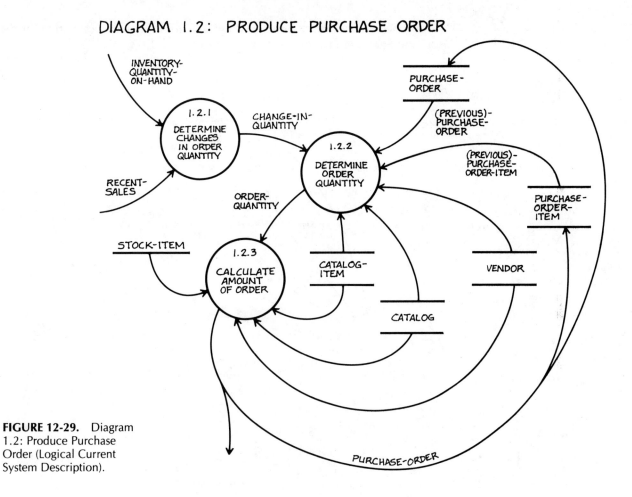

FIGURE 12-29. Diagram 1.2: Produce Purchase Order (Logical Current System Description).

DIAGRAM 2: SELL FOOD

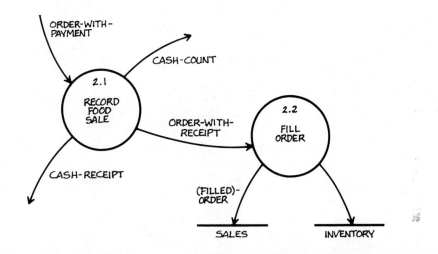

FIGURE 12-30. Diagram 2: Sell Food (Logical Current System Description).

We conclude our discussion of the Logical Current System Description for the FastFood Store with the Context Diagram of Figure 12-31.

CONTEXT DIAGRAM

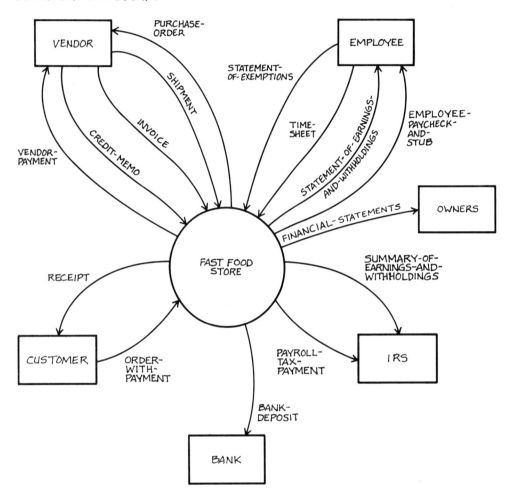

FIGURE 12-31. Context Diagram (Logical Current System Description).

SUMMARY

The process of structured analysis generally produces four system descriptions—a physical and a logical description of the current system and a logical and a physical description of the new system.

The process begins with a complete description of the current system. It concentrates on the areas that are expected to change and usually incorporates considerable physical detail. It consists of two activities: Describe the Current System (Transform 2.1 in Figure III-1) and Pare the Current System Description to Essentials (Transform 2.2).

First the current system is modeled as a leveled set of data flow diagrams with an accompanying system dictionary. This initial system description is then refined using the criteria of completeness, consistency, correctness, and communication. It is improved by minimizing the interfaces between transformations to make each transformation as independent as possible. Each transformation in the heirarchy is checked for its singleness of function and the closeness of its relationship to what is accomplished by its parent transform.

Next the logical current system description is derived from the physical current system description. As many implementation-dependent details as possible are removed. Nonessential transformations and data stores are eliminated. Developing a logical description of the data base is a major effort in producing the logical current system description. Objects and relationships in the data base must be identified and depicted in a data access diagram. Then the objects and relationships with their preconditions and postconditions are added to the system dictionary. The data flow diagrams are revised so that all the data stores shown there correspond to objects in the data base. The data dictionary is also pared to its essentials and the transform descriptions are prepared (or revised) for the primitive transformations. This logical current system description is then evaluated for overall completeness, correctness, consistency, and communication, and refined as necessary.

REVIEW

12-1. Why do we model the system as it is currently implemented?

12-2. What concern on the part of the analyst, with respect to modeling the current system, is answered by part of the Feasibility Report?

12-3. Identify two strategies for developing an initial Physical Current System Description. State the steps in each.

12-4. Distinguish between essential transforms and those that maintain data stores.

12-5. Why are data stores for objects identified in the DAD included in the DFDs? Where would you expect to find data stores for relationships referenced?

12-6. Why is derivable data eliminated from the DD descriptions for logical data flows and objects?

12-7. Name the tasks in constructing the Logical Current System Description.

12-8. When refining the current system description, why is it important to initially view all the diagrams at once?

12-9. What are the principal checks applied to the Context Diagram?

12-10. When factoring embedded objects from a complex object, why is the key of the encompassing object concatenated with the key of the factored object?

EXERCISES AND DISCUSSION

12-1. Why does the CPA appear suspicious in paragraph 17 of the dialogue? Do you think it is unusual for analysts to encounter this kind of reluctance in the analysis process?

12-2. Decompose the object Vendor-Statement. Discuss its role in the FastFood Store system. To what extent is the information in it essential? *NO*

12-3. State some problems with partitioning Diagram 0 in an initial physical current system description by department. Discuss any benefits of this approach.

12-4. Complete the data base description for the Accounts Payable subsystem by:
   a. Decomposing the objects: Purchase-Order, Invoice, Credit-Memo, and Vendor-Payment. Define the composition of each object you derive. Be sure to show its key (which may be a concatenation of several data elements).
   b. Modifying the data access diagram shown in Figure 12-23 to incorporate the objects and relationships discovered in a.
   c. Testing your DAD according to the procedure defined in Chapter 9 to be sure that the user views and data base accesses associated with the documents in Figure 12-18 are supported.

13 Modeling the New System

Following the study of the current system and the development of a logical description of it, the focus of analysis shifts to defining and modeling the requirements for a new information processing system. This chapter describes the activities leading to a description of the requirements for the new information processing system. As shown in Figure III-1, they are Activity 2.3: Establish the New System Requirements, Activity 2.4: Define Alternative Scopes of Automation, and Activity 2.5: Select the Best Scope of Automation.

Establishing the new system requirements uses the Logical Current System Description and the Feasibility Report to develop one or more Logical New System Descriptions. The Feasibility Report, produced during the initial step in system development, guides the description of these alternatives. As discussed in Chapter 17, the Feasibility Report states the objectives and constraints for the new system. It summarizes the most promising directions for system development, giving the expected advantages and disadvantages of each. The Feasibility Report also recommends one or more of these alternatives for the kind of detailed investigation described in Chapters 12 and 13. A Logical New System Description is prepared for each alternative that remains promising after the detailed study of the existing system.

Defining alternative scopes of automation for the new system is the next activity. It involves deciding what portions of the system are to be automated and what portions are to remain manual. The scope of automation is shown in a data flow diagram by drawing a boundary between the manual and the automated transformations. This boundary is called the **automation boundary** or **man-machine boundary.** Because the decision about where to locate this boundary is the basis for all the other implementation-related features of the automated system, it transforms a Logical New System Description into a Physical New System Description. Several possible automation alternatives may be considered for each Logical New System Description. There will be a Physical New System Description for each alternative, because each alternative proposes to automate different transformations.

Selecting the best scope of automation involves a comparison of the different Physical New System Descriptions. Their relative advantages and disadvantages are evaluated using cost/benefit analysis. Within the limits established by the constraints on the new system as well as on the development process, the best alternative is chosen. For this alternative, the complete structured specification is prepared, as described in Chapter 14.

ESTABLISHING THE NEW SYSTEM REQUIREMENTS

The transition from a logical description of the current system to a logical description of a new system is a pivotal point in the analysis process. To study the current system presupposes sufficient similarity between it and the new system to make the effort worthwhile. Yet the fact that users desire to change the system implies that the new system will be different.

What will be essentially different about the new system? This question drives the Logical New System Description. As befits a logical system description, the emphasis is on *what* data flows, data stores, and transformations are different in the new system, not *how* these differences are to be implemented.

Essential differences between the current system and a new one are primarily differences in scope or function. A more inclusive scope implies the addition of inputs, outputs, transformations, or data stores to the system. A reduced scope, on the other hand, suggests the retention of current functions within a tighter boundary. Even without enlarging the system boundary, users may wish to add new information or new transformations to the system.

The first task, therefore, is to identify what will be different in the new system. Some guidance as to these differences is provided by the Feasibility Report. It summarizes at the management level what changes are anticipated.

The next task is to identify the ***domain of change***—those parts of the current system that will be different in the new system. These differences might include changes to the system inputs and outputs, changes to the transformations, or changes to the data stores. In some situations portions of the current system may be eliminated; in others, modifications may be required to accommodate additions to the existing system.

A boundary can be drawn on the Logical Current Data Flow Diagram (as close to the top of the leveled set as possible) to separate the domain of change from the parts of the current system that will not be changed. This partitioning of the system permits us to concentrate on the domain of change. The data flow diagrams can then be redrawn to collect the domain of change into a single bubble at the top level. (Occasionally, there may be several bubbles for the domain of change, each depicting a contiguous portion of it.) If this is done, the data flows entering and leaving the domain of change constitute all the interfaces between the portions of the current system that will not change and the portions that will be affected by change.

The remaining task is to revise the data flow diagram within the domain of change to produce one or more Logical New Data Flow Diagrams. There are no hard-and-fast rules for how to do this, but there is knowledge of what must happen inside the domain of change. Information from the Feasibility Report will help. Making a list of new inputs, outputs, data stores, and transforms will help. Working top-down within the domain of change will help. So will working backward from new system outputs to trace information dependencies. Analysts and users must work together, drawing on both imagination and experience to invent the new system description. It may take numerous iterations.

It is often desirable to develop the Logical New System Description in two stages. The first stage models all the essential components of the new system by modifying the Logical Current System Description. In the second stage, other components are added to protect the system from errors and inconsistencies in the data and to detect or prevent fraud and dishonesty. These edits, checks, and audits adapt the essential system to the practical realities of the imperfect and error-prone world in which it must operate. Thus many system components that were removed during the derivation of the Logical Current System Description are reintroduced at this stage. Taking them out and putting them back in may seem like unnecessary effort. However, it reduces the complexity of the intermediate system descriptions so that users and analysts can concentrate on the essentials first, and it assures a complete review of the audit and control features of the new system. Some system controls may depend upon the choice of what part of the system is to be automated. In that case, further controls are added as each alternative for automating the new system is defined.*

A Logical New System Description is subject to the same evaluation criteria as the current system description. However, there is no real-world counterpart to compare it with to check for correctness. The users are responsible for deciding that the data flow diagrams depict a system that will be workable for their organization. At this point in the process, key users should be sufficiently familiar with data flow diagrams from the study of the current system that they can understand and contribute to the development and evaluation of the new system model.

Walkthroughs of the Logical New System Description will spot its flaws and inconsistencies.

The data dictionary for the new system is developed in parallel with the data flow diagrams. Entries for new inputs, outputs, and data stores should be added as soon as they are identified. The content of existing data flows and data stores is modified as required. Some stability at the upper levels of the data flow diagrams is desirable before lower-level data flows are included in the data dictionary and before the new transforms are added to the transform dictionary. The data base

---

*See Alan E. Brill, **Building Controls Into Structured Systems**, Yourdon Press, New York, N.Y., 1983, for a discussion of incorporated internal accounting controls during the structured system development process.

description must be kept logical by adding new attributes to logical objects, by adding new objects and relationships, and by modifying the data access diagram accordingly. When the new primitive transforms have been identified, transform descriptions for them can be written.

Note that it is very important to keep the current system description distinct from the new system description. It must always be possible to compare the two at any time during subsequent system development. When maintaining the system descriptions manually, make a copy of the entire Logical Current System Description and modify the copy. If automated aids are used, it may be possible without redundant data storage to identify which system components belong to which system description—Physical Current, Logical Current, Logical New, and Physical New.

In some cases, it may be desirable to develop alternative Logical New System Descriptions; such alternatives may even be required by the Feasibility Report. At other times, the Logical Current and Logical New System Descriptions may be the same. If so, look carefully at the current system model. As office-automation consultant Michael Hammer has put it, "Automating a mess yields an automated mess."* Be sure that the current system is depicted in its most logical form and cannot be improved further.

DEFINING ALTERNATIVE SCOPES OF AUTOMATION

After all the required components have been added to the new system description, several alternatives for automating the system are defined. These alternatives are enumerated in or implied by the Feasibility Report.

For each alternative, the scope of automation is defined. The **scope of automation** specifies the portion of the system that is to be automated. The scope of automation is depicted by drawing its boundary, called the **automation boundary** or **man-machine boundary,** on a set of Logical new Data Flow Diagrams. This automation boundary partitions the system into an automated part and a manual part; everything inside the boundary is automated. The set of data flow diagrams is expanded until a level is reached at which the automation boundary does not intersect any bubbles. This means that every transform in the system at that level or below has been specified as entirely manual or entirely automated.

The addition of the man-machine boundary to a Logical New System Description is sufficient to transform it into a Physical New System Description. That is why it is often said that the Physical New System Description is not very physical, at least when compared to a Physical Current System Description.

However, nearly all the important implementation-related features of the new system follow from the specification of a man-machine boundary. On the human side of the boundary (outside it), information is represented, transmitted, and stored in media that people can read, understand, and use. Inside the automation

*Quoted in Jonathan Schlefer, "Office Automation and Bureaucracy," **Technology Review,** vol. 86, no. 5, July 1983, p. 32. Reprinted with permission from **Technology Review,** © 1983.

boundary, information is represented, transmitted, stored, and transformed electronically; it is not directly accessible to or processible by people. As a result, there must be devices at the boundary for data capture and data display. A data-capture device physically transforms system inputs from human-readable form to machine-readable form; a data-display device transforms system outputs from an electronic information medium to a medium in which humans can understand information.

An important consequence of where the automation boundary is located is a decision about what generic types of input/output devices will be required to convert information as it crosses the boundary as well as the characteristics of these devices. Procedures for interacting with the automated system are necessary on the human side of the boundary. They will include data collection and data entry procedures as well as procedures for the distribution of system outputs. Unless the details of these procedures are prescribed in advance by the users, they are specified during system design. During analysis, it is usually sufficient to define the composition of the data flows crossing the automation boundary and to identify the generic types of required input/output devices.

Another decision related to the automated portion of the system is whether it is to be batch, on-line, or real-time. In a **batch** system, several instances of a data flow (a batch) entering the automated portion of the system are held for processing together as a group. Sometimes a batch is related to a processing cycle, as when time cards are collected daily or weekly, but employees are paid biweekly or monthly. Sometimes a batch is used to provide accounting controls, as when the number of dollars received in coins, currency, and checks is compared to the number of dollars associated with a bank deposit. Batching of data flows implies at least a temporary data store until it is time to process the group.

In an **on-line** system each data flow entering the automated portion of the system is processed as soon as it is received. The appropriate response and its timing depend on the specific system.

A **real-time** automated system responds to external stimuli, which may be either data flows or control flows. When it receives input, the system must produce the appropriate data or control flows, responding quickly enough to affect events in the external world. Real-time systems occur in communications, laboratory, and manufacturing applications.

If the system is widely dispersed geographically, there will be a need for telecommunication links and devices. Automation alternatives will also consider the degree to which the data base and processing power of the system are concentrated in a single location and the degree to which they are distributed throughout the system.

These tentative high-level decisions about the significant characteristics of each alternative for automating the system become the basis for determining the alternative hardware and system software environments. At this stage of system development, the concern is to be able to compare the implications of the different automation alternatives. Further hardware/software studies are a separate step in system development, as shown in Figure 3-1. Because these implications of automation affect the hardware and software configuration, they are sometimes referred to as **configuration-dependent** characteristics of a system. To show these configuration-dependent features may require little change to the data flow diagrams, perhaps only a few annotations. The important thing is to identify and depict data flows that will be critical interfaces at major hardware boundaries in the automated system.

Where the automation boundary is located may affect transforms inside the boundary. Some calculations, such as those required for extensive statistical studies, complex financial modeling, or mathematical optimization, are impossible or impractical to do manually. With automation, they become feasible. This means that automation offers the possibility of incorporating new transforms in a system or of replacing existing manual procedures with algorithms that are more complex and sophisticated. Indeed, this very possibility may have prompted the users' desire to automate. If not, such opportunities should at least be considered.

The result of this activity is several alternatives for automating the new system. For each alternative there is, in principle, a separate Physical New System Description. There is a set of data flow diagrams showing the location of the man-machine boundary for that alternative as well as any important configuration-dependent details. In practice, the detailed differences in data flows, data stores, and transforms that follow from differences in the data flow diagrams for each alternative are best described by exception in order to minimize redundancy.

Two to five alternatives are usually enough for the comparison that follows, although a number of minor variations on each alternative were probably considered in choosing the final set. They should be as different as possible within the limits set by the Feasibility Report. The choice of several alternatives should lead to a better decision. Making the alternatives dissimilar extends the comparison over a broader area and should also improve the result.

SELECTING THE BEST ALTERNATIVE

The next activity is to evaluate and compare the alternative Physical New System Descriptions and to select the best. The alternative chosen as the best will be the basis for the structured specification and continue through system design, hardware and system software selection, and the rest of the system life cycle.

In this context, **best** does not mean the best of all possible systems in the best of all possible worlds. It means best for this organization in these circumstances with all their inherent constraints and limitations. To the extent that these limitations are made explicit and understood, a realistic framework is established for making decisions. A review of the technical, economic, and organizational constraints contained in the Feasibility Report is in order to assure that the constraints remain valid and that the alternatives remain feasible.

The evaluation of alternatives entails assessing each of them with respect to a common set of criteria so that they can be compared on a consistent and objective basis. It should use a formal, systematic process for comparing the relative advantages and disadvantages of each alternative, generally known as *cost/benefit analysis.*

The costs and benefits considered in this analysis are not limited to the conventional economic considerations. They should include a variety of significant descriptors of expected system performance—both quantitative and qualitative—that the system owner can use to assess the relative desirability of the alternatives to the organization.

Cost/benefit analysis is hard because it requires estimates of future system performance or behavior based on a description of system requirements. Developing reliable estimates is the primary difficulty. Other difficulties, such as the lack of good data about past or current performance of similar systems, are secondary.

With estimates in hand, procedures for carrying out the comparison are straightforward. A variety of techniques is available. The comparison of dollar costs and benefits usually recognizes that they accrue over time. The anticipated life of the system (including the development time) is chosen as the period of comparison, and the costs and benefits occurring throughout the period are adjusted to make them comparable using present-value or discounted-cash-flow methods. Similar methods are used not only in information system development but also in making other kinds of capital budgeting decisions.

A detailed treatment of cost estimating and decision-making techniques is outside the scope of this book.*

The analyst presents the results of the comparison of alternatives to user management. Sometimes the cost/benefit analysis shows that one alternative is clearly superior. If two alternatives are close in their overall desirability, the analyst may recommend one, or may leave the choice for the users. Ultimately, the decision is the users' responsibility.

The other important decision to be made at this point is whether to continue development of the system. The cost/benefit study provides the best available estimates of the time and cost required to complete the system. These are reviewed to see whether the best alternative is still worth pursuing. If not, the project is canceled; otherwise, management must commit resources for the next steps in system development—the design, the selection of hardware and system software, and the planning of the system acceptance tests.

Note that the decision to continue or not precedes completing the structured specification. Most of the contents of the specification have been produced by now. However, if the project is canceled, it makes no sense to expend additional effort. If system development proceeds, the system requirements are completed and packaged, as discussed in Chapter 14.

THE FASTFOOD STORE—NEW SYSTEM DESCRIPTIONS

To illustrate some of the methods and issues involved in modeling a new information processing system, we continue our discussion of the FastFood Store system.

In talking with their system analyst, Mr. and Mrs. O established as their most important objective saving Mrs. O's time by relieving her of some of the paperwork.

*See Thomas H. Athey, **Systematic Systems Approach**, Prentice-Hall Inc., Englewood Cliffs, N.J., 1982, for a general method of evaluating alternatives. See Barry W. Boehm, **Software Engineering Economics**, Prentice-Hall, Inc., Englewood Cliffs, N.J., 1981, for estimating information system development costs. See DeMarco, **Controlling Software Development Projects**, Yourdon, Inc., New York, N.Y., 1984, for how to develop an estimating capability.

It became clear that the most promising area for improvement was the preparation of purchase orders and payment of the vendors. Much of Mrs. O's time is spent there. The accountant does most of the work to maintain the general ledger and prepare the financial statements, while payment of FastFood's few employees is done by the bank and would not be all that time consuming anyway, except for keeping current with the applicable government regulations. Thus priority for automation goes to purchasing and vendor accounts payable.

Note that in this case the new system requires automation of some of the functions of the current system, but no major change in scope and function. Thus the Logical New System Description, as far as we can tell, is the same as the Logical Current System Description. All the hard work to study the existing system has begun to pay off.

Identifying the Domain of Change

However, it is still important to identify the domain of change so that we can concentrate our efforts on the portions of the current system that will be affected by Mr. and Mrs. O's tentative decision to automate her most burdensome paperwork. A look at Diagram 0 in the Logical Current System Description (Figure 13-1) shows that Transform 1: Manage Inventory, which produces the Purchase-Orders, and Transform 4: Pay Accounts Due are involved in the change. The other transformations in Diagram 0 will be untouched by the proposed change.

We must now look at Diagrams 1 and 4 (Figures 13-2 and 13-3) to see whether the domain of change includes all the subordinates of Bubbles 1 and 4, or only a subset. The answer to this question is that Transform 1.1: Check-In Shipment and Transform 4.3: Pay Payroll Taxes need not be changed when the remaining transforms are automated. Thus the domain of change for the FastFood Store comprises Transforms 1.2, 4.1, and 4.2.

In many situations it is helpful to draw a data flow diagram like Figure 13-4, which presents the domain of change on a single diagram by combining Diagrams 1 and 4.

Diagram 0 may also may expanded and redrawn to show the domain of change as a single bubble, as in Figure 13-5.

This is usually helpful to the analyst, because it partitions the current system model explicitly into the parts that will be changed and the parts that will remain as they are. Whether or not these diagrams are drawn or presented to users, this partitioning is important conceptually because it is the first step in moving from the Logical Current System Description to the Logical New System Description.

DIAGRAM 0: THE FAST FOOD STORE

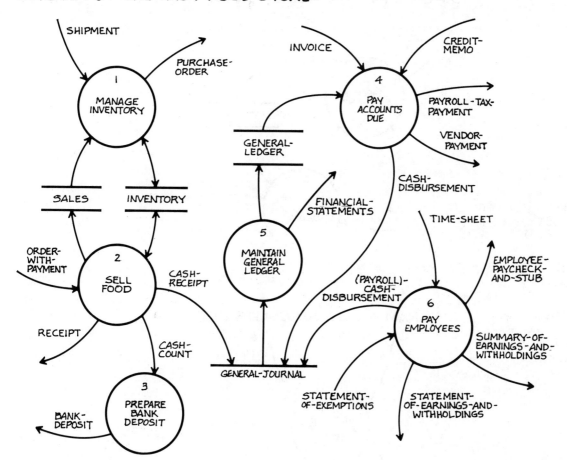

FIGURE 13-1. Diagram 4: Pay Accounts Due (Logical Current System Description).

DIAGRAM 1: MANAGE INVENTORY

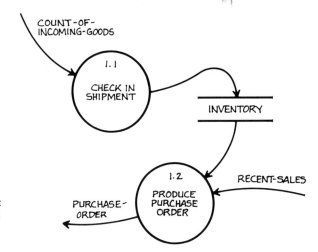

FIGURE 13-2. Diagram 0: The FastFood Store (Logical Current System Description).

FIGURE 13-3. Diagram 4.1: Maintain Accounts Payable Data Base (Logical Current System Description).

DIAGRAM 4: PAY ACCOUNTS DUE

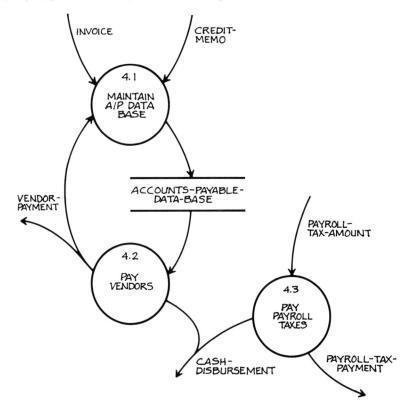

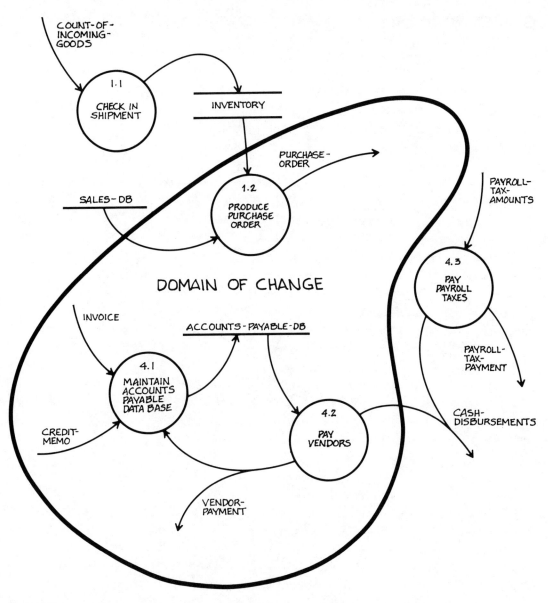

FIGURE 13-4. Depicting the domain of change.

DIAGRAM 0: FAST FOOD STORE

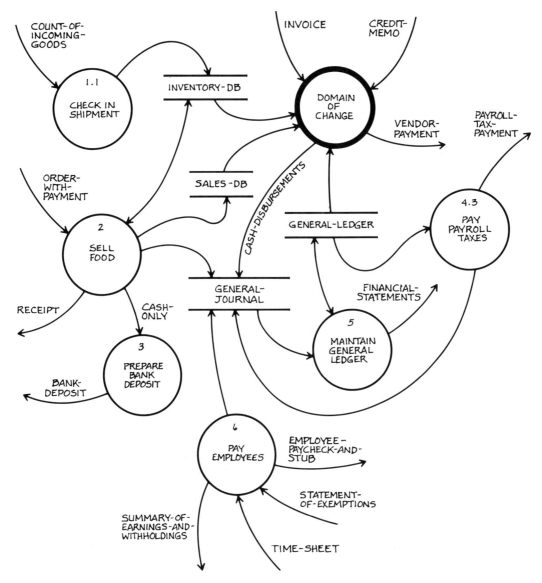

FIGURE 13-5. Diagram 0 showing the domain of change as a single bubble.

Modifying the System Description Inside the Domain of Change

For the FastFood Store, the new system is essentially the same inside the domain of change as the current system. Thus Figures 13-1 and 13-4 depict the essential components of the new system description without further modification. In other systems, revisions to the data flow diagrams within the domain of change might be necessary to achieve the users' objectives.

The next task is to review the entire Logical New System Description and add the appropriate edits and controls. Transform 1.1 is modified to include a check of the goods received against the packing slip that accompanies the shipment (Figure 13-6).

The reconciliation of the cash count with the total sales is reintroduced into Diagram 2, as shown in Figure 13-7.

Bank-Statements are reconciled with FastFood's record of deposits and checks written. The changes to Diagram 3 are shown in Figure 13-8.

DIAGRAM 1: MANAGE INVENTORY

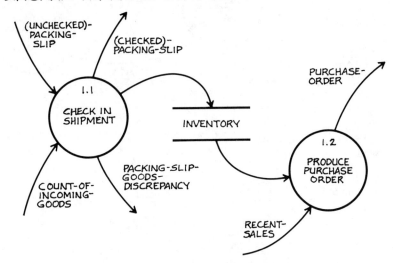

FIGURE 13-6. Revised Diagram 1: Manage Inventory (Logical New System Description).

Diagram 4 also changes (Figure 13-9). Before vendors are paid, checks are made for consistency among the purchase orders, packing slips, and invoices. Discrepancies between the goods received and the packing slips were detected in Transform 1.1, which sends this information to Transform 4. Each Vendor-Statement is also

DIAGRAM 2: SELL FOOD

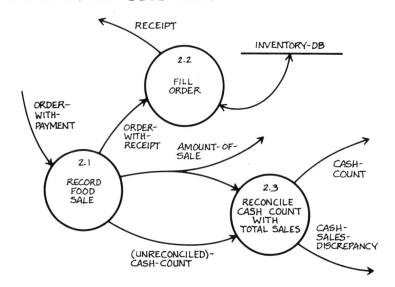

FIGURE 13-7. Revised Diagram 2: Sell Food (Logical New System Description).

FIGURE 13-8. Revised Diagram 3: Prepare Bank Deposit and Reconcile Bank Statement (Logical New System Description).

DIAGRAM 3: PREPARE BANK DEPOSIT

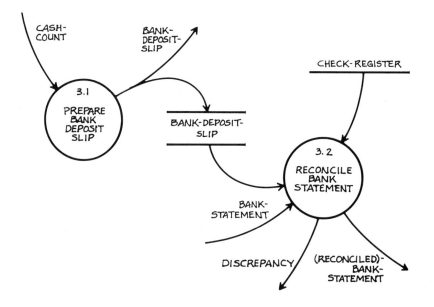

FIGURE 13-9. Revised Diagram 4: Pay Accounts Due (Logical New System Description).

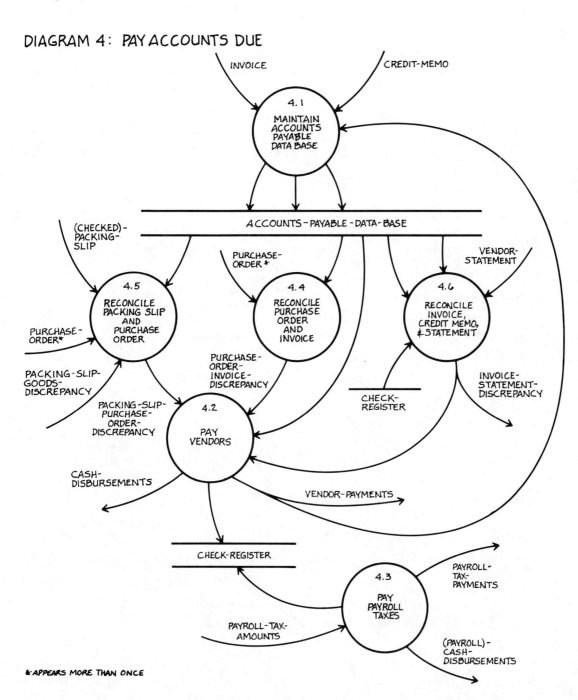

DIAGRAM 4: PAY ACCOUNTS DUE

* APPEARS MORE THAN ONCE

DIAGRAM O: FAST FOOD STORE

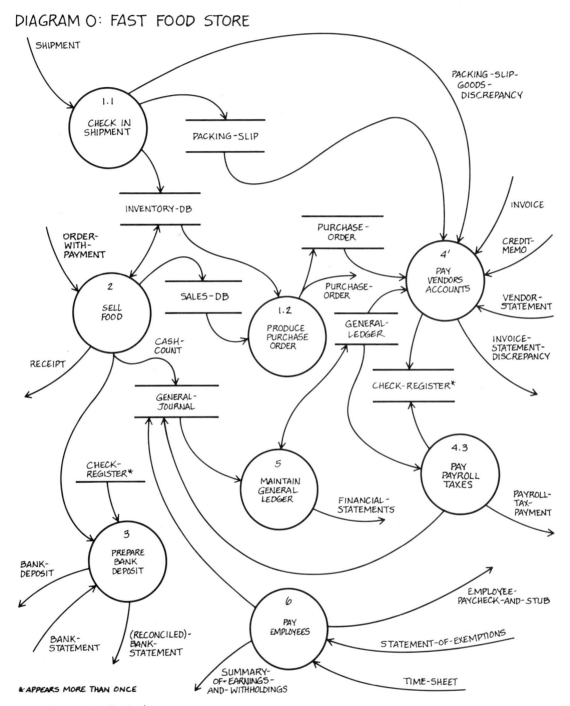

★ APPEARS MORE THAN ONCE

FIGURE 13-10. Revised Diagram 0: FastFood Store (Logical New System Description).

reviewed to see that it correctly shows all invoices and credit memoranda as well as all payments made by FastFood to that vendor.

These additional data flows and data stores are incorporated in the data dictionary and data base description, as well as being shown on the data flow diagrams. Diagram 0 (Figure 13-10) and the Context Diagram (Figure 13-11) are revised to balance with the lower-level diagrams. In Figure 13-10 Transforms 4.1, 4.2, and 4.4 through 4.6 are collected as Bubble 4'. This makes Figures 13-5 and 13-10 directly comparable without revising the transform numbers at this stage. This completes the Logical New System Description.

FIGURE 13-11. Revised Context Diagram. (Logical New System Description).

CONTEXT DIAGRAM

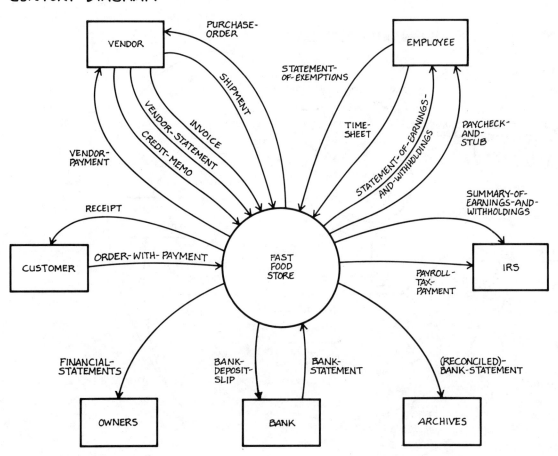

Defining Automation Alternatives

Since the primary objective of the new system is to automate information processing within the domain of change, the basic scope of automation for the FastFood Store is already known. In general, the man-machine boundary will coincide with or lie inside the boundary of the domain of change.

The next task is to define alternatives for automation by carefully considering details implied by the requirement to automate preparation of purchase orders and payment of vendors. Note that in examining these alternatives the focus is on the information requirements of each alternative. The discussion of how the procedures are carried out is a means of discovering what these information requirements are. At the same time analysts and users show that they are able to imagine at least one feasible way of implementing the requirements. The system designers may be able to provide a better way.

FIGURE 13-12. Diagram 1: Manage Inventory (Logical Current System Description).

Consider each of the transformations to be automated. Producing a purchase order (Figure 13-12) requires its inputs to be available in machine-processible form and yields its output, Purchase-Order, as an electronic data flow.

DIAGRAM 1.2: PRODUCE PURCHASE ORDER

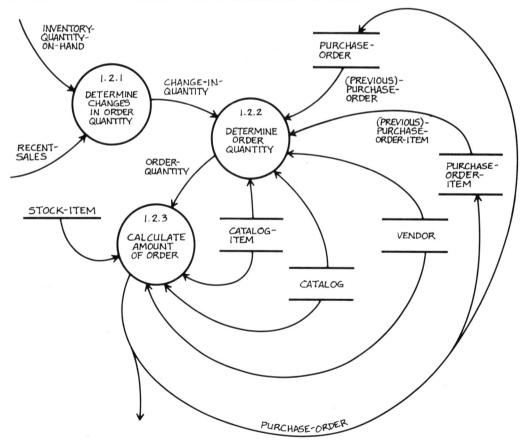

The Purchase-Order can be stored electronically, as well as being printed for Mrs. O to see and send to the vendor. The list of vendors and the portion of each vendor's catalog containing items purchased by the FastFood Store become automated data stores. The other inputs to Process 1.2, the quantity on hand for each item in inventory and the level of Recent-Sales, must also be available electronically.

How can this be accomplished? One alternative is to keep the Inventory and Sales data stores in the computer. Another is for Mrs. O to enter the current values of Inventory-Quantity-on-Hand and Recent-Sales whenever she is ready to order an item. Which alternative is preferable? Automating the Inventory and Sales data stores will add to the amount of file space required and the amount of data that must be entered to keep the data stores up to date.

A look at the policy for Transform 1.2.1: Determine Changes in Order Quantity, presented in Chapter 8, shows that a precise count is not required for either Inventory-Quantity-on-Hand or Recent-Sales. Letting Mrs. O talk to Mr. O and enter High, OK, or Low for Inventory-Quantity-on-Hand and Fast, Normal, or Slow for Recent-Sales seems to be much simpler. Note that the consideration of these two alternatives has been a decision as to where to put the man-machine boundary—whether the Inventory and Sales data stores are inside or outside the scope of automation.

Similar questions arise with respect to paying the vendors (Transform 4.2). Is the computer to produce the checks, or will it just generate a list of vendors to be paid and the amount due each so that Mrs. O can make out the checks? The trade-off is the cost of checks that can be mounted in a computer printer and the care that must be taken to see that they are properly aligned versus the time it takes Mrs. O to make out the checks. If Mrs. O is willing to enter the number of the next available check, then the computer can assign a check number to each vendor payment, producing in effect an automated check register for these payments. Again, the important concern here is not which alternative is best, but the information requirements of each: if Mrs. O wants a check register, and does not wish to maintain a separate account for paying vendors, then she must enter a beginning check number each time vendors are paid. That data flow must be shown on any data flow diagram depicting this alternative.

Examination of Transforms 4.4, 4.5, and 4.6 leads to additional alternatives for automation. In these transformations the Packing-Slips and Purchase-Orders are compared to see that the goods received were those ordered; the Purchase-Orders and Invoices are compared to see that the goods billed were those ordered; and the Vendor-Statement is compared to Invoices and Credit-Memos to see that they are also in agreement.

The following issues arise: to what level of detail will the comparison be made, and will the comparison be made by a computer or by Mrs. O? If Mrs. O does the reconciliations, all the data elements to be compared must be on the human side of the automation boundary, so she can read them. If the reconciliation is automated, all the necessary data must be in electronic form so that a computer program can make the comparison. The principle illustrated here, though seemingly obvious, is crucial in locating the man-machine boundary.

The following alternatives are identified:

Transform 4.4: Reconcile Packing-Slip and Purchase-Order.

Alternative 1: manual reconciliation. When a shipment comes in, Mrs. O will use a printed copy of the Purchase-Order and check off each line of the Packing-Slip against the corresponding item on the Purchase-Order. The data base description shows that vendors include the applicable Purchase-Order-Number on each line of a Packing-Slip.

Alternative 2: automated reconciliation. This requires Mrs. O. to enter at least the Packing-Slip-Line Quantity, Packing-Slip-Line-Stock-Number, and Purchase-Order-Number from each line of each Packing-Slip, as well as the Packing-Slip-Number. The other information on a packing slip is not essential to the reconciliation; requiring Mrs. O to enter it will contribute nothing to the process. Note that the description of each item is available from the vendor's catalog, since the stock number is known.

Transform 4.5: Reconcile Purchase-Order and Invoice.

Alternative 3: manual reconciliation. Mrs. O will use a printed copy of the Purchase-Order to check against each Invoice. Each Invoice-Line, like each line of a packing slip, includes the appropriate Purchase-Order-Number.

Alternative 4: Detailed automated reconciliation. Details of each purchase order are available because the preparation of purchase orders is to be automated. The minimum additional data elements required for each invoice are the Invoice-Number and, for each line, the Invoice-Line-Quantity, Invoice-Line-Stock-Number, Invoice-Line-Total, and Purchase-Order-Number.

Transform 4.6: Reconcile Invoice, Credit-Memo, and Vendor-Statement.

Alternative 5: manual reconciliation. Mrs. O will use the Invoices, Credit-Memos, and Statements received from a vendor to make the comparison. She will also need the list of Checks-Written to be sure that Vendor-Payments made since the last statement have been credited by the vendor to FastFood.

Alternative 6: automated reconciliation. In this alternative, information from the invoices, credit memos, and vendor statement will be compared by the computer to check their consistency. To make this possible, all the information the computer will need must be inside the automation boundary in machine-processible form. At a minimum, the following data elements must be available to the computer.

From the Vendor-Statement:

- Statement-Total
- Invoice-Number and Invoice-Total from each Invoice-Reference-Line
- Credit-Memo-Number and Credit-Amount for each Credit-Reference-Line
- Check-Number and Payment-Amount from each Payment-Line.

These will be compared to the:

- Invoice-Number and Invoice-Total for each Invoice
- Credit-Memo-Number and Credit-Amount for each Credit-Memo
- Check-Number and Amount for each check written to the vendor.

In either case, if a discrepancy is found, the reason will be identified manually.

These six alternatives produce eight possible combinations. For the purpose of this example, three of these are now selected for further consideration, disregarding minor variations for improving each of them. These three, along with the automation decisions mentioned previously, are the alternative Physical New System Descriptions referred to as the result of Activity 2.4: Define Alternative Scopes of Automation in Figure III-1. Note especially the impact that shifting the man-machine boundary has on data entry to the automated system and on the requirements for data display at the boundary. In practice, which alternatives to pursue for the FastFood Store will depend upon how frequently discrepancies are found in the reconciliation process, where they are most likely to occur, how serious they are, and how much time they take Mrs. O to pinpoint.

Alternative A, shown in Figure 13-13, combines Alternatives 1, 3, and 5. It is a manual reconciliation. Except for the fact that purchase-order data is stored electronically, it is the same as the current system and will provide a base line for comparison in the cost-benefit studies.

Alternative B (Figure 13-14) combines Alternatives 2, 4, and 6. This is the most fully automated version of the new system.

Alternative C (Figure 13-15) includes Alternatives 1, 4, and 6; Reconciliation of the packing slip and purchase order is manual; the other reconciliations are automated. The major input burden is entry of the detail from the invoices. Note that by automating Transform 4.2, the computer can pinpoint where the discrepancies are. Also note that a choice of Alternative 5 instead of 6 would permit the computer to print a report that should be identical to the vendor-statement except for any credit memos. This would greatly facilitate Mrs. O's manual reconciliation.

Moreover, automating Transform 4.2 makes it easy to automate writing the checks for the vendor payments, as discussed above, because all the necessary information is already inside the computer.

Selecting the Best Alternative

The next task is to evaluate and compare Alternatives A, B, and C, considering the resources required to develop each alternative and the costs and benefits with each in operation. Some of the advantages, disadvantages, and trade-offs have already been suggested, but there will be no further discussion of the cost/benefit studies here.

After reviewing the cost/benefit studies of these alternatives with the analyst, Mr. and Mrs. O select Alternative C as the best statement of their requirements for the new FastFood Store system.

The transforms are renumbered so that the man-machine boundary does not intersect any bubble on Diagram 0. Bubble 1.4 becomes Bubble 7; Bubble 4.3 becomes Bubble 8. The manual portion of Bubble 4' becomes Transform 9, and the automated portion remains as Transform 4, with the subsidiary transformations being renumbered accordingly. These changes to Diagram 0 are shown in Figure 13-16.

Mr. and Mrs. O and the analyst will now work together to resolve any other critical detailed requirements implied by Figure 13-15 and the rest of the Physical New System Description so that the structured specification can be completed.

DIAGRAM 4: PAY ACCOUNTS DUE

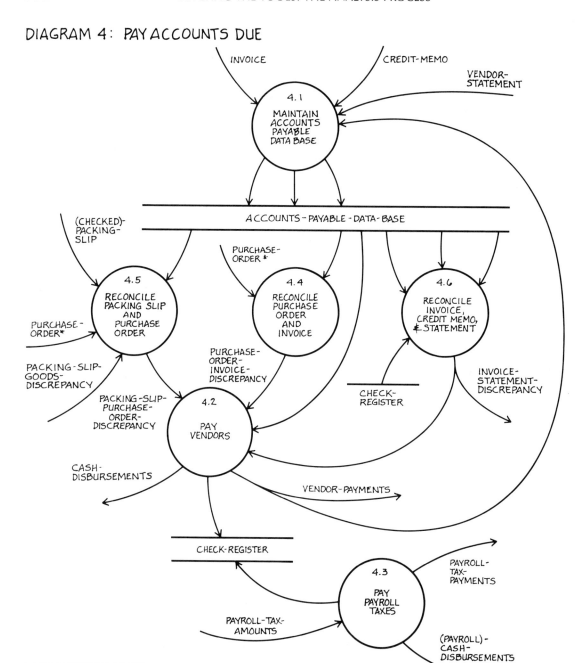

* APPEARS MORE THAN ONCE

FIGURE 13-13. Alternative A.

DIAGRAM 4: PAY ACCOUNTS DUE

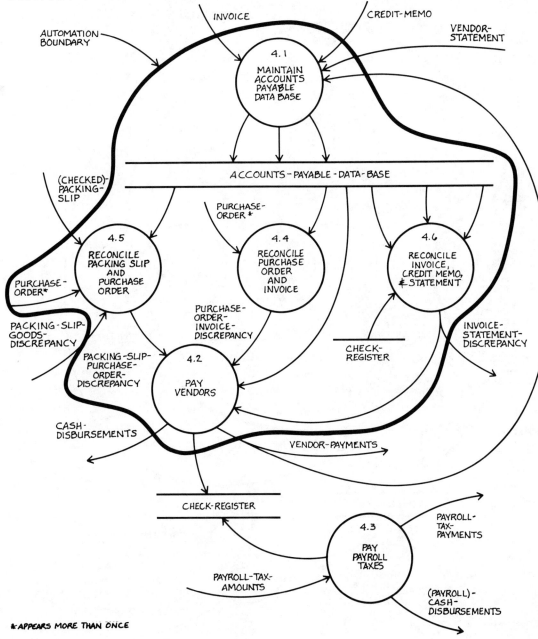

* APPEARS MORE THAN ONCE

FIGURE 13-14. Alternative B.

DIAGRAM 4: PAY ACCOUNTS DUE

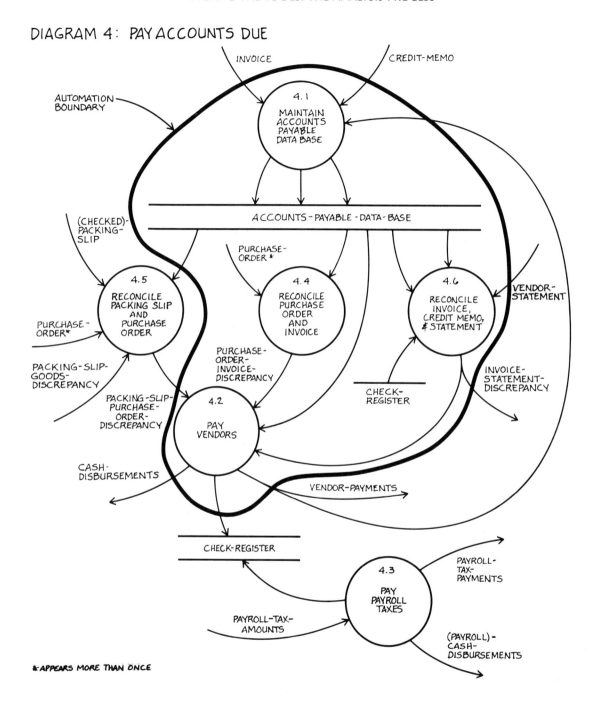

★ APPEARS MORE THAN ONCE

FIGURE 13-15. Alternative C.

DIAGRAM 0: FAST FOOD STORE

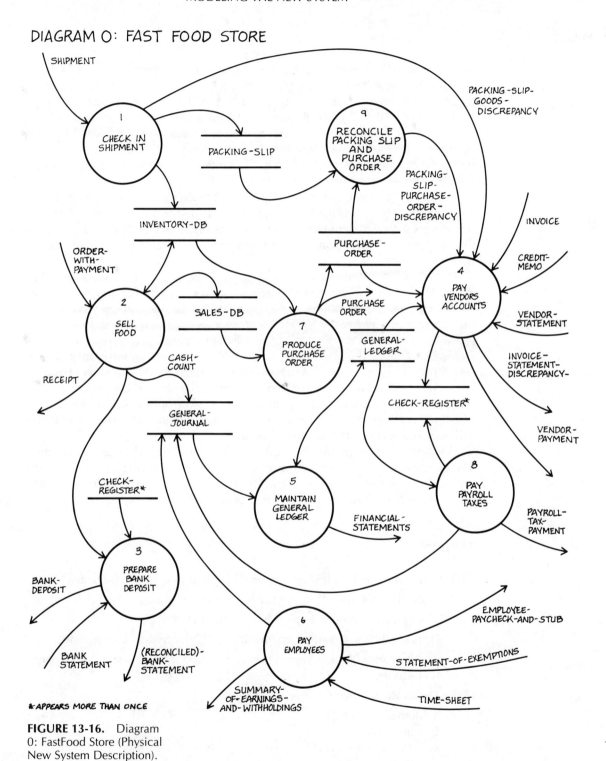

FIGURE 13-16. Diagram 0: FastFood Store (Physical New System Description).

*APPEARS MORE THAN ONCE

SUMMARY

Modeling users' requirements for a new system consists of three activities: Activity 2.3: Establish the New System Requirements, Activity 2.4: Define Alternative Scopes of Automation, and Activity 2.5: Select the Best Scope of Automation.

Guided by the Feasibility Report, users and analysts identify the domain of change—those portions of the current system to be changed. The Logical Current System Description is then altered to show the required changes. The emphasis is on what data flows, data stores, and transformations are different in the new system, not how these differences are to be implemented.

The Logical Current System Description is then altered; the data flow diagrams, system dictionary, and data base description are all modified as necessary. The result is the Logical New System Description. In some cases, users wish to consider more than one possibility for the new system, and alternative logical new system descriptions are prepared.

The crucial implementation-dependent decision in an information system is determining what portion of it will be automated. In a set of data flow diagrams the scope of automation is shown by drawing a man-machine boundary. It separates the system into a portion in which information is transmitted, stored, and transformed by a computer and a portion in which information is prepared and processed by people. Defining the scope of automation and drawing the automation boundary transforms a Logical New System Description into a Physical New System Description. The Physical New System Description may also show other implications of the automation decision, such as additional transformations that are practical if automated but impractical if manual, or such as data-capture and data-display devices and transformations at the automated boundary.

Usually several alternative Physical New System Descriptions are considered. These are compared using cost/benefit analysis to select the alternative that is best overall for the users. A complete structured specification for this alternative is the basis for the next phase of information system development—system design.

REVIEW

13-1. Name the activities involved in modeling new system requirements.

13-2. State the essentials of each of the following transformations:

    a. Logical Current to Logical New System Description

    b. Logical New to one alternative Physical New System Description.

    c. Alternative Physical System Descriptions to the best Physical New System Description.

13-3. Define the following terms:

    a. domain of change

    b. automation boundary

    c. scope of automation

    d. data-capture device

    e. data-display device

    f. batch system

    g. on-line system

    h. real-time system

    i. cost/benefit analysis

13-4. Why is it important to keep the Current System Description distinct from the Logical New and Physical New System Descriptions?

13-5. Why is it desirable to develop several relatively dissimilar Physical New System Descriptions?

13-6. Why is cost/benefit analysis difficult?

EXERCISES AND DISCUSSION

13-1. What is meant by

    a. present value

    b. discounted cash flow

    Why would they be used in comparing the dollar costs of new system alternatives?

13-2. What must happen to the FastFood Store data base when Mr. and Mrs. O receive a new catalog from a vendor? Modify the Physical New System Description (Figure 13-16 and related data dictionary and transform descriptions) to accommodate these additional functions. Note that this is an example of deferring some secondary detail, as mentioned in Chapters 6, 10, and 14.

13-3. Consider the conversion of the vendor catalog data stores from printed form to electronic storage when FastFood begins to use the system. Is it necessary to store every item in each vendor catalog? Is it necessary to begin the operation of the new automated system with a complete vendor catalog in it? What alternatives for building the automated vendor catalog can you suggest?

13-4. Redraw Figure 13-16 to show the entire automated portion of the Physical New System Description as a single bubble. What purpose does a diagram such as this serve?

14 Completing the Structured Specification

The Physical New System Description for the best alternative is now essentially complete. The final activity in the analysis phase is to add whatever is necessary to specify all the users' requirements for the new system and to package the entire set of requirements into the structured specification. Everything critical to the users' acceptance of the completed system must be included. This chapter describes some typical detailed requirements that may be appropriate for inclusion in the completed system specification and that should be considered now if they were deferred earlier. It also presents a suggested outline for the structured specification.

CRITICAL DETAILED REQUIREMENTS

The structured specification is the sole and definitive statement of the users' requirements for a new information processing system. As such, it must provide a basis for testing the performance of the system after it has been constructed. That is why DeMarco sometimes calls the structured specification the *Target Document*—it establishes performance targets for the system. We would like the Target Document to contain all the essential requirements. On one hand, this means eliminating nonessential system details, deferring as many of the implementation-related decisions to allow flexibility during system design and construction. On the other hand, any requirement, no matter how detailed, that is critical to the users' acceptance of the system must be in the structured specification. Otherwise, when the system is presented for acceptance, there are likely to be disappointed and angry users as well as frustrated and angry developers.

298

Some of the kinds of details that can be critical for user acceptance are listed and discussed below. They will not all be critical for every system development project, but analysts should at least review them to see which are crucial to each specific project. Many of them may already have been considered in defining the alternative Physical New System Descriptions and comparing their relative costs and benefits to select the best, but perhaps not in the same detail as is necessary for the final system specification. In practice, previous discussions with the users are likely to have uncovered those system features about which the users feel strongly, no matter how irrational those features may seem to the analysts and how difficult they may be to implement. In extreme cases, all the analysts and designers can do is to present to the users the cost of these idiosyncrasies and let them decide. After all, it is the users' system.

1. Editing and Error Checking

Most of the edits and checks of data flows required by the users' business policies should have been incorporated in the Physical New System Description. These assure the completeness, correctness, and consistency of the data being stored and transformed. The concern here is for checks at the conceptual level of the business functions being implemented. Low-level checks, such as whether a data element is numeric or alphabetic or has the correct number of characters, should not clutter up the data flow diagrams and transform descriptions. They are expected as a matter of course in a good system design.

2. Error Messages

The text of error messages for the critical checks and edits should also be included in the data dictionary. This serves as a checklist for the users and assures that they will understand the messages produced by the system and that the messages contain enough information for the users to take appropriate action as a result.

3. Transforms for System Start-Up and Shutdown

If the comparison of system alternatives was made by considering the system in its steady state, it may be necessary to add transforms, data flows, and data stores for system start-up and shutdown. For example, in a data base environment, if the system or application programs are not in continuous operation, transforms may be required to perform system initialization or to make the data base accessible to the application programs when the system is started. Similar operations may be needed when the system is shut down to close the data base, gather and report operating statistics, and terminate the application.

4. Building and Maintaining Stores for Reference Data

It may also be necessary to add transformations to load or alter the contents of data stores that are read-only in the steady state, such as tax tables used for calculating the amount of income tax to be withheld.

5. More Data Dictionary Detail

Data dictionary entries for the data elements should be completed to show the ranges of values for continuous elements and the set of values for discrete elements. If the number of characters in each element cannot be determined from the element values, or is critical, it should also be specified. Details of critical edits, such as consistency between zip code, state, and perhaps city, should be noted in the transform descriptions rather than in the data element entries. Quantitative information for data stores should also be added: the expected number of records, their volatility, and the anticipated rate of growth in the size of the data store during the life of the system. Similar information about data flows should be specified, at as high a level as possible, to minimize redundancy. This information consists of the volume of flow over a specified time period as well as the volume, time, and duration of peak flows. Figures 14-1, 14-2, and 14-3 show examples of completed data dictionary entries for a data store, a data flow, and a data element. Remember to limit the data dictionary (as well as the rest of the structured specification) to what is essential or required in the new system. Do not unnecessarily preempt the designers' decisions.

FIGURE 14-1.   Completed data dictionary entry for a data store.

Data Store Name:	Packing-Slip
Composition:	Vendor-Name & Packing-Slip-Number & Date
Number of Records:	25
Volatility:	Low
Rate of Growth:	0

FIGURE 14-2.   Completed data dictionary for a data flow.

Data Flow Name:	Order-with-Payment
Composition:	Order-Number & Date & 1 { Order-Line } & Order-Subtotal & Sales-Tax & Order-Total
Flow Volume:	400 per day
Peak Volume:	60 per hour
Time of Peak:	11 am. – 1 pm.

FIGURE 14-3.   Completed data dictionary entry for a data element.

Data Element Name:	Vendor-Name
Type of Value:	Alphabetical
Length:	36 characters

6. Critical Control Flows

Control flows critical to the users are often related to the timing or occurrence of events that activate transformations in the system. These were omitted from the data flow diagrams because the transformation required to produce a report or to pay employees, for example, is the same whether the report or the payroll is generated hourly, daily, weekly, quarterly, or annually. However, if the end of a cycle coincides for a large number of transformations, the peak load on the system can have a significant effect on the resources required to comply with critical response requirements. These critical control flows may be added as annotations to the Physical New Data Flow Diagram. They are depicted as named arrows drawn with a dashed line, as shown in Figure 14-4. The names are usually self-explanatory and

DIAGRAM O: FAST FOOD STORE

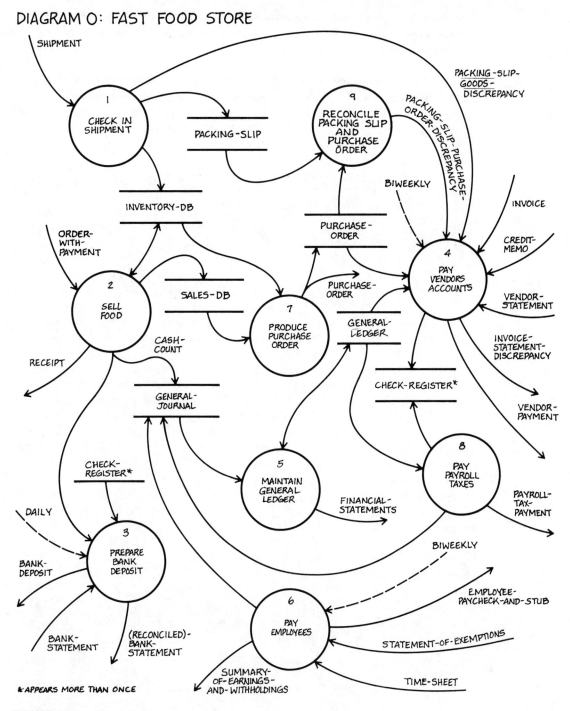

FIGURE 14-4. Diagram 0 annotated with control flows to show time cycles.

therefore do not require a data dictionary entry. Sometimes an appropriate alternative is to modify the name of a data flow to show time cycles, as in (Biweekly)-Payroll-Summary.

7. Critical Timing and Response Requirements

Other critical user requirements may state that system inputs are not available until a specific time, but that there is a deadline for receiving the resulting outputs. A manager may need reports at specific times during the year. In on-line systems, response time is usually critical to user acceptance. Response time depends on the interaction between the number of users, the volume of input and output, and the hardware/software environment as well as the system design. Real-time systems are characterized by the necessity to respond appropriately to external events as they occur.

8. Inflexible Input/Output Formats

The Logical New System Description focused on the essential information content of the data flows entering and leaving the system. Ideally, the formatting of the inputs and outputs will be left to the system design phase. At that time the designers will lay out the formats for the system inputs and outputs and present them to the users for review and approval. However, if the format of some of these data flows is prescribed by the users in advance and is essential to user acceptance of the system, they must be a part of the structured specification. These may include the layout of source documents, the arrangement of reports (perhaps on preprinted forms), special graphic symbols, and even screen layouts for video display terminals.

For example, the IRS Form 941-A must be produced using one of a limited number of type fonts, such as OCR-A, to expedite their processing. If the data is submitted on magnetic tape instead, the IRS prescribes the character codes, the record lengths and formats, and the blocking factor.

The best way to specify these requirements is to include a facsimile of each format in the specification, annotated if necessary. Figure 14-5 shows an example from the FastFood Store. It lists the payments made to vendors in the format that Mrs. O has used successfully for many years and does not wish to change. The vendors appear in alphabetical order. The payments to each vendor since the beginning of the year are listed chronologically. There are subtotals for each month and for each vendor.

For the benefit of the users, we recommend including typical data values, so that the facsimiles look like typical input or output. If form and printer layout charts are deemed necessary, they should be additions to, not replacements for, the input/output facsimiles.

In some cases, analysts may have developed reports or input forms to help the users visualize the new system. If these are included in the structured specification, it must be made clear that they are merely illustrative, not mandatory.

9. Conversion Requirements

Conversion from the current to the new system also deserves attention. If the change is from a manual to an automated system, or affects the interface between users and a currently automated system, user training is likely to be required. The extent of the required training and the kinds and details of users' manuals to be developed as part of the system design are outlined here. The major conversion

FIGURE 14-5. Required format for vendor payment report.

FastFood Store

VENDOR PAYMENTS REPORT

Date: 4/30/86   Page 1

Date Paid	Check Number	Amount of Payment	Vendor Total
A-1 Restaurant Supplies			
1/15/86	5097	$ 235.59	
2/10/86	5304	347.90	
3/20/86	5605	219.04	
			802.53
Brad's Wholesale Grocery			
1/16/86	5104	1097.18	
2/12/86	5314	998.27	
3/19/86	5603	1130.55	
4/18/86	5873	1235.93	
			4461.93
Brown Berry Coffee Company			
2/16/86	5329	507.00	
4/20/86	5902	675.50	
			1382.50
Chuck and Dave Meats			
1/15/86	5098	500.78	
1/29/86	5269	658.09	
	January	1158.87	
2/11/86	5309	726.08	
2/27/86	5398	567.94	
	February	1294.02	
3/13/86	5585	664.36	
3/27/86	5692	597.15	
	March	1261.51	
4/16/86	5869	795.58	
4/28/86	5921	529.47	
	April	1325.05	
			5039.45

effort usually is related to the data stores in the system. Careful planning can minimize the conversion effort. The choice of strategies includes three basic alternatives: operate the current and the new system in parallel, phase in the new system gradually, and change abruptly to the new system without phase-in or parallel operation.

It is often unnecessary to convert the entire existing data base to the new system. In many financial systems, an appropriately chosen end of cycle permits the new system to begin with balances forward as of that date. Subsequent transactions are processed by the new system. The end of the fiscal year is not always the best time, because the extra effort of learning a new system may coincide with the extra effort involved in the annual audit and unduly burden the users' staff. Often a data base may be converted by building or rebuilding it incrementally, adding each instance of an object or relationship as it is needed. Minimizing the conversion effort involves carefully analyzing what information must be stored in the new system before it can be used and planning how to make that information available.

10. Audit Requirements and System Controls

Some of the controls that provide auditability and guard against fraud, theft, and other losses may already have been incorporated in the Physical New System Description. If not, the DFDs should be extended to include these requirements. Requirements for suitable controls apply to both the manual and the automated portions of an information processing system. These requirements are concerned with the accuracy and completeness of the system's inputs, outputs, and transformations.

Some of the most common controls on application system inputs include, **authorization** of inputs, using **check digits** to test for the validity of data elements, and testing that a data element has a value from the allowable set (if discrete) or within the allowable range (if continuous). If inputs are processed in batches, sequential **batch numbers** assure that the loss of a batch is detected. The number of records or transactions in a batch can be counted, and this transaction count used as a control. Summing corresponding data element values, for example, hours worked or invoice amount, for each transaction in a batch yields a **control total.** The total is calculated manually before processing and compared with a value calculated by the computer during processing. If sums are calculated for data elements not usually added (such as part numbers or other identifiers), the result is called a **hash total,** but it is used like a control total.

In many situations, the transaction counts, control totals, or hash totals can be used throughout subsequent processing as well as to check the system outputs.

Other types of controls applicable to system outputs are verification controls and distribution controls. Verification controls are used to account for printed forms or documents. For example, in a payroll system, each employee's net pay is assigned a check number. This number is preprinted on the blank check and also printed by the computer program that writes the paycheck. The two numbers must agree. In this way the use of each check is accounted for. Distribution controls are procedures to assure that reports and other outputs from the system are delivered to the intended recipient. In some cases, the user may be required to sign a receipt.

Qualified EDP auditors should be consulted when the requirements for system controls are determined. The earlier they are involved in the preparation of the Logical New System Description, the better the opportunity for this aspect of system quality control. At the very least, the auditors should participate in final system walkthroughs and review the structured specification.*

---

*A fuller treatment of system controls may be found in Alan E. Brill, **Building Controls into Structured Systems**, Yourdon, Inc., New York, N.Y., 1983, upon which this discussion is partly based. See also **The Auditors Study and Evaluation of Internal Controls in EDP Systems**, American Institute of Certified Public Accountants, New York, N.Y., 1977.

11. Security and Backup	The auditability controls primarily involve functions within the information processing system to preserve the integrity and quality of the data and to provide cross-checks to detect errors. Other aspects of security are related to preventing unauthorized access to automated systems or to portions of the data base. Security of the physical environment of the information processing system seeks to protect the hardware from vandalism, sabotage, or natural hazards such as fire, flood, or earthquake. The ability to resume operation in the event of physical catastrophe and to recover from destruction of programs and data, whether deliberate or accidental, is another critical requirement for most systems. The financial exposure to loss for a business dependent on automated systems, such as an airline or a bank, can far exceed the cost of the hardware and software.

12. Quantitative and Testable Performance Targets	This is the area of the system specification that has historically been the weakest. In some cases, targets have been set that were arbitrary, unreasonable, or unnecessarily expensive for the area of application. In others, the targets were only qualitative or so vague that neither users nor developers had a clear and definitive test for whether or not a system was acceptable.

Users and analysts have much to learn about stating desired system performance appropriately, precisely, and quantitatively, just as designers have much to learn about how to predict system performance quantitatively.

PACKAGING THE SYSTEM DESCRIPTION

Packaging the Physical New System Description is the final step in the specification of users' requirements. The term *packaging* is used in two ways. In the most general sense it means the process of collecting and organizing the products of analysis into the structured specification. More specifically, it also means the process of partitioning a system description into implementation-related units.*

One packaging task is a repartitioning of the data flow diagrams to show the major aggregations in both the manual and the automated portions of the system. Aggregations in the manual part of the system will reflect the organizational structure of the new system. This is desirable only when the system is sufficiently complex to require reducing the number of bubbles on the top-level diagram or where it is important to communicate organizational changes in the new system. Otherwise it is preferable to retain logical names for the manual transformations when combining them at the upper levels of the DFDs.

Aggregations in the automated part of the system show hardware boundaries as well as boundaries between batch, on-line, and real-time portions of the system. In a distributed data processing system with a central computer and local computers or terminals, the hardware boundaries often correspond to geographic boundaries. In other systems, hardware boundaries will mark functions performed by special

---

*DeMarco, **Structured Analysis and System Specification**, p. 273, and Page-Jones, **A Practical Guide to Structured System Design**, Yourdon Press, New York, N.Y., 1980, p. 345.

processors, such as graphics workstations, photocomposers, data base machines, or telecommunications processors. In this repartitioning, Diagram 0 is organized so that the automation boundary does not split any of the bubbles. Separate high-level bubbles should also depict significant hardware boundaries. This may require renumbering the transforms and revising the data dictionary accordingly. Further partitioning of the DFD to show job and job-step boundaries also occurs either at the end of analysis or at the beginning of design, but it is not discussed in this book.

AN OUTLINE FOR THE COMPLETE STRUCTURED SPECIFICATION

A suggested outline for the structured specification is shown in Figure 14-6. Each section of the document is discussed briefly below.

FIGURE 14-6. Outline for Structured Specification.

System Requirements Specification

Introduction

System Summary

Project Summary
    Project Name
    Responsible User
    Responsible Analyst
    Scope
    Start Date
    Project Completion Date
    Completion Date for System Design Specification
    Project Budget Amount

System Constraints and Assumptions
    Interfaces to Other Systems
    Hardware Constraints
    Software Constraints
    Failure and Backup Requirements
    Security and Privacy

System Improvements
    Improvements in Service
    Increases in Revenue
    Reductions in Cost

System Impacts
    Hardware
    Software
    Organizational
    Operational

FIGURE 14-6.   *(cont.)*

Performance Requirements (Acceptance Constraints)
    Workload and Volume
    System Growth
    Response and Timing
    Accuracy and Validity of Data
    System Access

New System Description ~ *new physical*
    Overall System Description (Data Flow Diagrams)
    Data Base Description
        Data Access Diagram
        Integrity Constraints (Conditions on Objects and Relationships)
        Data Store Contents (Data Dictionary)
    Data Flow Descriptions (Data Dictionary)
    Data Element Descriptions (Data Dictionary)
    Transform Descriptions (Transform Dictionary and Transform Specification)

Conversion Requirements

Project Constraints
    Personnel
    Budget
    Facilities
    Schedule

Appendices
    How to Read the System Description
    Cost/Benefit Comparisons of Alternatives
    Project Cost Estimates and Schedule
    Detailed Activities, Schedule and Budget for Design Phase
    Detailed Activities, Schedule and Budget for Planning Acceptance Tests

---

The Introduction provides any background essential to an understanding of the specification as a whole. It may also include references to documents produced previously that contain material not in the specification. Figure 14-7 shows a possible Introduction for the FastFood Store specification.

The System Summary is a one-paragraph summary of the most important characteristics of the new system—the reason for system change; the scope and

FIG 14-7   Introduction for FastFood accounts payable system specification.

Introduction

The FastFood Store is a restaurant established 15 years ago by Mr. and Mrs. Owner. With the growth in their volume of business over the years, Mrs. Owner, who does most of the bookkeeping, has had increasing difficulty in carrying out her responsibilities. As a result of a study of the FastFood Store's operations, the Owners have decided to automate a portion of their accounts payable system.

FIG 14-8:   System summary for FastFood accounts payable system specification.

System Summary

The portion of the accounts payable system to be automated includes a data base of invoices, credit memos, and vendor information.

The system will use data from purchase orders and vendor statements for reconciliation with the invoices and credit memos. System outputs will be lists of any discrepancies found during the reconciliation, a check register, a list of vendor payments, and cash disbursement entries for the general journal. Packing slips and purchase orders will be reconciled manually, with any discrepancies entered into the automated system.

---

functions of the new system; its principal inputs, outputs, and data stores; the recommended type and scope of automation; and any critical impacts the implementation of the recommended alternative will have on the way the organization operates.

Figure 14-8 presents a possible system summary for the FastFood Store accounts payable system.

The Project Summary provides a synopsis of the important management aspects of continued system development.

The section on System Constraints and Assumptions lists specific requirements and conditions with which the new system must comply to be acceptable. These may include a specification of the interfaces to other systems, generic or specific hardware on which the system must operate or with which it must communicate, and the software environment in which the system must operate. Provisions for failure prevention and recovery as well as requirements for security and privacy are also covered. If these constraints or other system requirements contained in the specification are based upon assumptions, those assumptions should be stated explicitly so that their continuing validity can be reviewed as system development proceeds.

System Improvements summarizes the benefits expected from the new system. The specific benefits are system-dependent, but generally may be classified as improving service or performance, increasing revenues, or reducing costs. It may be appropriate to relate these benefits to deficiencies in the current system. Details are provided in an appendix to the structured specification containing the result of the cost/benefit studies. These studies were made in Activity 2.5: Select the Best Scope of Automation, as discussed in Chapter 13. System Impacts presents other significant effects on the existing system and its environment.

For example, the System Improvements section of the FastFood Store requirements specification would include the following benefits

- reduction in the time Mrs. O requires to reconcile the purchase orders and the vendor invoices
- reduction in the time Mrs. O requires to reconcile the invoices, credit memos, and statements from the vendors

These reductions will result from

- eliminating mistakes in calculation as a cause of error in the reconciliation process
- pinpointing the inconsistencies among the documents, thus focusing Mrs. O's attention on the problem areas

The Performance Requirements section states values of measurable and specific system characteristics that must be achieved for the completed system to be acceptable to its users. These requirements may apply to the workload on the system, the volume of information to be processed or stored, provisions for growth in the system capacity, response times to routine and critical inputs, timing or scheduling of system functions, requirements for the integrity, accuracy, and validity of the data, and required or prohibited access to the system or data. These performance requirements must be stated as specifically, precisely, and quantitatively as possible, because they are the sole basis of the acceptance tests that follow system construction.

For example, some performance requirements for the FastFood Store might be stated as follows:

- number of vendors: 50, to grow to 100 in three years
- number of invoices per month: 150
- average number of invoices per vendor per month: 3
- number of statements per month: 40
- number of inventory items: 1000
- average number of items per invoice: 5, maximum: 20

The system must provide overnight turnaround for producing the vendor payment and cash disbursement lists.

The New System Description is the Physical New System Description of the alternative selected for design, as discussed in Chapter 13. The basis for choosing this alternative is explained in the appendix to the structured specification that contains the cost/benefit comparisons. The New System Description consists of the leveled set of data flow diagrams, the logical data base description, data flow and data element descriptions, and transform descriptions. The data store, data flow, and data element entries appear alphabetically within each section; the transform entries are in numerical order. Some people include the specifications or descriptions for the primitive transforms in numerical sequence with the System Dictionary entries for the nonprimitive transforms. Others put them together with the data flow diagrams, so that the reader finds a transform description instead of another diagram at the lowest level.

Where appropriate, a section on Conversion Requirements is included to define constraints on the transition from the current to the new system. It may also recommend an approach to this transition as a basis for a subsequent detailed conversion plan—parallel operation, phased conversion, or abrupt switchover. Details of the current system that critically affect the conversion should be summarized here. However, the structured specification focuses on the new system. Thus, neither the current system descriptions nor the new system descriptions for the rejected alternatives should be routinely incorporated in the structured specification. Instead they should be packaged separately and made available if needed in later system development activities.

An appropriate conversion strategy for the FastFood Store accounts payable system would recognize the fact that the current system is manual and that most of the information in the data base changes rapidly. Only the vendor catalog information changes relatively slowly, with new catalogs issued once or twice a year for each vendor. However, vendors may make price changes more frequently. The vendor catalog data base can be built incrementally as orders are placed. Thus, it can be tailored to the needs of the FastFood Store and contain records for only those items regularly ordered by Mrs. O. The effort to include in the data base all the items sold by all the vendors is probably not warranted.

The summary of Project Constraints aids in project management. It presents the limits on personnel, budget, schedule, and facilities available for system development. This summary is supported by appendices estimating the cost and schedule for the remainder of the project, with more detailed breakdowns for designing the system and preparing the acceptance test plan.

Other appendices may include a short tutorial explaining how to read and understand the New System Description and the Cost/Benefit Comparisons. The tutorial will aid readers who have not been closely involved in the analysis of the system and thus are not familiar with data flow diagrams, data access diagrams, the data dictionary conventions, and transform descriptions. Writing this appendix the first time may be difficult, but it can be reused. It is certainly desirable to modify it each time to incorporate examples from the system specification to which it is appended.

SUMMARY

After the best alternative for the new computer information system has been selected, the final step in establishing the users' requirements is to complete and package the structured specification.

All details that are critical to user acceptance of the completed system must be specified. They may include requirements for editing and error checking, error messages, transforms for system start-up and shutdown, transforms for building and maintaining data stores of reference information, additional system dictionary detail, critical control flows, critical timing and response requirements, inflexible input or output formats, conversion requirements, audit requirements, system controls, and security and backup. All performance requirements should be stated quantitatively so that they can be measured and tested when the new system is delivered.

A complete structured specification includes an introduction, a system summary, and a project summary, as well as a statement of system constraints and assumptions. It contains a description of the new system, comprising data flow diagrams, a system dictionary, and data base description. Where appropriate, a section on conversion requirements is also included. A summary of project constraints assists the project management. Appendices may present additional supporting detail as well as a brief explanation of how to read and understand the new system description and the cost/benefit comparisons.

DISCUSSION

14-1. Why is the Physical New System Description of the selected system alternative (the data flow diagrams, system dictionary, and data base description) insufficient as a structured specification?

14-2. List 12 kinds of detailed requirements that may be critical in some systems. Give an example of each and explain why it is critical.

14-3. Why might these detailed requirements not have been made explicit earlier?

14-4. Explain what each of the following system controls is and how it is used to assure accuracy or completeness:
   a. authorization
   b. batch numbers
   c. check digit
   d. control total
   e. distribution control
   f. hash total
   g. output verification control
   h. transaction count
   i. value or range check

14-5. Explain the purpose of each section of the structured specification.

15 Coping with Changing Requirements

THE FACT OF CHANGE

Change is inherent in physical, biological, and social existence. Thus, business systems, like the rest of the world, are subject to change. This chapter describes how the fact of change affects the process of analysis.

Computer information system development would be much easier if we could guarantee that users' requirements would stay the same after the structured specification had been completed. In reality, changes continue to occur, for valid business reasons. How can the system development process respond appropriately to legitimate changes in users' requirements for a new system?

After the system requirements specification is complete, some organizations attempt to stifle, frustrate, and delay requests to modify the structured specification. The result of this intolerance of change is often a new system that no longer meets the real requirements of the organization, because the requirements changed while the system was being designed and constructed. As systems become larger and more ·complex, the penalty for ineffective systems becomes increasingly severe.

On the other hand, a system development project becomes unmanageable if anyone and everyone can make changes in a haphazard and uncoordinated way, with no consideration of their impact on either the development process or the completed system as a whole. There need to be orderly procedures for identifying necessary changes and modifying the system requirements to incorporate them.

If requirements have changed, it does no good to deny the fact and proceed with system development as if nothing had happened. It is in this context that DeMarco speaks of "The Myth of Change Control":

It is a fallacy to talk about controlling change. The best you can hope for is to keep track of change.*

The remainder of the chapter discusses some techniques for keeping track of change.

SOURCES OF CHANGE IN BUSINESS SYSTEMS

Changes in business systems arise externally, in the environment, as well as from within the organization.

In his ***Systematic Systems Approach,*** Thomas Athey provides a helpful framework for understanding system change. He views business as a cybernetic system, consisting of inputs, processes (transforms), and outputs, with measures of system effectiveness, as illustrated in Figure 15-1:[†]

Cybernetic systems are those systems that are affected by environmental shifts but have means through feedback control to continue to meet system objectives. Additionally, the system objectives are not rigidly fixed but are adaptable to changing conditions and responsive to new understanding.[‡]

The environment is subdivided into an immediate environment and a general environment, as illustrated in Figure 15-2.[•]

Together they account for the external influences on a business. The immediate environment is composed of suppliers, competitors, customers, and sociopolitical influences. The general environment is composed of resource forces, technological forces, political forces, and cultural forces. The distinction between the two lies in the effect of a change. A change in the immediate environment has a direct and immediate influence on system behavior. For example, a competitor alters his prices, a customer goes out of business, a supplier experiences a strike, the government imposes a new payroll tax. The general environment

includes forces that are more indirect in their effect on a system. While the general environment can have a tremendous impact on a system, these forces. . . are long range in scope and generally change very slowly. However, they are the underlying forces that eventually cause changes in the immediate environment.[§]

---

*Tom DeMarco, ***Structured Analysis and System Specification,*** Yourdon Press, New York, N.Y., © 1979, p. 295. Reprinted by permission of Yourdon Press.

[†]Thomas Athey, ***Systematic Systems Approach***, Prentice-Hall, Inc., Englewood Cliffs, N.J., © 1982, p. 31. Reprinted by permission of Prentice-Hall, Inc.

[‡]Thomas Athey, ***Systematic Systems Approach***, Prentice-Hall, Inc., Englewood Cliffs, N.J., 1982, p. 30.

[•]Thomas Athey, ***Systematic Systems Approach***, Prentice-Hall, Inc., Englewood Cliffs, N. J., © 1982, p. 32. Reprinted by permission of Prentice-Hall, Inc.

[§]Thomas Athey, ***Systematic Systems Approach***, Prentice-Hall, Englewood Cliffs, N.J., 1982, p. 30.

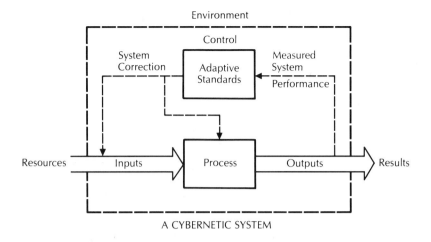

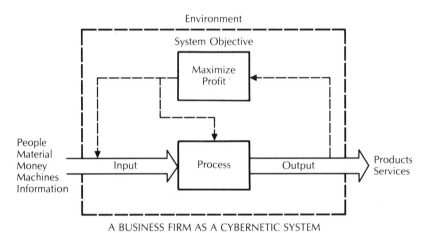

FIGURE 15-1. A business firm as a cybernetic system.

Changes arising from within an organization may be associated with three levels of decisionmaking—strategic, tactical, and operational. A change introduced at each of these levels also has a corresponding scope of effect. Strategic decision makers are concerned with what the overall organizational goals and objectives should be. A decision by an automobile manufacturer to enter the video game business is an example of a strategic decision. The scope of effect of such a change is quite comprehensive. At the tactical decision making level, the scope of effect is less. The decision maker here is concerned with how to best accomplish the overall system objectives. For example, should the auto maker acquire an established video game manufacturer or develop a production facility from scratch? At the operational level the scope of effect of change is much more limited. Here the manager makes the day-to-day decisions to operate the business in harmony with the goals and plans established at the strategic and tactical levels.

During analysis, the analyst needs to assess the stability of the various environmental factors that may affect the information processing system being developed as well as the business it will support. In addition, the analyst needs to be aware of

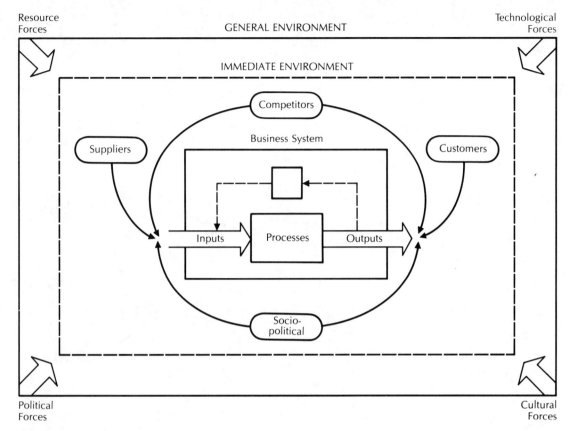

FIGURE 15-2. Forces in
the business environment.

decisions within the organization that may affect the system being developed. Identifying potential sources of change is important because it allows the analyst to cope with change by anticipating it instead of reacting to it. Figure 15-3 presents a decision tree for identifying and classifying the various sources of system change.

THE IMPACT OF BUSINESS CHANGE ON INFORMATION SYSTEM REQUIREMENTS

In the early stages of analysis, while modeling the current system, the analyst may discover that parts of an organization are in a state of flux. As discussed in the introduction to Part III, there may be strong pressure to move quickly to the description of the new system. In these situations it may be useful to identify the aspects of the users' current system that are changing.

Changes to the physical aspects of the users' organization suggest operational or possibly tactical decisions made to cope with a change in the immediate environment. In this situation, the analyst should proceed as quickly as possible to the Logical Current System Description.

Changes at the strategic and tactical levels may result in modifications to the Logical Current System Description. In *Managing the System Life Cycle,* Edward Yourdon says

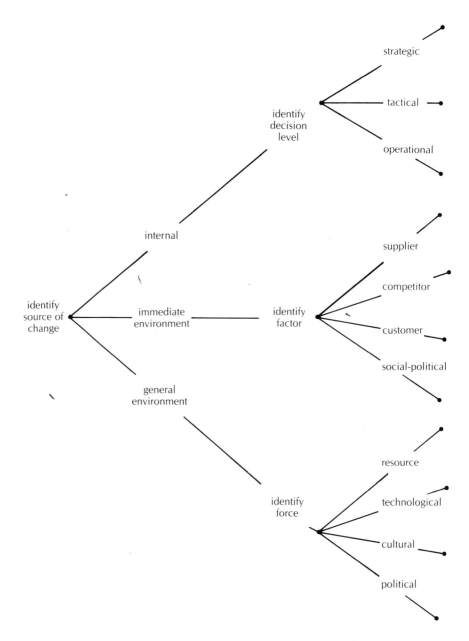

FIGURE 15-3. A decision tree for sources of system change.

If it is the ***logical*** model that is changing rapidly, then the users' whole business must be under great stress, for changes in the logical model mean that the organization is rapidly entering into new sorts of business, abandoning others, making fundamental changes to policy, and so on. Such changes, if they occur rapidly, will probably cause varying degrees of stress and disorientation at all levels of the user organization—and users may well feel that any attempt to model the business is futile. The analyst can take advantage of this situation to help the user to model his current and projected needs, thus enabling the user to feel more in control of things.*

Increased understanding is another reason to change a system description. In the FastFood Store example, there were many modifications to the Physical Current System Description first presented in Chapter 6 as it evolved into the Physical New System Description at the end of Chapter 13. At each stage, these changes resulted from increased understanding and, in turn, brought new insight into the system. This is a common experience in many projects as each of the four system descriptions is developed.

Thus, the number and magnitude of changes to a system description declines with time. However, once the structured specification has been produced, it cannot be expected to remain static and still provide an accurate target for the system designers and implementers. The following section describes revision of the structured specification by means of incremental modifications.

CHANGING THE STRUCTURED SPECIFICATION INCREMENTALLY

The partitioned structure of the structured specification encourages an incremental approach to keeping it up to date. Since more changes are likely to be proposed than will be accepted, the procedures for formulating and evaluating changes need to be efficient. Each decision about a proposal to change the specification can be viewed as a small-scale system study. The steps in this study are to describe the proposed change, analyze its impact, decide its disposition (accept, defer, or reject), and update the structured specification. A change to the structured specification is, of course, followed by a corresponding change in the system as the development process continues.

Four documents are used in this process—the Specification Change Request, the Specification Change Description, the Specification Change Cost-Benefit Analysis Findings, and the Specification Change Disposition. Each step in the process and each document is discussed below.

1. *Describe the proposed change.*

The user initiates the specification change procedure by submitting a Specification Change Request. A sample Specification Change Request is illustrated in Figure 15-4.

An analyst then prepares the Specification Change Description. The Specification Change Description shows the desired revisions to the structured specification. As appropriate, it consists of annotated data flow diagrams, modified data dictionary entries for affected data elements, data flows and data stores, a modified data base description, and revised transform descriptions.

Preparation of the Specification Change Description may involve decom-

---

*Edward Yourdon, *Managing the System Life Cycle*, Yourdon Press, New York, N.Y., © 1982, p. 74. Reprinted by permission of Yourdon Press.

Specification Change
Number ☐

Requested By: _____   Request Date: _____

Reason for change:

Attach affected Diagrams, Data Dictionary & Transform Descriptions

Attach Cost-Benefit Analysis

Disposition: ☐ Accept

☐ Defer until _____
Date

☐ Reject for following reason:

FIGURE 15-4. A specification change request form.

posing a complex change into several change requests. Generally, there is only one change per request. However, it may be necessary to indicate the related nature of clustered change requests. The analyst should select the level in the structured specification where the context for the proposed change is enclosed within one bubble. Appropriate modifications are then annotated on the child diagrams and affected data dictionary, data base description, and transform descriptions are modified. A sample Specification Change Description is given in Figure 15-5.

Preparation of the Specification Change Description will usually involve several iterations of refinement until all affected parties agree that it is an adequate description of the proposed change.

The evaluation criteria of completeness, consistency, correctness, and communication apply to the proposed change to the specification just as they did to the original requirements statement. For extensive changes, formal walkthroughs may be necessary.

FIGURE 15-5. An annotated change description.

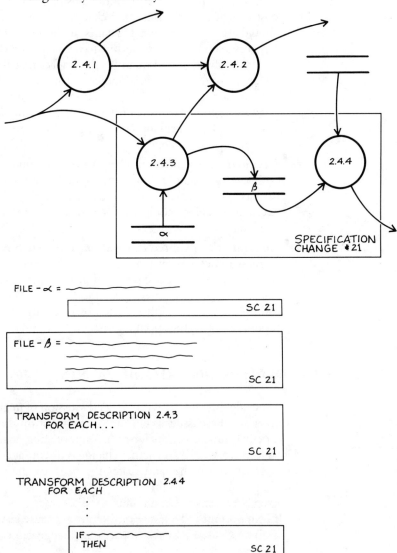

Note that only the affected parties need be involved in describing the change and then only in the limited context of the proposed change. Thus, the procedure can be tailored to each proposed change. Affected parties are readily identified through the partitioning of the structured specification. The task of preparing the Specification Change Description belongs to the analyst. The Specification Change Description is organized so that the change may be easily integrated into the structured specification upon acceptance.

2. *Analyze the impact of the proposed change.*

This step is similar to the cost-benefit analysis of the alternative Physical New System Descriptions, as described in Chapter 13, but on a smaller scale. The proposed change must be quantified using the same measures as in the previous cost-benefit analysis. Since the change is incremental, the cost-benefit analysis should also be incremental. That is, it should be concerned only with the differences in costs and benefits due to the proposed change. In addition, the cost-benefit analysis should quantify the impact that the proposed change is likely to have on the system development effort. This is very much dependent upon how far along the project is when the change is proposed. The cost-benefit analysis should at least quantify the impact of the change on the project budget and schedule, identifying any savings or additional effort directly attributable to the proposed change.

3. *Decide the disposition of the proposed change.*

The cost-benefit findings are weighed, and the change is accepted, deferred, or rejected. In the case of deferral or rejection, a brief justification is given. A decision to defer should be for a definite period of time to avoid *de facto* rejection.

4. *Integrate the accepted change into the structured specification.*

As stated earlier, this task is simple. The real work took place earlier during the preparation of the Specification Change Description. Now the obsolete portions of the structured specification are replaced as provided in the Specification Change Description. The obsolete parts are retained as an audit trail of modifications to the specification.

MANAGEMENT ISSUES IN CHANGING REQUIREMENTS

We conclude this chapter with a brief discussion of management concerns. Detailed treatment of these issues is beyond the scope of this book.

An environment conducive to accommodating change is extremely important. Users learn very quickly whether change is discouraged, tolerated, or actively solicited. An approach that anticipates the need to change goes a long way toward building such an environment.

Keep bureaucracy to a minimum.

It is important to be aware of the turmoil caused by a rapidly changing system. We have glossed over the impact that change can have on project management.

Estimating resources and schedules is difficult enough in a stable environment. Thus, the importance of attaching the cost of implementation to proposed changes cannot be overstressed.

SUMMARY

Businesses are systems that can adapt their goals and objectives to respond to change. The need to change a business may arise from factors in the environment as well as for internal reasons. Because the environment in which a business operates is by definition outside the organization's control, changes in the environment are uncontrollable. The best that can be done is to anticipate environmental change and respond to it in an orderly fashion. Management will have greater control over the scope and timing of internally initiated change. Whether of internal or external origin, changes may occur at operational, tactical, and strategic levels and often affect the information systems that support the business at each of these levels.

As a consequence, the process of computer information system development must accommodate changes to system requirements after the structured specification has been completed. Each proposed change must be described and its impact analyzed. Then a decision as to whether to incorporate the change in the system requirements must be made. If the change is accepted, the structured specification must be modified accordingly.

A proposed change is submitted on a Specification Change Request, then described precisely on a Specification Change Description. The Specification Change Cost-Benefit Analysis Findings present the effect of the proposed change on the schedule and budget for the remainder of the system development process as well as on the costs and benefits of the completed information processing system. Based on these findings a decision to accept, defer, or reject the proposed change is made and recorded as a Specification Change Disposition.

REVIEW

15-1. What is meant by the immediate environment of a business? Give some examples of components of the immediate environment.

15-2. What is meant by the general environment of a business? Give some examples of components of the general environment.

15-3. State the purpose of the following parts of a Specification Change:
   a. Specification Change Request
   b. Specification Change Description
   c. Specification Change Cost-Benefit Analysis Findings
   d. Specification Change Disposition

15-4. Why is the Specification Change said to be incremental?

DISCUSSION

15-1. Discuss the effect of each of the following changes in the immediate environment of the FastFood Store. Which of them would affect the information processing systems? Which would affect the structured specification for the accounts payable system?

a. A vendor goes out of business because his prices were too low.

b. The quality of locally available produce declines.

c. A local payroll tax is imposed.

d. A major computer vendor offers substantial discounts to small businesses that purchase a new line of microcomputers.

15-2. One of the most popular items on the FastFood Store menu is "Mr. O's Special Burger (with Secret Sauce)." Discuss the possible impact of a long-term shift in consumer preferences from beef to fish and poultry. To what extent would this affect the accounts payable subsystem?

15-3. Discuss the importance of considering the cost implications of altering the project budget and schedule when a change in the structured specification is proposed.

16 Alternative Processes for Systems Analysis

Chapters 12 through 15 have dealt with the process and products of structured analysis. The discussion assumed four relatively complete system descriptions, culminating in a detailed model of the new system requirements that is presented in the structured specification. This structured specification contains everything necessary for custom design and construction of the required information processing system. An understanding of this process provides the necessary conceptual framework for the most comprehensive kind of system development, which starts from scratch and takes no shortcuts. Not every system development project needs to follow this process in its entirety.

The conceptual framework of Figures III-1 through III-4 is idealized. It provides a sequential description of the models used in analysis, allowing us to understand the purpose of each system model. In reality, this idealized analysis process must must be tailored to each specific project. The process is usually iterative or cyclical rather than linear, and the preparation of the various models may overlap. The following section explores some of the factors influencing the analysis process. Subsequent sections describe some alternatives to the complete, detailed system development strategy presented in Chapters 12 through 15.

FACTORS INFLUENCING THE ANALYSIS PROCESS

When formulating the analysis and development strategy for a particular system an astute analyst will consider a variety of factors.

Early in a project, the analyst can identify these factors by preparing profiles of the users, the analysts, and the project.

The Users' Profiles   Factors contributing to users' profiles include the types of users, the users' knowledge of the application, the users' prior experience with a systems analysis, and the users' risk profile.

In Chapter 4 users were classified as system owners, responsible users, hands-on users, or beneficial users. The initial step in producing users' profiles is to determine an individual user's type.

When evaluating the users' knowledge of the application, consider not only years of experience but also the variety of experience. Remember, 10 years of the same experience are not the same as a sequence of 10 years of different experience.

The users' prior experience with systems analysis will have a bearing on how much time should be allotted for educating users about how a system study is conducted. Be sure to consider whether or not the users and analysts have worked together previously. Also try to determine if the users know what their roles will be during the course of analysis.

Knowledge of the extent to which users are willing to take risks can also have an impact on the choice of a systems analysis strategy. One concern here is whether users can base the decision to adopt the proposed system upon a relatively abstract system description or whether a more concrete presentation of the proposed system is required.

The Analyst's Profile   Factors contributing to a profile of analysts include the analysts' knowledge of the application and their experience in systems analysis.

An analyst's knowledge of the application will have a tremendous bearing on the analysis process. If this is the analyst's first exposure to a particular application area, not only will the analyst have to spend more effort learning about the application, but he will be more dependent on the users. If, on the contrary, the analyst has considerable experience, less effort will have to be spent in learning for this specific project. A similar argument holds for the analyst's prior experience in systems analysis.

The Project Profile   The project profile is concerned with factors that are largely uncontrollable by either the analysts or users. These factors include the stability of the project environment, the pressure to produce immediate tangible results, the pressure to produce accurate schedules and budgets, and the complexity of the system.

Environmental stability includes such external factors as the competitive situation, legislative requirements, and considerations of social responsibility. Influences generated within the organization include pressures to produce results and the need for accurate schedules or budgets.

Much research is being conducted to find reliable measures of system complexity. Until the results of this research are available, complexity estimates will continue to be based upon the number of transforms and interfaces anticipated in the new system.

Each of the profiles mentioned is intended to give the analyst a gross picture or feeling about critical factors affecting the choice of the analysis process. To a very large extent, these are subjective evaluations based upon experience.

Here we are reminded of an anecdote about experience. A successful businessman was asked to what he attributed his experience. He replied, "Good decisions." When queried about how one can make good decisions, he answered, "Experience." When pressed further about how one can gain experience, he replied, "Bad decisions." (It seems the businessman was also a master of short replies.)

By no means have we given an exhaustive list of factors. Rather, the reader should be aware that just as we can systematically describe and decompose systems, we can also systematically describe and decompose the factors that influence the systems analysis process. The following sections briefly explore some alternatives to the process of analysis suggested in Figures III-1 through III-4. We also mention some factors that favor the selection of each alternative.

WHEN NOT TO LOOK AT THE CURRENT SYSTEM

Consider the following situation:

- the users and analysts have extensive experience
- the users and analysts have worked together previously
- the users are willing to take some risks
- the environment is stable
- there is strong pressure to move to the new system.

Given this situation, it may not be cost effective to do an extensive investigation of the existing system. Suppose we add one more element to the scenario:

- the current system description was produced by these analysts and users in an earlier system development project.

The description remains complete, consistent, and correct. Clearly, it would be redundant to re-derive this current system description.

There are also situations where it may be unnecessary to examine the current system.

An obvious situation is where there is no current system to be considered. This may be the case when a new application is developed. Analysts involved with the development of the first word-processing or computer-graphics system would have been hard put to examine an existing system. In these research and development environments, analysts try to characterize the way things are currently done, pointing out deficiencies that should be remedied and identifying the essential inputs, processes, and outputs. Thus, some consideration of the current mode of operation is necessary. The question becomes whether or not a detailed study of the current environment is justified.

Another situation is where the new system is anticipated to be substantially different from the existing system. This may render detailed study of the current system unproductive. Thus, it may be beneficial to study some other system that more closely resembles the new system. Implicit in the discussion thus far is the supposition that we consider modeling only an organization's own current system. There may be too many deficiencies or difficulties with the in-house system to model it effectively. It may thus be more helpful to examine another organization's current system that is representative of the new system. Care must be taken to ensure that the other system is truly representative.

Another alternative in these situations is to develop the Logical New System Description from scratch. As stated in the introduction to Part III, it is often difficult to achieve success with this approach. Prototyping, discussed later in this chapter, holds some promise as an effective way to develop a new system description without modeling the existing system.

SHORTCUTS IN PRODUCING THE CURRENT SYSTEM DESCRIPTION

Rapid consideration of the current system is made possible by the availability and accuracy of the Logical and Physical New System Descriptions from the previous major system change. This was the case in one situation discussed in the previous section. However, suppose the user and analyst cannot state with confidence that the existing system description is complete, correct, and consistent. Care must be taken to ensure that it continues to reflect the way things are currently done. However, it is still much easier to check the accuracy of a system description than to develop it in the first place.

Another effective shortcut is selective deployment of resources during the study of the current system. This strategy may be an effective way to respond to pressures to change the system quickly. It can maximize tangible results while minimizing resources expended. Some analysts practice "blitzing"—applying extra resources intensively to study complex or important areas of the system in detail.* Less criticial areas may be given short shrift. This may be accomplished by waiting until after the production of the Logical Current DFDs to produce transform descriptions and data element definitions. By deferring these details, the analyst capitalizes on the elimination of nonessential data flows, data stores, and transforms. Another technique is to leave some of the edits, audits, and approvals in the Logical Current and New System Descriptions in anticipation of their necessity in the Physical New System Description. This calls for judgement based upon experience and is not suggested for novice analysts.

STRATEGIES FOR PRODUCING THE NEW SYSTEM DESCRIPTION

There are two alternative strategies for producing the new system description. The first is related to the use of packaged software for the new system. The other strategy relies on prototypes to aid in discovering the new system requirements.

---

*Brian Dickinson, **Developing Structured Systems: A Methodology Using Structure Techniques**, Yourdon Press, New York, N.Y., 1981, p. 337.

System Requirements for Packaged Software

In general, consideration should be given to the purchase of packaged software as an alternative to custom design and development. After the users and analysts have arrived at an appropriate scope of automation for the new system, it can serve as the basis for a request for proposal (RFP). The request for proposal asks vendors of application software packages to propose solutions to the system requirements.

This approach still requires a statement of system requirements as a basis for evaluating and comparing the vendors' proposals. However, the requirements statement can often be simpler than one intended for custom development. In addition to top-level DFDs, their associated data dictionary entries, and the data base description, an RFP may include suggested or required algorithms for certain critical transforms, specific interface or performance requirements, and budget constraints. In response to the RFP, the vendor is required to state the extent to which his package satisfies the new system requirements, whether it can be modified to meet the requirements, and what the cost of the necessary changes will be. An unfortunate ramification of this approach is that, in general, the vendor is free to use **any** system description technique when answering an RFP. Some organizations require responses to RFPs to adhere to stringent guidelines in order to facilitate comparison of the various vendors' system descriptions.

Prototyping

Some users find it difficult to envision what a proposed system will look like when it is described as abstractly as in a structured specification. This may be especially true for users with little experience as participants in a system development effort. As a consequence, these users may be unable to determine whether their needs will be met by the proposed system.

Prototyping attempts to bridge the gap between the users' view and the analysts' view of the system specification. Prototyping is an approach in which the abstract system description is quickly realized in a concrete form that simulates the proposed system. In this way, users can "play" with a mock-up of the new system and suggest changes. Users can see the proposed input and output formats and can evaluate the proposed scenarios for interaction with the automated system. Using prototype development tools, this can be done at less expense than in traditional approaches to system development.

There are two basic strategies in prototyping: build a mock-up of the new system that is discarded when construction of the actual system is completed, or let the prototype system evolve into the new system. Both strategies are facilitated by software development tools that allow and encourage the rapid development and integration of system components. Prototyping may use fourth-generation languages as well as off-the-shelf system components such as terminal-conversation managers, report generators, and data base management systems.*

Although currently restricted in distribution and use, Anthony Wasserman's

---

*Richard L. Wexelblat, "Nth Generation Languages," **Datamation**, September 1, 1984, Vol. 30, No. 14, pp. 111–117.

User Software Engineering (USE) is representative of the prototyping approach to system development.* USE assists in the rapid construction of partial systems and/or prototypes as an aid to systems analysis and specification.

In USE, diagrams describing the interaction of user and computer, such as that of Figure 16-1, are entered into a computer relatively easily. After the system has interpreted the diagrams, a user may carry on a conversation at a computer terminal simulating interaction with a real system.

We close this discussion of prototyping with some advice from Fred Brooks, the director of IBM's OS/360 project. In his classic book, **The Mythical Man-Month,** Brooks advises system developers to "plan to throw one away; you will anyhow."[†]

---

*Anthony I. Wasserman, "The Unified Support Environment: Tool Support for the User Software Engineering Methodology," **Proceedings of Softfair: A Conference on Software Development Tools, Techniques, and Alternatives**, Arlington, Va., July 25–28, 1983, IEEE Computer Society Press, Silver Spring, Md.

[†]Frederick P. Brooks, **The Mythical Man-Month**, Addison-Wesley, Reading, Mass., 1975, p. 116.

FIGURE 16-1. Diagram for a user-computer dialogue.

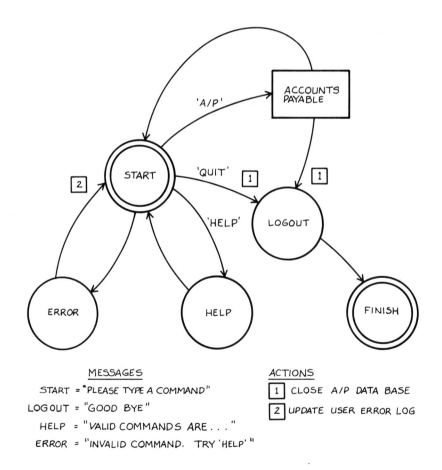

MESSAGES

START = "PLEASE TYPE A COMMAND"

LOGOUT = "GOOD BYE"

HELP = "VALID COMMANDS ARE . . ."

ERROR = "INVALID COMMAND. TRY 'HELP'"

ACTIONS

1   CLOSE A/P DATA BASE

2   UPDATE USER ERROR LOG

He warns that if this advice is not followed, the first version of a system will turn out to be a prototype instead of a production system. Sooner or later it will be replaced by a system that corrects the shortcomings of the first system.

SUMMARY

Not every system development project requires custom analysis, design, and construction. The four system models used in custom software development—Physical Current, Logical Current, Logical New, and Physical New—provide a starting point for tailoring the analysis phase of system development to the needs of each specific project as well as a standard against which simplified alternatives can be measured.

If no current system exists, or if it requires extensive change, detailed study of the current system may be impossible or unproductive. Blitzing (intensive study of selected critical areas) is often an effective alternative to an investigation of the entire current system. In some cases, the current system is studied to determine requirements for packaged software rather than for custom development. Profiles of the users, analysts, and project provide a framework for assessing the factors that may influence the analysis process.

The Logical New System Description can be developed from scratch, instead of by modifying the current system description. If the proposed computer application is breaking new ground, of if users cannot communicate their requirements clearly, prototyping may be a preferable way of determining the new system requirements. The prototype may become the requirements definition or may lead to a more conventional user specification after the requirements are better understood. In either case, the prototype will be replaced by a production version of the computer information system.

REVIEW AND DISCUSSION

16-1. Discuss the potential impact of each of the following factors on the analysis process:
   a. Type of user: system owner, responsible user, hands-on user, beneficial user
   b. Extensive vs. limited user knowledge of application
   c. Extensive vs. minimal user experience with systems analysis
   d. Risk averse vs. risk-taking users
   e. Extensive vs. limited analyst knowledge of application
   f. Extensive vs. minimal analyst experience with systems analysis
   g. Stable vs. unstable organizational environment
   h. Complex vs. less complex system

16-2. Discuss circumstances under which modeling the current system may be ineffective.

16-3. Chapter 12 identified some benefits of modeling the current system. If the analysis team chooses to skip modeling the current system, what measures may be taken to recoup any benefits that may have been lost?

16-4. Discuss the practice of "blitzing." What are its benefits to analysts? To users? Identify any potential pitfalls.

16-5. In order to facilitate comparison of vendor proposals, what steps should be taken to ensure that responses to an RFP adhere to specified guidelines for system description?

16-6. Considering Brooks' advice to "throw one away," discuss the prototyping approach with respect to each of the following:
   a. Sophisticated versus naive users
   b. Evolutionary versus expendable prototypes
   c. Availability of prototype development tools

PART IV

THE CONTEXT FOR STRUCTURED ANALYSIS

Parts II and III have dealt in some detail with the specification of user requirements, emphasizing techniques for modeling information systems as well as the sequence of models that leads to the complete structured specification. Part IV is concerned with the relation of structured analysis to the activities that precede, support, and follow it in the information system processing life cycle. Chapter 17 discusses the first phase of the life cycle, Identifying User Needs. Chapter 18 presents techniques for information gathering, management, and reporting, which support analysts as they develop the various system models. Chapter 19 is an overview of computer information system design, which follows the specification of user requirements and takes as its starting point the data flow diagrams developed during analysis. Chapter 20 surveys the remaining phases of the system life cycle, including the roles of users and analysts. Chapter 21 concludes this introduction to structured analysis methods with a brief survey of system description methods other than data flow models and makes some observations about the future of information systems analysis.

17 Identifying User Needs

It should be clear by now that the activities of the analysis phase, especially the New System Descriptions, are based upon and guided by decisions made earlier in the system life cycle. This chapter discusses the initial phase of the life cycle—Identifying User Needs. (See Figure 3-1.)

Various authors exhibit considerable differences in naming and describing the step or steps that precede the specification of user requirements. This variability corresponds to the variability of the process, as the early activities in system development more than any others need to be tailored to the circumstances of each specific project. For this reason, we present as a single major step what others may divide into several steps.

QUESTIONS ADDRESSED

The detailed effort involved in producing the structured specification is evident. Before this effort is undertaken, a number of questions ought to be answered to be sure the effort is worthwhile and, if so, to guide the activities of analysis.

Every system development project results from a need to change the system. Sometimes a request for system change is initiated directly by the system's users; at other times the change is a response to pressures on the business organization exerted from outside and sensed by the users. This leads to questions such as:

What are the reasons for changing the system?

There may be a problem with the system as it is currently operating. Or the system may be adequate at present but not able to serve the anticipated growth of the business or provide satisfactory service in the future. On the other hand, the organization may recognize an opportunity to serve new markets, to add new services, to enhance its competitive advantage. In other cases, new government regulations or intensified competition may necessitate the system change.

Is a user's request for system change well considered?

Not every desire for system change expressed by users is worth pursuing. An organization needs a procedure for screening requests for new systems to see that it makes sense to try to implement them. Is a request frivolous or impractical? Is it consistent with the objectives, policies, and strategies of the organization? Is there a genuine need for the proposed changes, or does the request represent a wish list of features that would be an unwise use of the organization's resources? Perhaps some of the suggestions are even impossible to achieve in the foreseeable future.

What are the specific problems or opportunities to which the proposed system will respond?

Sometimes a precise definition of the problem or opportunity requires considerable effort in itself. Knowing that something is wrong is easier than pinpointing the problem, especially if the real difficulty is being masked by some of its symptoms. For this reason, many authors, especially in the context of general problem solving rather than information system development, treat problem definition as a separate step in the process.

What are the specific objectives to be achieved by the proposed system?

There are three general objectives for new systems: to reduce cost, to increase revenue, and to improve the quality of products or services. These all may lead to increased profit; how this profit is used varies with the organization. Nonprofit groups may choose to reduce revenues or expand services; a profit-making organization may invest in research or increase its dividend. The three general objectives must be made as specific as possible for each project to establish useful targets for system development. Which costs are to be reduced, and by how much, for the new system to be considered acceptable? What features of the system will be used to evaluate qualitative improvements? How many orders per day is the system expected to process next year? Five years from now? How many years do we expect the new system to operate before making further significant changes or replacing it?

The purpose of such questions as these is to make explicit the **constraints** on the system and its development—**the conditions that must be satisfied for the system to be acceptable.** A system that satisfies all these constraints is said to be **feasible.** Also made explicit are the **criteria—characteristics that are used to compare new system alternatives.** (Note that criteria is a plural noun form; the singular form is criterion.)

Are there systems that will satisfy the stated constraints?

Before undertaking the effort of detailed system specification, there needs to be a reasonable probability that a desired new system can be realized. If similar systems already exist, the chance of success is high. If the proposed system or significant subsystems are being constructed for the first time, the risk is considerably higher. Determining whether feasible solutions exist or can be developed is the focus of this activity. Its importance leads many authors to call the initial system development phase the Feasibility Study or, as in this book, to call the result of this phase the Feasibility Report.

Finding and listing feasible alternatives entails a quick investigation that makes a high-level sweep through the generation and evaluation of alternatives like that done in detail during Step 2.5: Select the Best Scope of Automation. This is in keeping with the iterative character of the system development process. There are some crucial differences between Step 2.5 and the earlier feasibility study. The feasibility study is limited and high-level. It should consume only about 5 to 10 percent of the resources expended for the entire system development. Only gross differences among alternatives are identified, and the comparison is highly selective. The quality of information available is usually low, and therefore the variability of estimates of costs, benefits, and performance, as well as of system development, is high. The outcome of the comparison is not a decision as to which alternative will be implemented. (This error has been the downfall of many projects.) Rather, it is a decision as to which alternatives, if any, are sufficiently promising to be pursued through the kind of detailed evaluation made possible by a Physical New System Description, as discussed in Chapter 13. At that time, the additional detail and specificity will improve the estimates.

> *Should the project proceed to the structured specification phase, and, if so, which system alternatives should be considered further?*

This decision depends not only on the list of feasible alternatives but also on estimates of the resources necessary for system development, especially those resources committed to requirements definition before the next cost/benefit analysis takes place in Step 2.5: Select the Best Scope of Automation.

THE ACTIVITIES OF IDENTIFYING USER NEEDS

An appropriate organization of activities to answer the above questions is shown in Figure 17-1. It is based on Yourdon's *Managing the System Life Cycle.* The discussion examines each activity, emphasizing the information it produces.

It may seem paradoxical that Identify User Needs begins with a statement of user needs. At this stage user needs are expressed informally, loosely, and often verbally. The system requirements they imply are similarly unstructured. They must be organized into an explicit statement of the most promising direction for system development to take.

1.1: Identify Any Current Deficiencies

As mentioned above, not every new system is motivated by deficiencies in the current way of doing things. However, a statement of these deficiencies, when they exist, is a good starting point for establishing objectives of the new system.

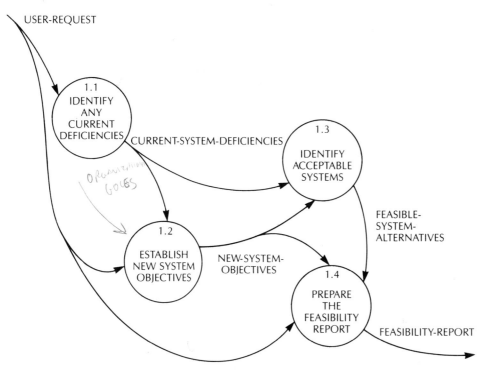

FIGURE 17-1. Activities of identifying user needs.

An examination of a user request for a new system draws upon the stated user needs as well as a similarly informal statement of user operating policies. Those deficiencies that can be identified are listed in order of the seriousness of their consequences.

1.2: Establish New System Objectives

This activity is based on the initial statement of user needs and the opportunities perceived for the new system as well as the list of current deficiencies. Analysts and users review the reasons for system change, check them for compatibility with organizational goals and policies, and prepare a brief, high-level, yet specific, statement of objectives for the new system.

1.3: Identify Acceptable Systems

This activity generates or identifies feasible alternative systems. Yourdon calls them **scenarios** to emphasize the fact that they, too, are high-level. For example, different system implementations considered at this level may be batch versus on-line or centralized versus decentralized. The use of a commercial software package may also be considered, as discussed in Chapter 16. Generic types of input/output devices—key entry, optical scanners, color displays, plasma panels, etc.—may be compared and evaluated. The feasibility of a new system also depends on management constraints on the development process—limits on the schedule and manpower available—as well as other constraints on the new system.

It is important to propose several alternatives, preferably rather different ones. One alternative is usually the current system, even if it is a total disaster, as a point of comparison for the other options. The explicit comparison increases the likelihood that the alternatives selected for further study will be better. A choice also requires users to participate in and be responsible for a conscious decision on an alternative. In some cases, it turns out that no alternative can be found that is better than the current system.

In this activity a brief summary, perhaps incorporating a high-level data flow diagram, is produced for each alternative along with a summary of its benefits and costs, both quantitative and qualitative.

1.4: Prepare the Feasibility Report

The alternative (or alternatives) selected as the basis for detailed study leading to the structured specification is described in the Feasibility Report. It presents the results of the activities carried out and the tentative decisions made during the identification of user needs. Conceptually, the Feasibility Report consists of two parts—one to provide technical guidelines for the project (the part emphasized here) and one to assist in project management. The technical guidelines incorporate the system scope and functions, its principal inputs and outputs, its objectives and constraints, a summary of each alternative to be studied in detail, and any critical assumptions used in selecting these alternatives. The project management portions of the report include the project budget, schedule and deliverables, perhaps with a more detailed breakdown of the activities and deadlines for the analysis phase, the project leader and responsible user representative, and personnel assignments for users and analysts. Cost/benefit estimates for alternatives are also included in or appended to the Feasibility Report.

THE FEASIBILITY REPORT

One way to organize the Feasibility Report is shown in Figure 17-2. Each section of the report is discussed briefly below.

The System Summary is a one-paragraph summary of the most important characteristics of the new system—the reason for system change; the scope and functions of the new system; its principal inputs, outputs, and data stores; the general type and scope of automation contemplated; the alternative(s) selected for detailed study; and any critical impacts the implementation of the alternative(s) will have on the operations of the organization. (This system summary is what DeMarco calls the Charter for Change.)

The Project Summary provides a synopsis of the important management aspects of continued system development. The Background section is a brief description of any significant events leading up to the need for a new system or an explanation of the reasons for the new system. Project Objectives states as specifically as possible the standards for evaluating the new system. Quantitative statements are preferable to qualitative statements, but all important objectives should be included, whether or not they are quantifiable.

FIGURE 17-2. Outline for Feasibility Report.

FEASIBILITY REPORT

System Summary *Short*

Project Summary
    Project Name
    Responsible User
    Responsible Analyst
    Scope   *Quarterly → monthly*
    Start Date
    Project Completion Date
    Completion Date for Structured Specification
    Project Budget Amount

Background

Project Objectives

Current System Summary
    System Deficiencies
    System Description

New system Constraints and Assumptions
    Specific Performance Requirements
    Interfaces to Other Systems
    Hardware Constraints
    Software Constraints

New System Alternatives
    Summary of Alternatives
    Description of Each Alternative
        Essential Transformations
        Principal Required Inputs and Outputs
        Required Data Stores
    Expected Improvements (for each alternative)
        Improvements in Service
        Increases in Revenue
        Reductions in Cost
    Expected Impacts (of each alternative)
        Hardware
        Software
        Organizational
        Operational

Project Constraints
    Personnel
    Budget

Appendices
    Cost/Benefit Comparisons of Alternatives
    Project Cost Estimates and Schedule
    Detailed Activities, Schedule and Budget for Specification of User Requirements Phase

The Current System Summary describes the existing system in high-level data flow diagrams, narrative, or other means, as most appropriate for the intended users of the Feasibility Report. Known deficiencies of the current system are highlighted. System Constraints and Assumptions list specific requirements and conditions with which the new system must comply to be acceptable.

The New System Alternatives presents a description of each alternative to be studied in further detail during analysis, along with a summary of the consequences of implementing it. If there are several alternatives, it is desirable to present the first alternative in full and then to discuss only the differences between it and the remaining alternatives. Details of and support for the cost/benefit analyses appear in an appendix to the Feasibility Report.

The other portions of the report deal with project management issues. They are the Project Constraints listed in the body of the report and the appendices containing the cost estimates and schedule for the remainder of the development effort and the more detailed plans for the activities required to produce the structured specification.

SUMMARY

The initial phase of the system life cycle—Identifying User Needs—precedes establishing users' requirements for an information processing system. Often called the feasibility study, this phase examines the desirability of a proposed system change and makes explicit the conditions that must be satisfied for a new system to be acceptable to the using organization.

Identifying user needs consists of four activities:

1. **Identify any current deficiencies.** Shortcomings of the current system are listed in order of their seriousness.

2. **Establish new system objectives.** Objectives of the new system that are consistent with the goals and policies of the using organization are stated.

3. **Identify acceptable systems.** In the next activity, feasible alternatives for the new system are identified and their implications investigated. One or more of these may be recommended for further study in the next phase of system development.

4. **Prepare the feasibility report.** The alternative (or alternatives) selected as the basis for detailed study leading to the structured specification is described in the Feasibility Report.

The Feasibility Report consists of two parts—one to provide technical guidelines for the project and one to assist in project management. The technical guidelines incorporate the system scope and functions, its principal inputs and outputs, its objectives and constraints, a summary of each alternative to be studied in detail, and any critical assumptions used in selecting these alternatives. The project man-

agement portions of the report include the project budget, schedule and deliverables, the project leader, responsible user representative, and personnel assignments
for users and analysts. Cost/benefit estimates for alternatives are also included in or
appended to the Feasibility Report.

REVIEW

17-1. What step in the system life cycle precedes the specification of user
requirements?

17-2. Explain some of the reasons for identifying user needs and summarizing the
findings in a Feasibility Report.

17-3. List five questions addressed in a feasibility study.

17-4. Define the following:
   a. constraint
   b. criterion
   c. feasible

17-5. Describe the content and purpose of each section of the Feasibility Report.

DISCUSSION

17-1. Distinguish between the use of the terms *system* and *project* in the outline
for the Feasibility Report (Figure 17-2).

17-2. Why is it important to consider several alternatives for a new system?

17-3. Why is the current system usually desirable as one alternative? Under what
circumstances could the current system not be considered?

17-4. Information processing system development is said to be an iterative process.
Support this assertion by discussing it in terms of the relationship of the
feasibility study (Chapter 17) to the specification of user requirements
(Chapters 12, 13, and 14).

18 Information Gathering, Management, and Reporting

Chapters 12 through 15 have presented the processes for determining the feasibility of information processing system development and for modeling the current and new systems to produce the structured specification. These activities require large quantities of information to be collected, evaluated, managed, and communicated. Although the quantity of information and its details vary from project to project, it should be clear *what* information is needed during the feasibility study and system requirements definition.

The information requirements for the first two steps of the sytem life cycle are outlined for the Feasibility Report and structured specification in Figures 17-2 and 14-5 respectively. This chapter focuses on *how* to gather, manage, and present the information needed by users, analysts, and other participants in the process. It describes sources of information, techniques for collecting it, and criteria for evaluating it. Aids to manual and automated management of changing information throughout the iterative process of system analysis are also described. The important considerations in communicating information and presenting the results of the feasibility and analysis activities are discussed. Throughout this chapter the emphasis is on the principles for collecting, maintaining, and reporting information. The skills and techniques summarized here are applicable in many other contexts besides systems analysis, which is the principal concern of this book. For more details, consult the references at the end of the chapter.

GATHERING INFORMATION

Information Sources

Some of the sources of information for analyzing and specifying system requirements have been mentioned in previous chapters. These include documents and information displays that are part of the current system—input forms, graphic displays, reports and other system outputs. They include written statements of procedures, organizational goals, and policies, as well as applicable laws and regulations. If the existing system is automated, there may be some system documentation—system flowcharts, module hierarchy charts, program flowcharts, file and record descriptions, system reference manuals, or users' manuals. There may also be portions of a data dictionary and perhaps, for newer systems, even data flow diagrams and data access diagrams or system descriptions using alternative methods described in Chapter 21.

Other sources may also be helpful. Among these are annual reports and brochures, which often contain summaries of an organization's goals, objectives, principal activities, and functions, as well as graphs of trends and projections for the future. Business- and industry-oriented publications deal with the general business climate and with trends in the larger environment of the new system. These sources as well as books and periodicals about information systems describe a variety of applications, the state of the art in hardware and software, and commercially available application software packages and turnkey systems.

People are the other major information sources. They provide information not available in photographs, drawings, and writing, and can suggest other information sources. People not only describe the current system; they identify its problems and deficiencies. They state future opportunities; they imagine how a future system could operate; they express hopes and fears, state policies, articulate goals and objectives. People also explain, interpret, and give perspective to the information they provide. They can withhold information and, as they choose, express or conceal their interests in system change. Users of the system, as described in Chapter 4, are the most important people to talk to during system development. However, it is often beneficial to consult users and analysts outside the organization who have had experience with using or developing similar information processing systems.

Information Gathering Methods and Techniques

In addition to collecting information from the written sources listed above, the most common methods by which analysts gather information are interviews, observation, and questionnaires. These may be guided by sampling techniques. In some situations with very ill-defined system requirements, it may be helpful to present users with simulated or actual systems to obtain their reactions. One such approach, prototyping, is discussed in Chapter 16.

Interviews.   An *interview* is a conversation in which questions are asked in order to gather information. An interview may involve only two people or a small group. It is preferable for interviews to be conducted face to face, to permit both interviewer and interviewee to communicate nonverbally. However, it may sometimes be necessary to interview by telephone. A successful interview builds rapport between analyst and user as well as collecting information. Perhaps the most important requirement for an interview is that the participants *listen* to each other. In preparing for and conducting an interview, an analyst should make every effort to assure that listening takes place. Evaluation of the information must wait.

Preparing for an Interview.   Preparation for an interview includes selecting the people to be interviewed, scheduling time for one or more sessions, and planning the questions to be asked. The top-down iterative process of identifying users' needs and specifying users' requirements makes top-down iterative interviewing effective.

During the feasibility study, the system owner will be interviewed to obtain an understanding of the relation of the proposed system changes to the goals and policies of the organization, the principal reasons for system change, and constraints on the system or its development. System owners can provide a long-range view of the future of the organization. They may be able to suggest some alternatives for the new system and why each alternative is desirable or undesirable. System owners may also identify specific people in the organization whom they wish the analyst to interview.

Responsible users have a more detailed view of the system's operation, but their perspective is still that of a manager. If the responsible users have sufficient expertise, there may be no need to talk to hands-on users in the early stages. However, discussions with hands-on users may be necessary to pinpoint deficiencies in the current system. Later, when the current and new systems are being modeled, interviews primarily involve the responsible and hands-on users (and perhaps an occasional beneficial user). System owners may need to be interviewed again only if the objectives, constraints, and approaches defined during the feasibility study are called into question by the detailed information gathered later.

A memorandum from user management announcing the interviews and explaining their purpose often precedes scheduling them. Otherwise, when an analyst makes an appointment, a brief statement of the purpose of the interview is a good idea.

Individual sessions should be kept short—under 30 minutes to an hour—with the shorter time limit for managers and the longer sessions where details of system operation are being investigated. If possible, leave enough time between interviews to summarize the information collected in the first and review the plan for the second.

Identifying whom to interview and obtaining an appointment are often some of the most difficult aspects of interviewing.

It is important to make a list of the questions to be asked during the interview. This frees you to listen instead of thinking about what you are going to say next. It also helps you carry out your responsibility for directing the conversation. Try to arrange the questions in a logical sequence and estimate the time it will take to answer them so that you can keep to the allotted time. Allow some time for unanticipated questions. Some questions will restrict the responses to "Yes," "No," or a limited number of specified alternatives. Others will provide an opportunity for freer response and reactions.

Conducting an Interview.   Courtesy, tact, and a businesslike attitude during an interview will demonstrate that the analyst is a professional. A period of light conversation at the beginning may encourage relaxation and rapport, but it should be appropriate to the position and personality of the person being interviewed and should be kept brief to avoid wasting time. Follow the questions prepared in advance and make notes of the essentials of the answers. Record attitudes and opinions as well as facts. Ask if documents are available to keep your notes from duplicating the information they contain. Listen carefully to sense what may be left unsaid. Be alert to unanticipated responses and new ideas and follow them up with additional questions. Look for nonverbal cues such as tone of voice, facial expression, and body language to interpret what is being said. If you believe a question was not understood, paraphrase it. If you do not understand an answer, say so, or restate it for confirmation. Provide feedback to demonstrate that you have understood. Often a neutral restatement of the answer to one question can introduce the next.

Sometimes your presence itself may be felt as a threat. In that case, you may have to spend part of the time, especially during an initial interview, to explain its purpose and reassure the interviewee that the information is being collected to help the organization. Unless a primary purpose of the interview is to obtain reactions to system changes or develop a new system description, do not propose solutions to system deficiencies. Do not attack the credibility or evaluate the opinions of the interviewee.

Conclude the interview with a brief summary of what was covered and express thanks for the other person's assistance or cooperation. State whether another session is necessary, and arrange to schedule it. After the interview is formally concluded, the person interviewed may volunteer further insights. For this reason, many analysts plan to end an interview a few minutes early, particularly if it is with a person whose time is tightly scheduled.

As soon as possible, review your interview notes, transcribe them if necessary for legibility, and add impressions and observations that you were unable to record at the time or that have resulted from reflecting on the interview. Occasionally, it may be appropriate to send a written summary of the interview to the interviewee for written confirmation, if the discussion was significant, and you need to make it a matter of record.

Questionnaires.   A *questionnaire* is a list of questions to which written answers are requested from several respondents. If the questions are asked orally, the technique is called a *survey.* A questionnaire is used in situations in which it is impractical or undesirable to hold individual interviews. Some of these situations are when there are many people to be surveyed, when they are located far away, or are geographically dispersed. Perhaps there is a lack of suitable interviewers or of enough time to schedule interviews before the answers are needed.

An interview usually focuses on information that a person is uniquely qualified to provide because of position, expertise, or experience. The intent of a questionnaire, on the other hand, is to ask everyone surveyed the same questions. Those chosen to receive the questionnaire may be selected from a larger population so that their answers will be representative of the entire group. For example, a business might mail or distribute questionnaires to its customers who will be beneficial users of a new system.

In comparison to an interview, a questionnaire is shorter and more highly structured. A long list of questions discourages people from giving thoughtful answers and sometimes from responding at all. Most often, answers to a questionnaire are true/false or multiple choice. Some open-ended questions may be included, but they solicit a brief, focused comment, not an essay.

The design of a questionnaire requires great care if the results are to be valid. The questions must be stated clearly so that they are self-interpreting. The questions and the choice of responses must not bias the answers. Respondents must be encouraged to ignore questions that they feel incompetent to answer, perhaps by offering choices such as "Don't Know," "No Opinion," or "Insufficient Information to Answer."

If the recipients of a questionnaire are chosen to represent a larger class of people, a statistician should prescribe the selection procedure to assure a valid sample. A statistician may also be helpful in interpreting the significance of the replies after they have been tabulated.

Whether or not the respondents are anonymous, there is no way to assure that their answers are honest. A questionnaire provides none of the nonverbal cues inherent in an interview. However, it is possible to include in a questionnaire of moderate length some questions to detect inconsistent responses.

Since few questionnaires achieve a 100 percent response, there is always uncertainty as to whether the answers of those who failed to reply would have differed significantly from the replies that were received.

When using a questionnaire, be sure to allow enough time for preparing the questions, identifying the respondents, distributing the questionnaires, receiving the replies, and tabulating and interpreting the results. Seek expert advice and assistance if reliable answers are crucial to the system development project.

Observation.   Observation is a natural and direct way of gathering information about an existing system. It is a good way to learn about manual information-processing procedures by following the flow and transformation of information. Observation of the interface between manual and automated portions of a system is also beneficial.

Watching what happens is a good way of confirming or correcting information collected by other techniques. A current system description should show what occurs, even if that differs from what official policies or procedures say should occur. In most cases, the discrepancies involve physical aspects of the system. A difference that affects the logical system description is more important to look for.

An observer without some knowledge of a system may miss important features or may misinterpret what is seen. Another problem with observation is that, when people know they are being watched, their behavior may change. This possibility must be considered when interpreting what is observed.

Preparation is important to observation as well as to interviews and questionnaires. Knowing what you want to look for or confirm will help you focus on the appropriate situations.

Observations are recorded by notes or sketches. It may also be appropriate to take photographs or use film or videotape. In that case, advance arrangements should be made.

An analyst should also be alert to informal opportunities for observing what happens in an information processing system and the organization it supports.

Evaluating Information

Regardless of its source or the methods used to collect it, information must always be evaluated. The analyst's expertise, the time available, and the project budget constrain the selection of the evaluation procedures. In evaluating information, the following questions must be answered, whether the procedures are formal, sophisticated, and quantitative, or less formal and more qualitative.

What does the information say?

The information content must be known before it can be analyzed, evaluated, and used. The ability to write a summary or paraphrase is a good test of an analyst's or user's comprehension of information, especially when technical terms have been used. The summary should try to maintain the precision of expression of technical language while eliminating jargon.

What does the information mean?

Here the emphasis shifts from content to understanding. What are the implications of this information for the organization and system being studied? How is it related to other information? To what extent is it reinforced or contradicted by other information? Is it primarily a statement of fact, opinion, perceptions, or feelings?

Is the information relevant?

This question is concerned with how the information can help in identifying user needs and specifying system requirements. Some information may contribute to formulating the system's objectives, to defining performance requirements, to identifying, generating, or comparing system alternatives, or to modeling the current or new systems. Other information may reveal people's attitudes toward changing the system, thereby suggesting problems and opportunities asociated with integrating the system into the operations of the organization. Sometimes information can be immediatedly discarded as extraneous; at other times the relevance of informtion becomes apparent only later.

Is the information reliable?

Information is not always reliable. Systems and procedures may have been changed without revising documentation and policy statements. Information gathered directly from people may also be unreliable. It may be intentionally misleading or unconsciously biased by its source or the methods used to gather it. It may be colored by emotion or self-serving motives. It may be only partially true, or it may omit significant information.

Generally, the information on which people make decisions and then act is incomplete, contradictory, partly false, and partly unreliable. The best we can usually do is to be aware of this situation and try to compensate for deficiencies in the information we use. Clues to the reliability of information can be found in the competence and credibility of the information sources as well as in the number of independent sources supporting it. That is why it is desirable to use more than one method of gathering information, to corroborate by observation when possible, and to cross-check written and oral sources of information with each other.

INFORMATION MANAGEMENT

For even a modest system development project, a large amount of information is gathered, evaluated, and used. Because some of the information collected is irrelevant, and because information from several sources is desirable to assess its reliability, analysts must manage more information than that contained in the Feasibility Report and structured specification. Moreover, it is usually impossible to control the information gathering process so closely that each stage is limited to obtaining precisely the information needed next. Some means of organizing and managing this information during system development is essential.

Information management is necessary for other reasons than the quantity of information. The complexity of the information to be managed is related to the complexity of an information processing system and its development. In addition, since development is an iterative process, the amount of information grows, and the content usually changes at each iteration. The documents and models produced during system development must keep their identity and be retained for use in later stages of the process. The additional understanding of a system and additional detail developed in modeling the new system cause the tentative decisions made during the feasibility study as well as the estimates of costs, benefits, and project schedules to be reviewed and revised. The result is increased reliability. Current and new system descriptions are retained so the extent of the changes can be seen. The new system description is kept as a statement of user requirements so that it can be compared to the system as designed and delivered for user acceptance.

Clearly the means of maintaining information must facilitate changes. This is in addition to orderly procedures for changing system requirements after the structured specification is finished, as discussed in Chapter 15.

Manual Methods of Information Management

Information management during systems analysis is concerned with organizing the information collected, as well as maintaining and modifying the various system descriptions produced in the process. Organizing the notes, documents, and other material is mostly a matter of setting up and keeping the discipline established by a good filing system. It also involves taking notes in a legible hand, transcribing them if necessary so that you and others can read them later. In this time of ubiquitous xerography, everything committed to paper, such as notes, data flow diagrams and other drawings, and reports should be recorded with sufficient contrast to yield clear, clean copies.

The format of a set of data flow diagrams, with limited information on each page, is designed to make it easy to modify. While the data flow diagrams are changing rapidly, it may be desirable to use translucent paper. This makes it easy to copy portions of a drawing underneath and change the rest. Dark pencil or ink will make it easy to make copies as they are needed. It is not usually worth the effort of using a template for the graphic symbols or of typing or mechanically annotating the diagrams until they are relatively stable (unless the additional appearance of quality is required for a management presentation). Learn to make clear, professional-quality sketches.

To keep the data dictionary in alphabetical order as it evolves, use index cards, one for each entry. The cards can be laid flat several to a page and copied for walkthroughs and intermediate system descriptions. An alternative is to use a loose-leaf notebook with a page for each entry. This is bulkier, but may be easier to store and transport than boxes of cards.

In a manual system, the elimination of redundancy is especially important. Every duplication of information, every cross-reference, is one additional item that must be maintained and one additional opportunity for an inconsistent system description.

Automated Aids to Information Management

Systems analysts, like the proverbial shoemaker's children, have failed to benefit from their own technology. However, with the widespread availability of time-shared computers, word-processing software, and microcomputers, some kind of automated aid to analysts' information management is usually accessible. A word processor or even a line-oriented text editor can facilitate changes to the documents produced during system development as well as the data dictionaries and transform descriptions for the various system models. As long as there is some ability to rearrange information, the data dictionary can be kept in alphabetical order. If a data base management system is available, it may also help in maintaining the data dictionary and transform descriptions, especially if it can be used to modify text as well as simpler forms of numeric and alphabetical information. Combined with selective query and report-generating capabilities, such a data base management system can help in monitoring project progress and in presenting the results of system studies. If a data base management system is not in use in the organization, the cost of acquiring one must be evaluated in relation to the size and cost of the project.

Proprietary data dictionary software packages were developed primarily to assist a data base administrator maintain a data base description for a completed system. An organization using one of these packages should evaluate its suitability for system requirements, but acquiring this kind of data dictionary software specifically for use in analysis may not be advisable.

Software developed specifically for systems analysis and development is likely to become increasingly available. One example is PSL/PSA, a product of the ISDOS Project at the University of Michigan, under the leadership of Professor Daniel Teichroew. PSL/PSA (Problem Statement Language/Problem Statement Analyzer) allows its users to create, maintain, and modify a model of an information processing system. It maintains a data base containing the system model, including data flows, data stores, transforms, and transform descriptions. The model encompasses a hierarchy of data structures and transforms. Output includes each component of the system model, the values of its defined attributes, and those attributes for which no values have been defined. Printer plots of data flow diagrams as well as of other objects and relationships in the PSL/PSA data base can be generated. PSL/PSA takes advantage of complementary relationships between objects, as described in Chapter 9, to provide cross-references and redundant output without requiring redundant input or additional redundancy in the model. Some other system description methods, discussed in Chapter 21, also provide software aids.

REPORTING INFORMATION

The importance of communicating to and interacting with users, analysts, and others throughout system development makes it imperative for systems analysts to become skilled at reporting and presenting information as well as at acquiring and maintaining it. Written, oral, and graphic communication skills are all necessary. Though they are distinguished below, in practice they are combined to produce the best results.

For the most part, the primary purpose of reports and presentations made by systems analysts is to inform—to present findings and recommendations objectively, relying on the intrinsic merit of what is reported. Occasionally it is important for analysts to become advocates of a recommendation—to persuade as well as to present. See Chapters 13 through 15 of Athey for a discussion of the issues involved.

Modes of Communication

The principal modes of communicating information are written, oral, and graphic.

Written Reports. The most common written reports prepared by analysts during system development are the Feasibility Report and structured specification, the subjects of the two preceding chapters, and the documents recording walkthrough findings, as discussed in Chapter 11. The Physical and Logical Current System Descriptions as well as the New Logical System Descriptions should also be available as separate documents. There may also be a document containing details of alternative Physical New System Descriptions that are not included in the cost/benefit analyses in the structured specification. This provides a record of all the system descriptions but eliminates nonessential detail from the statement of system requirements. In addition, there will be a variety of memoranda, both technical and those oriented to project management.

Oral Reports. Although oral communication is also involved in information gathering as well as throughout system development, oral presentations consist primarily of walkthroughs (Chapter 11) and management briefings or reports.

Graphic Communication. The effectiveness and desirability of graphic communication has already been emphasized. Advertisers exploit the visual impact of television and use photographs and strong graphics in newspapers and magazines. Even radio, a nonvisual medium, frequently tries to create visual images for its listeners. This power of the image is the reason for using data flow diagrams and data base diagrams for information processing system description. Decision trees, as we have seen, depict the policies for transformations. Similarly, charts and graphs should be used in preference to narrative for summarizing information, reporting trends, comparing costs and benefits of system alternatives, and presenting project budgets and schedules.

Principles of Communication

Understanding and putting into practice the following principles should result in clear, effective communication. Some of them are valid for all kinds of communication; others are directed toward specific media or situations.

Know your audience. To whom are you speaking or writing? What is their general background and level of education? How much expertise do they have in the subjects being dealt with? It is insulting to be talked down to, yet too high a level of discussion impedes or prevents communication. Sometimes only a little extra background or introduction will allow the reader or listener to follow the presentation. Try to put yourself in the place of your audience. Determine what you would want to know if you were they.

Answer the Basic Questions. These questions are summarized in verse by Kipling:

> I keep six honest serving men
> (They taught me all I knew);
> Their names are What and Why and When
> And How and Where and Who.*

Though Kipling recommends them for information gathering, these questions are also basic for reporting information. Perhaps in a business context, we should also add, At what cost?, but providing answers to these questions is fundamental.

Seek Simplicity. Clarity, directness, and simplicity reinforce each other. Writing and speaking clearly, directly, and simply require effort; making simplicity habitual requires years of discipline.

Use paraphrase or everyday language instead of technical language for a nontechnical audience. For example, a data flow diagram is a "picture of what is happening in the system"; a data access diagram shows "what information is stored in the system and how it can be retrieved." Otherwise, define technical or other unfamiliar terms and the concepts behind them in simple language.

Control the rate of information transfer. This is especially important in oral presentations, where listeners must rely on their memories of what was said earlier. The accompanying visual material should also be kept simple. To put so much information on a slide that it becomes too small to read is self-defeating. When time is limited, limit the number of topics to be covered or points to be made accordingly.

*"The Elephant's Child" by Rudyard Kipling from *Rudyard Kipling's Verse: Definitive Edition,* Doubleday & Co., Inc., Garden City, N.Y., p. 607. Reprinted by permission of the National Trust and Doubleday & Co., Inc.

Organize and Plan the Report or Presentation. Organizing a presentation or report gives it a structure, defining its components and arranging them in an appropriate sequence.

Top-down decomposition can help in organizing a presentation as well as in describing an information processing system. This involves beginning with an overview that states the principal points or topics and then developing them in detail. Reports or presentations to management often begin with a concise summary, which includes recommended decisions or actions when appropriate. Details supporting the recommendation or summary follow, perhaps at several levels. This permits the reader of a report to choose the areas and levels of detail to be investigated further. A table of contents with the outline made explicit throughout the document facilitates skimming and selective reading. Similar principles apply to oral presentations.

It is also important to arrange the components of the report in a logical sequence, to be sure that new concepts and terms are defined as they are introduced, and that each lays the groundwork for the next.

The plan should also consider the selection of examples, the coordination of written and graphic material, and appeals to the senses and to the imagination. It may require making a schedule to allow time for typing documents, preparing charts and slides, and reproducing reports or presentation aids before the deadline.

Prepare and Rehearse the Presentation. An effective oral presentation requires careful preparation and rehearsal. The more powerful the audience and the more critical the outcome, the more important it is to spend time getting ready. This means that visual aids must be ready early. If possible, practice in the place where the presentation will be given. Have others listen and comment on both the strengths and weaknesses, but remember that the purpose of a run-through is to identify and improve the weaknesses. If a videotape or recorder is available, use it to monitor the rehearsal. Practice until you feel comfortable. Be sure that you can stay within the alloted time; plan what to leave out if time runs short. Try to anticipate questions and prepare an answer for them.

Make a Final Check. Take the time for a final check, but provide enough leeway to correct deficiencies. Are there enough seats located where their occupants can see? Is all the equipment in place? Will the cords reach the power source? Are there spare lamps for the projectors? Are the slides, transparencies, charts, and handouts at hand, along with pens if you plan to write? A few minutes spent in this way can prevent last-minute panic and allow you to concentrate on what you are saying rather than worrying about what has gone wrong.

Watch for Feedback. In oral presentations it is important to notice the clues your listeners give as to how well they are receiving what you are saying to them. Watch their expressions and their body language. These clues can help you decide whether you are moving too rapidly or too slowly for your audience and to adjust your pace accordingly. Provide an opportunity for them to ask questions. If you repeat or paraphrase a question, you have additional time to understand it and to prepare your answer. This also assures that everyone has heard the question, particularly if the group is large. In some cases, you may be able to ask questions of the audience.

SUMMARY

Effective information gathering, management, and communication skills complement skills in system modeling—both are required for successful information system development.

The skilled analyst gathers information from a variety of sources, including people and documents. Principal methods of information gathering consist of interviews, questionnaires, and observation.

Regardless of its source or the methods used to collect it, information must always be evaluated for content, meaning, relevance, and reliability.

The amount of information to be managed during systems analysis is so large, and its rate of change so frequent, that automated aids are highly desirable.

In reporting information, written, oral, and graphic communication skills are all necessary. Important principles to remember in communicating information include: know your audience; answer their basic questions; seek simplicity; organize, plan, prepare, and rehearse presentations; make a final check before issuing a document or giving a presentation; and watch for feedback.

REVIEW AND DISCUSSION

18-1. Identify some sources of information for systems analysis and requirements specification.

18-2. Name and give a definition for three information gathering techniques used in systems analysis.

18-3. Why are a variety of sources and techniques desirable in collecting information?

18-4. What kind of information would you expect to learn from each of the following types of users (as defined in Chapter 4):
  a. System owner
  b. Responsible user
  c. Hands-on user
  d. Beneficial user

18-5. List what you would do to prepare for an interview.

18-6. Some authors suggest the use of a tape recorder during some interviews. What are the advantages and disadvantages? Under what circumstances might a tape recorder be appropriate?

18-7. In what circumstances is a questionnaire an appropriate technique for gathering information?

18-8. What role does a statistician play when questionnaires are used?

18-9. Discuss reliability and relevance in relation to information collected through:
a. Interviews
b. Questionnaires
c. Observation

18-10. Describe manual methods for maintaining the data dictionary.

18-11. What advantages to maintaining a system description can automated aids provide?

18-12. List three basic forms of communicating information. Give important features of each.

18-13. Briefly explain each of the following principles of communication:
a. Know the audience
b. Answer the basic questions
c. Seek simplicity
d. Organize and plan the report or presentation
e. Prepare and rehearse the presentation
f. Make a final check
g. Watch for feedback

19 Computer Information System Design

WHAT IS DESIGN?

In developing complex systems, it is extremely difficult to proceed directly from a specification of user requirements to the construction of a system that satisfies those requirements. In most cases there is an intermediate step—designing the system.

According to Herbert Simon, winner of the 1978 Nobel Prize for Economics:

> Design . . . is concerned with how things ought to be, with devising artifacts to attain goals.*

This brief description of design introduces us to some important aspects of system design. Design is a quintessentially human activity that serves human purposes. The artifacts devised by designers are the products of human creativity and invention, as diverse as tools, vehicles, clothing, movies, paintings, books, legal and governmental systems, and, of course, information systems.

Each of these artifacts serves to accomplish a variety of objectives. Some of these goals are those of the users of the artifacts, as discussed in Chapter 1. Others are objectives of the designer and are related to the activity of design.

Design is necessary because the desired artifact or system does not exist, and therefore must be devised. Traditionally the creation of new artifacts has been a slow evolutionary response to slowly changing human needs. The result of this

---

*Herbert A. Simon, *The Sciences of the Artificial*, 2nd edition, The MIT Press, Cambridge, Mass., © 1981, p. 133. Reprinted by permission of The MIT Press.

process is exemplified by the Slovakian peasant shawls mentioned in Chapter 1 in the quotation from Alexander. In the past century or two the pace of social and technological change has accelerated, and the complexity of man's artifacts has increased significantly. As a result, invention has become a deliberate human enterprise, and design is now an activity consciously pursued and a process explicitly taught.

The process of design begins with a statement of the objectives of the desired object or system. These are made as explicit as possible so that any artifact devised by the designers can be tested to determine whether it meets the objectives of the people who will use it. When the artifact can be shown to satisfy its objectives, the design process may terminate.

Design is often viewed as a problem-solving process. The statement of objectives defines the problem, and the tests characterize the criteria for deciding whether a solution has been found.

We could try to solve design problems adaptively, by trial and error, much as Thomas Edison invented the electric light after more than 1,000 attempts. But the process of constructing a series of artifacts and testing them is often undesirable or impractical, particularly in the case of large, complex, and expensive systems. (Can you explain why this is so? See discussion question 19-1.)

Instead designers work with models of the artifacts they are trying to create. For example, an architect makes a scale model of a building; a painter makes preliminary sketches before starting a painting; an engineer uses a set of equations to calculate the movement of a bridge carrying traffic. As discussed in Chapter 1, these models are abstractions. They select important characteristics of the system components and represent significant relationships among them.

Using models, designers are able to simulate or predict the performance of a new system. They can change the model and determine a resulting change in system performance. They can test the model to find out whether the system it represents will have the desired behavior. In this way, the design process produces a description of an artifact that will satisfy its objectives, rather than the artifact itself. In fact, most designers generate a number of alternatives, trying to produce better solutions to their problem.

But the final purpose of the design process is the creation of a new object or system. Designers must assure that their description is sufficiently complete and detailed for the construction of the artifact. Architects and engineers make detailed construction drawings supplemented by extensive descriptions of the materials and techniques to be used in construction. Composers convert their musical sketches into a full score containing every note played by every instrument.

Thus design is an intermediate step between a problem statement and the construction of a solution or between a statement of requirements and the construction of a system satisfying those requirements.

Design is a process that creates descriptions of a newly devised artifact. The product of the design process is a description that is sufficiently complete and detailed to assure that the artifact can be built.

We use the word **design** to refer to both the **process** and the **description(s)** it produces.

Thus far, we have used the term *designer* more often than *designers.* In fact, for most large systems, including large information processing systems, design is a team effort. The team comprises specialists in various aspects of the system being designed. Careful coordination and management of the design team is vital to their success, as is communication among the team members. In design as in analysis, system models are working tools essential for understanding and communication.

THE GOALS OF COMPUTER INFORMATION SYSTEM DESIGN

The design of computer information systems is a specialized process adapted to the types of artifacts its practitioners devise. Its models represent the significant features of a working system of computer programs, showing the interfaces between the various program units. The design of an automated information processing system may be characterized by its goals, processes, tools and techniques. (Chapter 5 presented a similar characterization of information processing systems analysis.)

The principal goals in creating a design model of a computer information system include the following:

1. In general, solving the problem posed in the structured specification.
2. More specifically, satisfying the performance requirements determined during systems analysis.

    These include the data flows, data stores, and transformations depicted in the data flow diagrams and defined precisely in the data dictionary and transform descriptions as well as a variety of details contained in the complete structured specification as described in Chapter 14.
3. Deriving an automated system whose structure fits the structure of the problem and that is integrated smoothly with the people who use it and the organizaton whose business it supports.
4. Considering alternative system designs to select the one most suitable for the organization within the limits of the time, resources, and information available.
5. Matching the system design to the hardware and system software environment in which it will operate.
6. Creating a system whose structure makes it easy to understand, construct, and modify, even if it requires more hardware and takes longer to execute.

THE DESIGN PHASE OF THE SYSTEM LIFE CYCLE

Chapter 3 presented an overview of the entire information processing system life cycle (see especially Figure 3-21). This section describes the activities of the design phase. Remember that design occurs concurrently with determining the hardware and system software environment.

Figure 19-1 shows the major steps in the design of a computer information system:

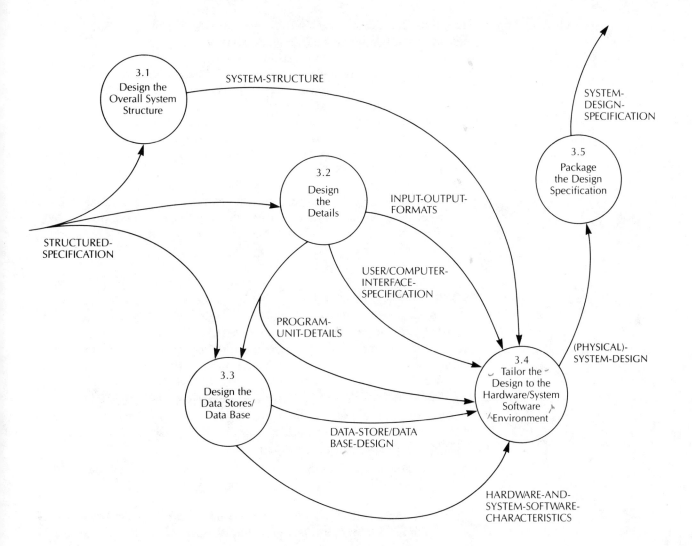

FIGURE 19-1. Activities of designing a computer information system.

1. Design the Overall System Structure

In principle, information processing system design can proceed by either aggregation or decomposition. In either case, the difficulties arise most often in finding an appropriate system structure—a good relationship among the components. In practice, this means that a top-down strategy, concerned first with the most important system interfaces, is likely to be the most fruitful, especially in the case of complex systems. The overall system structure has a greater impact on satisfying the system objectives than the constituent program units.

Designing the system structure means defining its components (the ***program units*** or ***modules***), the relationships between them (the ***connections*** by which control is passed from one module to another at execution time), and the ***information flows*** between pairs of modules.

2. Design the Details

Detailed system design involves several activities:

a. Design the details of the program units.

This involves a complete specification of the algorithm for each module in the system, preferably using a description that is independent of the language in which the program units will be coded.

b. Design the formats of all the system inputs and outputs.

These include the layouts of all the source documents and reports as well as other kinds of information displays. Some of these may have been defined as part of the system specification.

c. Design the user interfaces with the automated system.

This may include manual procedures for collecting and distributing system outputs. For interactive systems, user interface design includes the detailed procedures for entering data, requesting output, and initiating other system functions.

3. Design the Data Stores or Data Base

This entails defining physical storage structures for the logical data structures described in the analysis data dictionary. It also involves identifying a method of accessing the physical files and implementing the access paths specified in the logical data base description. In many cases, the file access software will be provided as part of the operating system and related data management facilities or will be part of a data base management system. In this case data base design is primarily a matter of providing interfaces to the system software environment.

4. Tailor the Design to the Hardware and Software Environment

The design is then adjusted to the hardware and system software environment in which it will be executing, as that environment becomes better defined.

5. Package the Design Specification

In this step, the portions of the design model are collected in final form into a document that will guide the construction of the system.

Computer information system design may be viewed as a sequence of decisions about how the users' requirements will be implemented. The first of these decisions—which portions of the system will be automated—is the essence of the Physical New System Description developed during analysis. Each subsequent decision defines the implementation in further detail, moving the system description from one that is almost entirely implementation-independent, or logical, to one that specifies details of the implementation.

TOOLS AND TECHNIQUES OF STRUCTURED DESIGN

Design Methods   There are a variety of design methods for computer information systems. As in the case of analysis methods, some are based primarily on data flow; others on data structure. Among the better known design methods in current use are the Warnier-

Orr and the Jackson design methods and Structured Design, developed by Yourdon, Constantine, and Myers. The method presented here is Structured Design, which uses the data flow system model from structured analysis as its starting point.

Design Tools

The principal tools for describing a system design are structure charts, a data dictionary, algorithm description languages, facsimiles of system inputs and outputs, interactive dialogue models, and drafts of user and reference manuals.

FIGURE 19-2. Example of a structure chart.

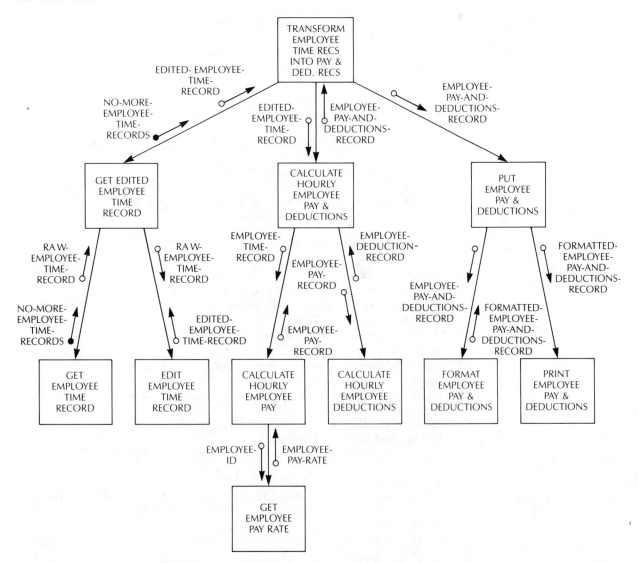

FIGURE 19-3. An example of program design language (PDL).

Module: Calculate Hourly Employee Pay

Input: Employee-Time-Record

Output: Employee-Pay-Record

Constants: Hourly-Employee-Overtime-Threshold = 40 hours
           Hourly-Employee-Overtime-Factor = 1.5

Accumulate hours by type.
  Set Regular-Hours, Sick-Hours, Vacation-Hours of Employee-Pay-Record to Zero.
  For each Day of Employee-Time-Record do:
    Case of Type-Of-Hours
      Work: Add Regular-Hours of Employee-Time-Record
                  to Regular-Hours of Employee-Pay-Record.
      Sick: Add Sick-Hours of Employee-Time-Record
                  to Sick-Hours of Employee-Pay-Record.
      Vacation: Add Vacation-Hours of Employee-Time-Record
                  to Vacation-Hours of Employee-Pay-Record.
    End-Case.
  End-For.
Determine overtime hours.
  If Regular-Hours of Employee-Pay-Record > Hourly-Employee-Overtime-Threshold
    Then
      Overtime-Hours of Employee-Pay-Record =
        Regular-Hours of Employee-Pay-Record − Hourly-Employee-Overtime-Threshold.
      Set Regular-Hours of Employee-Pay-Record
        to Hourly-Employee-Overtime-Threshold.
    Else
      Set Overtime-Hours of Employee-Pay-Record to Zero.
  End-If.
Get employee pay rate.
  Retrieve Employee-Pay-Rate from Employee-Payroll-File
    using Employee-Id of Employee-Time-Record.
Calculate pay for each type of hours.
  Overtime-Pay of Employee-Pay-Record =
    Overtime-Hours of Employee-Pay-Record *
    (Hourly-Employee-Overtime-Factor * Employee-Pay-Rate).
  Regular-Pay of Employee-Pay-Record =
    Regular-Hours of Employee-Pay-Record * Employee-Pay-Rate.
  Sick-Pay of Employee-Pay-Record =
    Sick-Hours of Employee-Pay-Record * Employee-Pay-Rate.
  Vacation-Pay of Employee-Pay-Record =
    Vacation-Hours of Employee-Pay-Record * Employee-Pay-Rate.

Complete employee pay record.
  Set Employee-Id of Employee-Pay-Record to Employee-Id of Employee-Time-Record.

A *structure chart* depicts the overall system structure, as shown in Figure 19-2. Each rectangle represents a *module*—a named, bounded, contiguous set of executable program statements. The name of a module sums up the transformation performed by that module and all its subordinates. All the internal details of the modules are suppressed. Each is treated as a black box. Only the interfaces between the modules are shown. Invocation of one module by another (a call from a superior to a subordinate) is represented by a directed arrow known as a *connection.* An information flow across the connection is known as a *couple* and is represented by a small labeled arrow with a circle at the tail. A *data couple* is shown by an open circle; a *control couple* (or *flag*) is shown by a solid circle. Additional symbols appearing on structure charts are not discussed here.

Several tools are used in detailed design. Details of the modules are specified in an algorithmic language such as *pseudocode* or a *program design language.* (See Figure 19-3 for examples.) The transform specification tools of structured analysis—Structured English, decision trees, and decision tables—may also be used. These are collected into a *module dictionary.* The composition of the information flows (both data and control couples) are defined in a *design phase data dictionary.*

The design phase data dictionary may be built by copying and extending the analysis phase data dictionary, but the two must retain their distinctive identities—one as a requirement, the other as the satisfaction of that requirement. If it is necessary to change the system requirements during design, then the procedures discussed in Chapter 15 should be followed. Figures 19-4 through 19-6 present examples of *report, screen, and source document layouts.*

Scenarios and *dialogue diagrams* as well as *screen layouts* are used to study and communicate the human-computer interface for an interactive system. The design of this interface is outside the scope of this book. These design decisions are also

FIGURE 19-4. Example of a report layout.

Report Date 12/12/86		Hourly Employee Earnings Register For Pay Period From 12/03/86 to 12/10/86									Page 1
Employee Identification	Regular		Overtime		Sick		Vacation		Total	
	Hours	Earn.	Hours	Earn.	Hours	Earn.	Hours	Earn.	Hours	Earn.
ADAMS–347	40.0	200.00	0.0	0.00	0.0	0.00	0.0	0.00	40.0	200.00
BAXTER–524	30.0	300.00	0.0	0.00	10.0	100.00	0.0	0.00	40.0	400.00
CASEY–128	0.0	0.00	0.0	0.00	0.0	0.00	40.0	280.00	40.0	280.00
DELANEY–111	40.0	200.00	5.0	37.50	0.0	0.00	0.0	0.00	45.0	237.50
EVANS–321	32.5	325.00	0.0	0.00	0.0	0.00	0.0	0.00	32.5	325.00
Totals	142.5		5.0		10.0		40.0		197.5	
		1025.00		37.50		100.00		280.00		1442.50

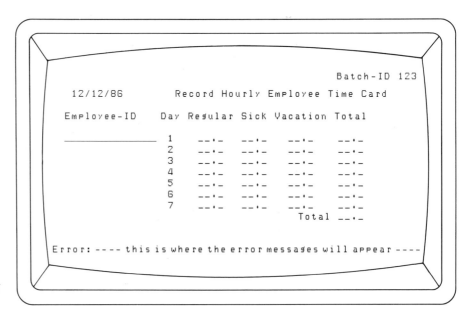

FIGURE 19-5. Example of a screen layout.

Hourly Employee Time Card

From: ________ /________ /________ To: ________ /________ /________
Month Day Year Month Day Year

Employee ID: ADAMS–347

Employee Signature

Day	Regular Hours	Sick Hours	Vacation Hours	Day's Total Hours
1	______.___	______.___	______.___	______.___
2	______.___	______.___	______.___	______.___
3	______.___	______.___	______.___	______.___
4	______.___	______.___	______.___	______.___
5	______.___	______.___	______.___	______.___
6	______.___	______.___	______.___	______.___
7	______.___	______.___	______.___	______.___

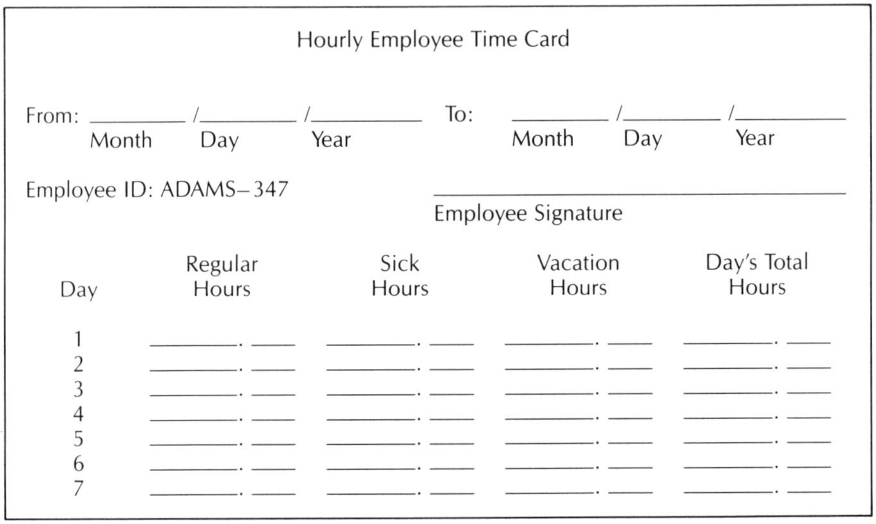

FIGURE 19-6. Example of a source document layout.

recorded and communicated to the system builders in the form of ***draft users' and reference manuals.***

Data store and data base design use the ***data dictionary*** and ***data structure diagrams.***

Techniques for Generating an Initial Design

Structured design begins with the product of structured analysis—a set of data flow diagrams. It has two principal strategies for producing a structure chart that matches the structure of the design to the structure of the requirements shown in the data flow diagrams—***transform analysis*** and ***transaction analysis.***

The resulting structure chart will contain a module for each primitive transform in the data flow diagrams as well as a data couple for each data flow. It will contain other components not shown in the data flow diagrams—modules to manage the flow of information as well as control couples. Modules that access data stores are shown, but the data stores themselves do not appear in a structure chart.

Transform analysis is used for a data flow diagram (or portion of a data flow diagram) in which a single important transformation can be identified. This transformation is accomplished by modules placed in the structure chart at the center of the system. A chain of modules called an ***afferent branch,*** provides the incoming data flow to the ***central transform*** or ***transform center;*** another module chain, called an ***efferent branch*** of the system, takes the output of the central transform and formats it for presentation to the world outside the automation boundary. Figure 19-7 shows a portion of a structure chart for the FastFood Store accounts

FIGURE 19-7. Structure chart for portion of system derived using transform analysis.

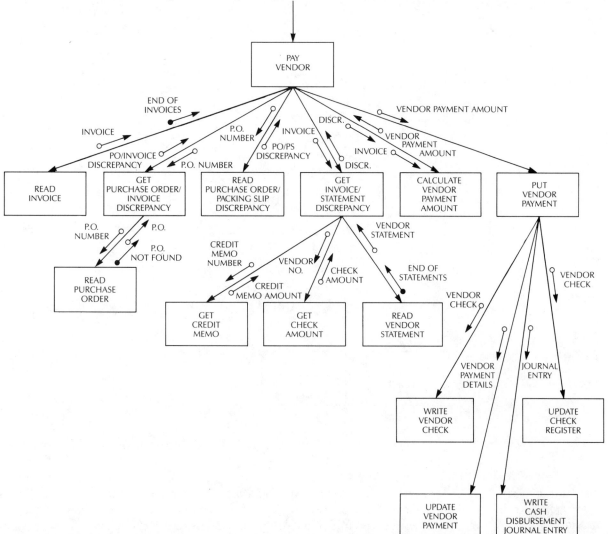

payable system. Here Transform 4.2: Pay Vendors in Figure 13-15 has been taken as the central transform.

Transaction analysis is used for a data flow diagram (or portion of a data flow diagram) in which there are several independent information flows. Each of these data or control flows is regarded as a stimulus—called a **transaction**—that triggers a separate response by the system. The structure of the system contains a **transaction center,** which identifies each type of transaction, and a separate subordinate module for each transaction type, which carries out the appropriate system response. Figure 19-8 shows a structure chart for a portion of the FastFood Store accounts payable system (See Figure 13-15). Here three transactions are the basis for the

FIGURE 19-8. Structure chart for portion of system derived using transaction analysis.

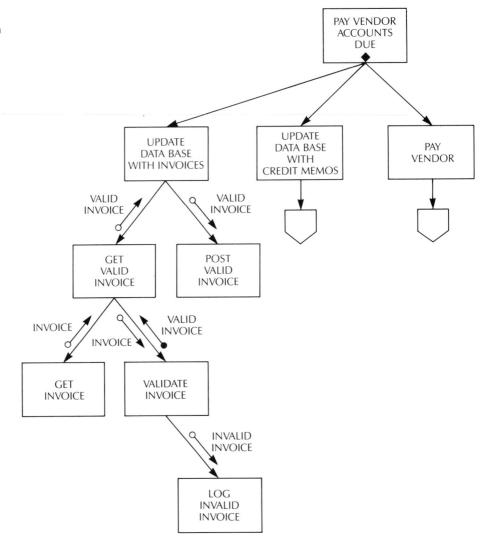

system structure—adding an invoice to the data base, adding a credit memo to the data base, and paying the vendors.

Many systems have structures that combine transform and transaction centers, as shown by Figures 19-7 and 19-8.

ITERATION AND REFINEMENT

Structured information system design, like structured analysis, is an iterative process. Each successive system description is evaluated. It is reviewed to be sure that it satisfies all the users' requirements. It is also judged using known criteria for a good information system design. These criteria, not specified here, emphasize a system structure that is easy to understand, maintain, and modify and that localizes the effects of change. Walkthroughs are also held to identify defects and to guide where the design can be improved at the next iteration.

ROLES OF USERS AND ANALYSTS DURING SYSTEM DESIGN

During system design, users and analysts may be called upon to help plan the continued system development. They may be asked to set priorities for delivering the various portions of the automated system. Analysts and users may also serve as consultants to the development team by interpreting the structured specification. This role is essential when ambiguities, omissions, or errors remain in the specification in spite of everyone's best efforts. These must be corrected as they are discovered.

The interface between the manual and the automated parts of the system is critical. Responsibility for designing this interface is often given to the part of the development team that designs the manual part of the system. Since this is such a critical interface, all the design groups, the users, and the analysts must agree on its design. Recently, this has become such an important consideration that some analysts include a logical specification of the man-machine interface as part of the structured specification.

Other aspects of the design of the manual system involving both users and analysts include restructuring the users' organization, the reassignment of tasks to personnel, the reorganization of work and document flow, and the development of documentation and training aids.

THE STRUCTURED SPECIFICATION AS THE BASIS FOR DESIGN

In the preceding chapters the structured specification has been viewed primarily as a statement of users' requirements for a new information processing system. The perspective has been that of the analysis phase of the system life cycle.

To the designers of an information processing system, the requirements stated in the structured specification are a problem to be solved. As indicated earlier in this chapter, a goal of the designers is to understand the problem structure in order to devise an information processing system whose structure is appropriate to the structure of the requirements.

The movement from problem structure to solution structure is facilitated by the data-flow approach to system modeling presented in this book. It provides a top-down decomposition of the system in terms of a small number of fundamental types of components. It minimizes the complexity of the interfaces among these components. It minimizes redundancy and nonessential detail. It also organizes the description of the data base into relationships among a minimal set of logical objects.

These aspects of the new system model are characteristic of any effective specification of requirements, perhaps using one of the alternative approaches to system description surveyed in Chapter 21. In addition, the techniques of transform analysis and transaction analysis illustrated in Figures 19-7 and 19-8 provide a link between the data flow diagrams in the requirements specification and an initial design for a computer information system. Although this link is direct, the application of transform and transaction analysis is not a mechanistic or mindless process. Its existence, however, is an important feature of structured analysis. Moreover, the criteria discussed in Chapter 11 for improving a system requirements model also lead to improved system designs.

This compatibility between analysis and design models and the availability of methods for easing the transition from analysis to design are important reasons for our choosing the data-flow approach in this introduction to structured systems analysis. The connections will become clearer when the reader acquires a deeper understanding of structured design accompanied by experience. This chapter has provided an overview of information system design and a background for further understanding of the relationship between structured design and the analysis of requirements for computer information processing systems.

SUMMARY

Design is a process that creates descriptions of a new artifact. The product of the design process is a description that is sufficiently complete and detailed to assure that the artifact can be built.

The principal design goal for a computer information system is to satisfy the users' performance requirements stated in the structured specification.

The major activities in the design of a computer information system are

1. **Design the overall system structure.** This means defining the program units, the connections by which control is passed from one to another at execution time, and the information flows between pairs of modules.
2. **Design the details of the program units.** This activity involves specification of each algorithm, the formats of all system inputs and outputs, and the user interface with the automated system.

3. **Design the data stores or data base.** This entails defining physical storage structures for the data and implementing the access paths specified in the logical data base description.

4. **Tailor the design to the hardware and software environment.** The design is then adjusted to the hardware and system software environment in which the system will be executing.

5. **Package the design specification.** In this activity, the design model is collected in a document that will guide the construction of the system.

In structured design, the structure chart is used as the model for the overall system structure. A program design language is used to specify the details of each program unit. Final input and output formats are specified. Scenarios, dialogue diagrams, and screen layouts are used to study and communicate the design of the human-computer interface for an interactive system. Other design decisions are presented in users' and reference manuals.

Transform analysis and transaction analysis are structured design techniques for generating an initial system design from a data flow model of users' requirements. This initial design is iteratively evaluated and refined, using walkthroughs to identify defects and guide improvement of the design.

REVIEW

19-1. What is an artifact?

19-2. Define design:
   a. as a process
   b. as a product

19-3. How do models aid the design process?

19-4. State six goals of computer information system design.

19-5. Briefly describe the five major steps in the design of a computer information system.

19-6. Name the components of a structure chart, showing the symbol for each.

19-7. Name some tools used for module specification.

19-8. Name two major strategies for generating an initial system design.

DISCUSSION

19-1. What are some of the difficulties of constructing a solution to a problem without first designing one or more solutions? What if the tentative solution must be tested to determine whether it satisfies the problem requirements?

19-2. Why is it important for the analysis phase data dictionary to be kept distinct from the data dictionary developed during system design?

19-3. Discuss the relationship between the components of a data flow diagram as a model of an information processing system and the components of a structure chart as a model of a computer information system.

20 System Construction, Testing, Installation, and Operation

The analysts' and users' participation in information system development does not end with the analysis phase of the life cycle, though their involvement in system design is often rather limited. This chapter describes the remaining phases of the life cycle in somewhat less detail than the overview of system design presented in Chapter 19. It also discusses the continuing roles of users and analysts at these stages.

The analysts' most important role is as a liaison between the users and the system developers. The users ultimately decide whether to integrate the new system into their operations. As system development proceeds, they should be involved wherever further important decisions are made that will affect the operation of their business.

Figures 20-1 and 20-2 present the information life cycle as outlined in Chapter 3, the former as a diagram, the latter as a table. They provide the framework for this chapter's discussion of developing the system acceptance tests; constructing the system; integrating the system with the using organization; and operating, modifying, and enhancing the system.

DEVELOP THE SYSTEM ACCEPTANCE TESTS

The structured specification is the definitive statement of users' requirements for a new information processing system. As such, it describes what performance standards the new system must meet after the system has been designed and constructed. However, these performance requirements must be expressed as a specific set of

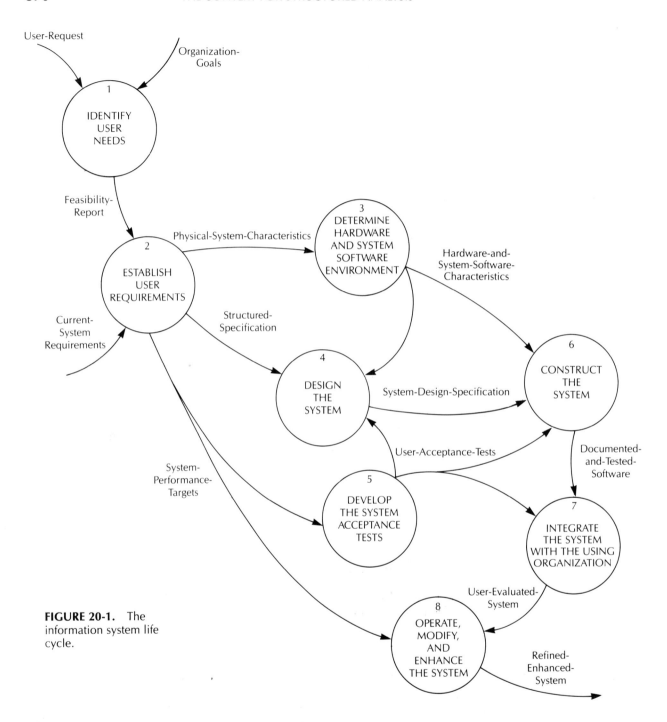

FIGURE 20-1. The information system life cycle.

FIGURE 20-2. Phases, participants, products, and decisions in the information system life cycle.

PHASE	PARTICIPANTS	PRODUCTS	DECISIONS
1. Identify User Needs	USERS ANALYSTS	Feasibility Report	Is the proposed system desirable to the users? Is it technically and organizationally feasible? Are the estimated benefits worth the expected costs?
2. Establish User Requirements	USERS ANALYSTS	Structured Specification (includes: Physical System Characteristics and System Performance Targets)	Have the user requirements been stated completely and accurately?
3. Determine Hardware and System Software Environment	DESIGNERS analysts	Hardware Characteristics System Software Characteristics	Are the proposed hardware and system software environment adequate to support the specified user requirements?
4. Design the System	DESIGNERS programmers	System Design Specification with Preliminary User Manuals	Does the proposed system design meet the specified requirements? Is it still technically, operationally, and economically feasible?
5. Develop the System Acceptance Tests	QUALITY ASSURANCE analysts	User Acceptance Tests	Do the acceptance tests measure compliance with all the stated performance targets with sufficient precision?
6. Construct the System	PROGRAMMERS designers analysts evaluators users	Documented and Test System with User and Operations Manuals Report of Completed User Acceptance Tests	Are the documentation and manuals for the completed system adequate? Has the system passed all the user acceptance tests?
7. Integrate the System with the Using Organization	USERS analysts designers evaluators	Operational System Postimplementation Evaluation Report	Has the new system been smoothly integrated with the operations of the organization? Is the system still meeting the specified performance targets?
8. Operate, Modify and Enhance the System	USERS ANALYSTS DESIGNERS PROGRAMMERS EVALUATORS	Debugged, Refined, Enhanced System Performance Requirements for System Modifications and Enhancements	Have all the known bugs in the system been corrected? Is there a need to change parts of the system? Is there a need to enhance the capabilities of the system?

tests that will decide whether the system is acceptable. If the system passes all these tests, it is considered to be acceptable by the users.

The acceptance tests establish measurable standards of system performance. In principle, they are passed or failed as a whole, although in practice some tests may be considered as less important and therefore may be relaxed somewhat if their requirements are not satisfied exactly.

The acceptance tests may be generated before system design or in parallel with system design. The designers' task is to anticipate and achieve the required system performance. It is important that the designers know what the tests will be, because the tests provide the standards for evaluating the system design.

Acceptance tests make an important contribution to quality control in system development. As we have seen, to be effective, quality control must begin in the early stages of system development. Quality control involves a continuing critique of the system requirements and of what is being produced in response to those requirements.

Ideally, the acceptance tests for a system should be defined by people who have neither analysis, design, nor construction responsibilities for that system. In this way there is an independent interpretation of the system requirements as stated by the analysts. In addition, systems designers and constructors are prevented from setting the standards by which their own work will be evaluated. In many situations, however, analysts prepare or help prepare the acceptance tests because they are most familiar with the users' requirements. Users are consulted if necessary to clarify the requirement statements.

CONSTRUCT THE SYSTEM

Following system design, the system is constructed.*

The program units of which the system is made are coded, tested, and debugged individually. Then they are combined incrementally. That is, only one program unit at a time is added to the previous combination, and its interfaces to the rest of the system are tested. If there is a problem, its location is in the newly added program unit or in the interfaces.

Users' manuals, operations manuals, and reference manuals are produced for both the manual and automated portions of the new system.

The products of system design—the structure chart, module specifications, and dialogue designs—can be used to plan a sequence for constructing the system. Principles of decomposition can be used to divide the construction activities among the programmers and other personnel. Construction may proceed top-down, bottom-up, or in some other combination. Large systems are often constructed, tested,

---

*The phase of the life cycle in which the system is fabricated is widely called "implementation." However, in other contexts "implementation" is used to include integrating the new system into the using organization. For this reason, we prefer to use language that specifies more precisely what is happening at each of these phases of system development.

and delivered in versions. The initial version furnishes users with an important subset of the system's ultimate capabilities; each subsequent version adds a group of additional capabilities.

During system construction, users and analysts observe the system (or its various versions) as it evolves. In this manner they have tangible evidence of progress. It is also possible that they will catch design and construction errors. Depending on their technical abilities, some of the analysts and users may be invited to participate in development walkthroughs.

When construction of the system is complete, the acceptance tests are performed. Passing the tests marks the completion of this phase of the life cycle. Although minor adjustments may be made to correct system deficiencies, this is much too late to be correcting major flaws. In complex systems, designers must design out errors, while designing in ease of future modification. On the other hand, in the worst case, acceptance tests will prevent defective or inadequate systems from being placed in use.

If the system requirements are satisfied by software packages or by turnkey systems, there may not be a construction phase at all. Or perhaps the construction phase will be limited to tailoring the system to the specific user organization or its computing environment. In either case, acceptance tests are necessary for the same reasons as in a custom-designed and custom-built system.

INTEGRATE WITH THE USING ORGANIZATION

Several activities are involved in integrating the tested and accepted system into the users' organization. The hands-on and responsible users need to be trained to operate the system. They need to understand the interface between the manual and automated procedures for the new system. Other activities required for this phase may be dependent on the approach taken to system testing. In some large or geographically dispersed systems, initial system installation and testing may be limited to users in one region or area. In such situations, the installation of the new hardware or software in the other regions may follow the initial tests.

Conversion

Conversion to the new system also occurs during this phase if there is to be a period of parallel operation of the current and new systems. Data base conversion could also occur at this time if acceptance tests do not require building the entire new data base.

Certainly there may be appreciable technical problems when moving to a new computer information system. Data base conversion alone can be overwhelming when converting from a manual, paper-based system to an on-line interactive data base management system. If the conversion is spread out over a period of time, users who were promised that the new system would be much more efficient and productive are likely to be dismayed at the extra effort required. What these users fail to realize, whether they were told or not, is that the benefits of efficiency and productivity come after data base conversion.

Preparation for System Change

Because integrating the new system into the using organization involves a transition from one system to another, users should anticipate some difficulties during system start-up as they learn their new tasks and procedures. The more these difficulties are minimized through well-designed system interfaces, proper user training, and cooperation among all the participants, the better the transition will be.

Information system change often introduces significant organizational change. Users and analysts need to work together to formulate a plan that provides incentives for both types of change. In discussing how to plan for organizational change, Athey points out that acceptance of a new system depends upon the attitudes of the group of people affected by it. He summarizes the major factors affecting the acceptance of a new system as:

> The more **pressure** on the acceptance group to make a decision, the more likely the group will be to try something new.
> The greater the **relative advantage** the recommended solution has over the present system solution, the more likely it will be accepted.
> The greater the **goal congruence** among the acceptance groups, the greater the likelihood of the new system acceptance.
> The better the match in the **behavioral change** required to that which is desired, the more likely the new solution will be implemented.*

When planning for system conversion and integration, these factors can be used to increase the likelihood of success. The plan for system change can induce pressure to change, demonstrate the relative advantage of the new system, try to promote common goals among the acceptance groups, and minimize the amount of behavioral change required.

By addressing these organizational behavior issues as well as the technical issues, users and analysts can increase the probability of a smooth transition from the existing system to the new one.

OPERATE, MODIFY, AND ENHANCE THE SYSTEM

When the period of integrating the new system into the organization is over, routine system operation begins. Both manual and automated procedures are working smoothly and normally. The anticipated benefits of the new system are available to the organization on an ongoing basis. It is now time for a **postimplementation** (or postinstallation) **review.**

Postimplementation Review

This review assesses the successes and shortcomings of the system development. Two specific areas are included in the review—system performance and project performance.

In evaluating system performance, two questions are addressed.

Have the operational expectations for the system been met?

*Thomas H. Athey, **Systematic Systems Approach**, Prentice-Hall, Englewood Cliffs, N.J., 1982, p. 261.

Have the predicted benefits been achieved?

These questions deal with whether or not the users got what they expected and whether the specified system requirements are still being met. This is especially important if acceptance tests were not made with the system at full load or with the full data base. The review helps identify any bottlenecks in the system and determines whether it is desirable to reduce the execution time or response time of the automated system. Even if the stated requirements are satisfied, unanticipated requirements or problems may be discovered by this review.

The project review assesses the technical performance of the system designers and builders. It also gathers data for estimating resource requirements for future projects. In judging project performance, the following questions are addressed.

Were the stated project goals and objectives achieved?

What is the relationship between estimated and actual development costs and schedules? What are the explanations for any variances?

These questions deal with what happened in the project and why it happened.

If the system was constructed and delivered in several versions, there may be a postimplementation review for each version.

Care must be taken to keep the review from turning into a witch hunt. The emphasis should be on giving an honest, accurate, and dispassionate evaluation of successes and failures. In this manner, the documented successes can be repeated and the failures avoided on future projects. This review process provides critical feedback so that both users and analysts can learn from the effort. Otherwise, it will be easy to commit the same errors in the future.

Defining Requirements for System Enhancement

A good job of system development often leads to requests for system enhancement. Satisfied users have an increased understanding of the potential of computer information systems. As a result, they may ask to add "just this one more thing." The *ad hoc* incorporation of many new features can compromise a system design that did not anticipate them. Major system enhancements should be treated as a new development project.

SUMMARY

Analysts and users continue to participate in information system development after the structured specification has been completed and approved.

The performance requirements stated in the structured specification must be expressed as a specific set of system acceptance tests. These tests are prepared by quality control personnel, with analysts and users helping to interpret the system requirements. After the system has been constructed, debugged, and tested by its developers, users and analysts participate in performing the acceptance tests.

The accepted system is then integrated into the users' organization. Users are trained to operate and use the system, any remaining hardware and software are installed, and conversion from the old system takes place. The plan for system change can incorporate pressure to change, demonstrate the advantages of the new system, promote common goals among the people affected by the change, and minimize the amount of behavioral change required. By addressing these organizational behavior issues as well as the technical issues, users and analysts can increase the probability of a smooth transition from the existing system to the new one.

At the end of this transition, both manual and automated procedures in the new system should be working smoothly. A postimplementation review evaluates the system to see whether its performance requirements are being met. It also evaluates the technical performance of the system designers and builders to provide feedback for future system development projects.

It takes the combined effort of users, analysts, designers, and programmers to develop today's complex computer information systems for business. In the process each group has essential responsibilities and contributes to success or failure through its decisions and actions.

REVIEW

20-1. What is the purpose of system acceptance tests? From what are they derived?

20-2. What is the role of users and analysts during system construction?

20-3. Who has the primary responsibility for integrating the new information system into the using organization? What steps can be taken to encourage a smooth transition to the new system?

20-4. What is the purpose of the postimplementation evaluation? What two aspects of system development are assessed?

21 The Future of Structured System Development

In previous chapters, this book has described structured analysis methods for computer information systems in the context of the system life cycle. It has emphasized data flow diagrams, the system dictionary, and logical data base description, with Structured English, decision trees, and decision tables for specifying transformations. The approach is applicable to specifying users' requirements for custom software development, or, as discussed in Chapter 16, may be tailored to suit projects for which custom development is inappropriate.

This concluding chapter offers a preview of what may be in store for analysts as they master these methods and wish to extend their expertise in information system development. There are three major topics—alternative system modeling techniques, trends in system development, and the impact of these trends on the participants, especially on users and analysts.

ALTERNATIVE SYSTEM MODELING TECHNIQUES*

As an introduction to structured analysis methods, this book has emphasized a single principal technique for modeling the relationships among components of an information processing system—a method that uses data flow diagrams as its basis. However, there are alternative approaches to system modeling. In his "Manifesto for Systems Analysts," information systems consultant James Martin exhorts analysts to "understand the benefits and limitations of the different structured techniques."*

*James Martin, *An Information Systems Manifesto,* Prentice-Hall, Inc., Englewood Cliffs, N.J., © 1984, p. 278. Reprinted by permission of Prentice-Hall, Inc.

This section provides a brief overview of some alternative system description techniques.

Approaches to information processing system description may be classified into three categories: data-flow based, data-structure based, and control-flow based approaches. The discussion presents one or two techniques that are representative of each category. The interested reader is referred to the references cited for more information.

Data-Flow Techniques   Data flow diagrams, of course, are one of the data-flow approaches. Another data-flow approach is Structured Analysis and Design Technique (SADT), a proprietary product of SofTech, a firm that specializes in software engineering.*

Systems are described in SADT with complementary diagrams for processes and data. The process models are called *actigrams* (as in action diagrams). They are used to represent data and processes (transformations) in much the same way that DFDs are used. The basic building block for describing data and processes is illustrated in Figure 21-1. On an actigram (Figure 21-2) processes are connected in networks.

FIGURE 21-1. Basic building blocks for transformation and data in SADT.

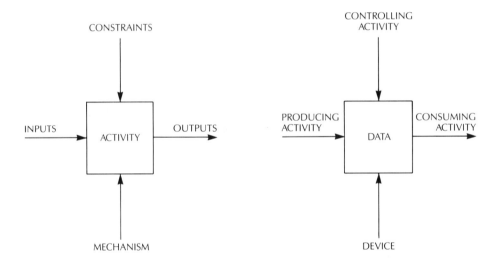

Processes are decomposed on lower-level diagrams. Data is modeled with *datagrams,* which show the hierarchical structure of the data and the processes that create and use the data. (See Figure 21-3.)

SADT provides stringent guidelines for the production of actigrams and datagrams. SADT also stresses the importance of project management and recommends a specific, distinctive project organization.

*Douglas T. Ross, "Structured Analysis: A Language for Communicating Ideas," *IEEE Transactions on Software Engineering,* Vol. SE-3, No. 1, January 1977, pp. 16–34.

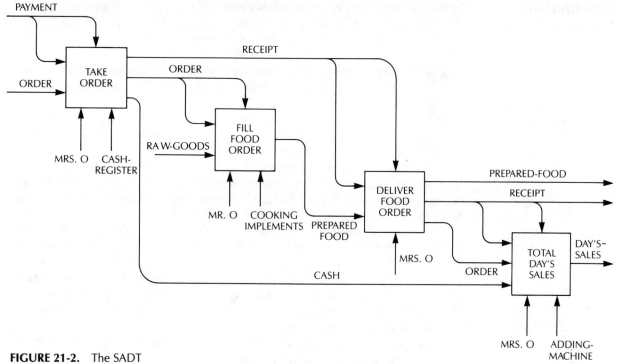

FIGURE 21-2. The SADT actigram equivalent to Diagram 2: Sales (FIGURE 6-3).

FIGURE 21-3. An SADT datagram.

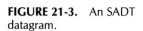

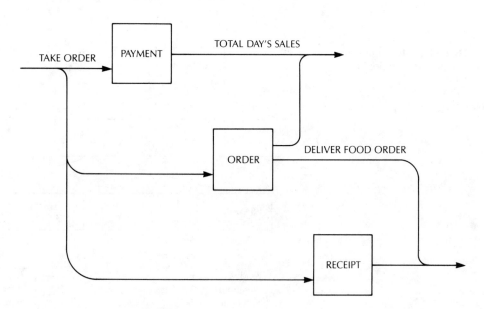

Data-Structure Techniques

Jackson System Development and Warnier-Orr techniques are representative of data-structure based approaches to system description. Each emphasizes computer-based systems and says very little about nonautomated systems.

Jackson System Development.   Jackson System Development (JSD) is an outgrowth of JSP, a program design method proposed by the Englishman Michael Jackson.*

In JSD, system development is viewed as consisting of three steps. The first two steps are concerned with the user and with abstracting a system model. The third step is concerned with realizing the system on a computer.

Step one is to describe the system in terms of **entity actions** (external entities, their actions, and attributes) and **entity structure** (data structure.) Entity actions are independent of time. Entity structure is time-dependent and is described using hierarchical box structures, as illustrated in Figure 21-4.

Step two is to realize these abstract descriptions as a process model composed of known functions. The initial model is described using a System Specification Diagram. A System Specification Diagram shows what the processes are, how they are connected to one another, and their inputs and outputs at the system boundary. The detailed specification of each process is given in a form similar to Structured English. The description of system timing varies from informal to formal and is used to constrain the implementation.

Step three, implementation, consists of converting the specification into a set of executable programs that are constrained by response time requirements and by the number and power of available processors. In this step, a Specification Imple-

---

*Michael A. Jackson, **System Development,** Prentice-Hall International, Inc., London, 1983.
Michael A. Jackson, **Principles of Program Design,** Academic Press, New York, N.Y., 1975.

FIGURE 21-4.   Entity structure description in JSD.

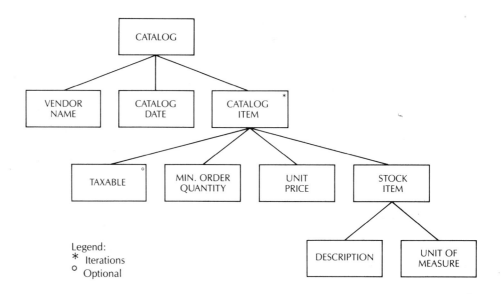

FIGURE 21-5. A Warnier diagram with equivalent data dictionary definitions.

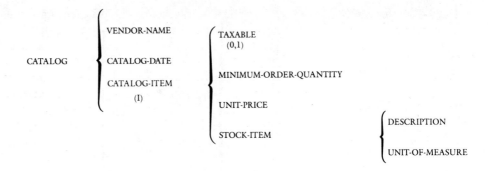

CATALOG = VENDOR-NAME & CATALOG-DATE & {CATALOG-ITEM}

CATALOG-ITEM = (TAXABLE) & MINIMUM-ORDER-QUANTITY
                 & UNIT-PRICE & STOCK-ITEM

STOCK-ITEM = DESCRIPTION & UNIT-OF-MEASURE

mentation Diagram is derived from the System Specification Diagram. It reflects implementation details and timing considerations.

Though described sequentially here, JSD is intended to be an iterative procedure for system development.

Structured Systems Development (Warnier-Orr).   Kenneth Orr's Structured Systems Development is based upon earlier work by a Frenchman, Jean-Dominique Warnier.* Warnier proposed a method for system development that grew out of his method for program development through synthesis.

In the Warnier-Orr methods, data structure plays an important role. It is modeled with a ***Warnier diagram.*** Figure 21-5 presents a Warnier diagram and its equivalent in the data dictionary form used throughout this book.

Warnier diagrams may also be used to represent actions, as in Figure 21-6.

*Jean-Dominique Warnier, ***Logical Construction of Programs,*** Van Nostrand Reinhold Company, New York, N.Y., 1974.
Jean-Dominique Warnier, ***Logical Construction of Systems,*** Van Nostrand Reinhold Company, New York, N.Y., 1981.

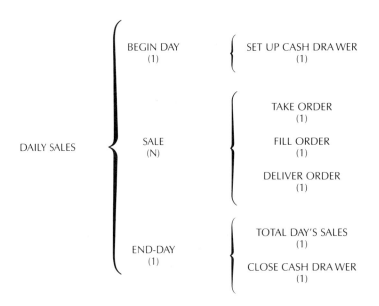

FIGURE 21-6. A Warnier diagram representing actions.

Orr extended the use of Warnier diagrams to include updating the data base, as illustrated in Figure 21-7.*

The steps in Structured Systems Development are

1. Use the Warnier diagram to sketch the overall problem.
2. Identify the structure and frequency of the system outputs.
3. Identify the logical data base.
4. Place the system requirements into a basic system-flow hierarchy.
5. Look to see if the data required already exists.
6. Identify events in the real world that affect the data base.
7. Place the actions that update the data base into the basic system hierarchy.†

Thus the Warnier-Orr approach works from outputs backward to the inputs.

**Control-Flow
Techniques**

Software Requirements Engineering Methodology (SREM) is intended for the specification of requirements for real-time systems. SREM is the result of efforts by the Department of Defense to meet the following three goals:‡

- a structured medium or language for the statement of requirements, addressing the properties of unambiguity, design freedom, testability, and modularity and communicability.

---

*Kenneth T. Orr, **Structured Systems Development,** Yourdon Press, New York, N.Y., © 1977, p. 41. Used by permission of Yourdon Press.

†Kenneth T. Orr, **Structured Systems Development,** Yourdon Press, New York, N.Y., © 1977, p. 81–102. Reprinted by permission of Yourdon Press.

‡Mack W. Alford, "A Requirements Engineering Methodology for Real-Time Processing Requirements," in **Advanced System Development/Feasibility Techniques,** J. Daniel Couger, Mel A. Coulter, and Robert W. Knapp, Eds., John Wiley and Sons, New York, N.Y., © 1982, p. 349. Reprinted by permission of John Wiley and Sons and Mack W. Alford.

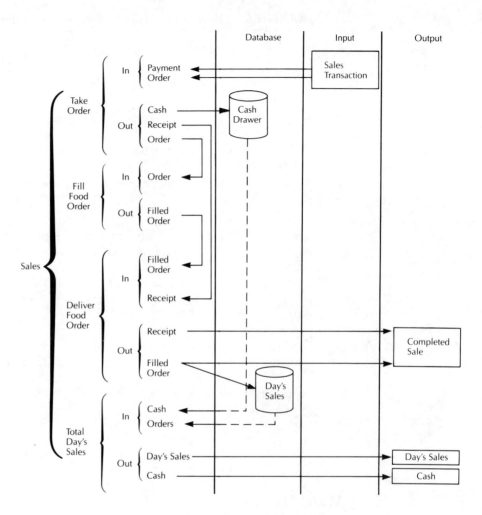

FIGURE 21-7. An extended Warnier diagram.

- an integrated set of computer aided tools to assure consistency, completeness, auto-matability, correctness, and

- a structured approach for developing the requirements in this language, and for validating them using the tools.

SREM is based on the observation that real-time software is tested by applying a stimulus (input message) to the system and observing the results of processing the stimulus—the response of the software and the contents of the computer memory. Thus, testable requirements must be specified in terms of stimulus, response, and memory.

SREM employs a formal language, Requirements Statement Language (RSL), for specifying requirements. The requirements are maintained in a centralized data

FIGURE 21-8.   Characteristics of requirements statement language (RSL).

SREM Requirements Statement Language Components

1. Elements (Nouns)—identify classes, "things" with names, e.g., source, destination, data, file.
2. Attributes (Adjectives)—properties of the classes of "things" attached to the elements, e.g., description, units.
3. Relationships (Verbs)—identify relations between element classes, e.g., inputs, outputs, passes, makes.
4. Structures (Sequence)—provide the mechanism for defining sequences, e.g., R–Nets.

---

base. A flexible extraction facility provides a means for automatically generating documentation. RSL is used to describe the required flow of processing to meet system objectives, including timing, accuracy, and response for computations. A summary of the characteristics of RSL is given in Figure 21-8. In the statement of performance requirements, a test is defined in terms of variables measured on processing paths. The Requirements Network (R-Nets) language is used to describe processing paths.

SREM relies heavily on automated tools. These tools are integrated in the Requirements Engineering and Validation System (REVS), which accepts RSL as input. The tools perform completeness and consistency checks. They also provide traceability of requirements and generate simulations to validate the correctness of the requirements.

TRENDS IN SYSTEM DEVELOPMENT

Structured analysis methods are widely used and have proved effective in the development of information processing systems for a wide variety of applications. They have led to better communication among users, analysts, designers, and other participants in system development, producing improvements in system quality. They have most often been used to develop applications that support day-to-day business operation, applications that are relatively routine and may be large in scope.

Yet many business have other, unmet information processing needs. The advent of low-cost personal computers has given users their own processing power to met these needs, and they have become impatient to put that power to work. New means of system development are providing users with the software they need, especially users who are managers and need decision-support systems. Three aspects of these changing approaches to system development are discussed—user-driven computing, fourth-generation languages, and the information center.

FIGURE 21-9.   Characteristics of prespecified and user-driven computing.

CHARACTERISTICS OF PRESPECIFIED COMPUTING

1. Formal specification of requirements, design, and programs.
2. Typically follows a system development life-cycle.
3. Documentation of requirements, design, programs, and user instructions.
4. Development time may be measured in months or years.
5. Maintenance may be formal, time consuming, and expensive.

CHARACTERISTICS OF USER-DRIVEN COMPUTING

1. Users do not know ahead of time what they want in detail. After using a prototype, users typically want to modify it.
2. Applications may be generated by the user alone (or with the assistance of a consultant), by interacting with an automated development tool.
3. Incremental changes are made by the user (or with the assistance of a consultant), thus maintenance is continuous.
4. The automated development tool may also generate (or assist in the generation of) documentation as a natural by-product of application generation.

User-Driven Computing

James Martin has done much to popularize the term "user-driven computing." The users to whom Martin refers are primarily decision-makers. Their information requirements stem from the kinds of questions these users typically ask—"what if" questions. Typically, there is some key decision pending. It must be addressed in a timely fashion under considerable pressure. There are many interrelated factors impinging on the decision. As alternative solutions are explored by changing the values of some of the factors, the decision-maker becomes aware of new factors and may alter the problem, requiring additional information. Thus information systems to support decision-makers need a high degree of flexibility. They arise from user needs and must be controlled by the user in a rather direct way.*

Martin contrasts user-driven computing with what he calls "prespecified computing," systems developed from formal requirements statements. This contrast is shown in Figure 21-9.

User-driven computing suggests the following requirements for the process of application development:

1. Responsiveness: Applications must be capable of being created quickly (on the order of days).

---

*James Martin, *Application Program Development Without Programmers,* Prentice-Hall, Inc., Englewood Cliffs, N.J., © 1982, p. 178. Adapted by permission of Prentice-Hall, Inc.

†James Martin, *An Information Systems Manifesto,* Prentice-Hall, Inc., Englewood Cliffs, N.J., © 1984, p. 44. Adapted by permission of Prentice-Hall, Inc.

2. Modifiability: Ongoing applications must be capable of evolving as the user comes to understand more about the problem.

3. Economy: Applications must be cost-effective, even though some applications will be used only once.

With user-driven computing, a decision-maker, who is the responsible user, perhaps also the system owner, becomes not only a hands-on user, but a program developer as well.

What sort of language will a user who is not a professional computer programmer use to communicate with the computer? With a procedural language, such as COBOL or Pascal, the programmer gives a detailed, step-by-step specification of ***how*** a transformation is accomplished. Users want ***nonprocedural languages,*** which allow them to describe ***what*** should be accomplished without becoming bogged down in specifying the details of ***how*** it is to be accomplished. Their desires can be fulfilled by fourth-generation languages.

Fourth-Generation Languages

Early programming languages forced the programmer to understand the computer at a very low level of detail. First-generation computer languages were called ***machine languages*** because the programmer communicates by zeroes and ones that are the codes for the hardware instructions—the machine operations and the addresses of the data for those operations. Second-generation languages are characterized as ***symbolic assembly languages,*** because the programmer communicates via symbols for operations and variables. COBOL, FORTRAN, and BASIC are examples of third-generation, or ***procedural languages,*** also known as ***higher-level languages.*** Though fourth-generation languages are less easily definable, they characteristically result in significantly greater productivity in application development. Often they are nonprocedural.

Some important characteristics of ***fourth-generation languages*** are listed below.

1. They are interactive.
2. The novice can be productive quickly, yet the expert is not hindered, since the more advanced language features may be learned incrementally.
3. They provide on-line interactive help, instruction, trace, and debugging facilities.
4. They can be used to access existing as well as specially created data bases.
5. They make intelligent assumptions when not specifically instructed by the user.
6. In addition to the traditional tabular report, they provide a rich variety of alternative output options.

Some examples of fourth-generation languages are RAMIS, Focus, and NOMAD.

The Information Center

The proliferation of personal computers, their ability to communicate with other computers, and the advent of fourth-generation languages have presented users with an alternative to application development by the corporate data processing department. In spite of its name, an information center is not a repository of corporate information, where users satisfy their information needs. Rather, the **information center** is a group within an organization that serves users' application development needs promptly and directly. The information center has emerged as a result of the backlog of demand on system developers, the ever-increasing demand for new applications, the difficulty of prespecifying these new applications, and the availability of new system development technology.

The information center is staffed with technical support specialists who maintain the various types of system hardware and software and with consultants who interact directly with hands-on users. The information center consultant works with the hands-on user to create applications. The consultant may suggest appropriate application development tools, train users in the proper use of those tools, tap into existing data bases, or create new ones where necessary. The information center consultant may also recommend prespecified computing for an application when that is more appropriate. The information center consultant spends a great deal of time training users and handholding while they learn new systems.

The information center is intended to manage and support user-driven computing. Martin believes that "the overriding objective of information center management is to greatly speed up the creation of applications that end users require."[*] He summarizes the reasons that the information center needs management as:[†]

- To ensure that data entered or maintained by the users are employed to their full potential rather than being in isolated personal filing cabinets.
- To assist users so they develop applications as efficiently as possible.
- To encourage the rapid spread of user-driven computing.
- To ensure that adequate accuracy controls on data are used.
- To avoid redundancy in application creation.
- To avoid integrity problems caused by multiple updating of data.
- To ensure that the systems built are auditable and secure, where necessary.

Automating System Development

The need for higher productivity in information system development implies a need for increasing automation of the process of systems analysis, design, and construction. This section presents an overview of current efforts and areas of research in automating system development.

The Information Systems Design and Optimization System (ISDOS) project at the University of Michigan is representative of systems for automated system development.[‡] The aim of this project, under the direction of Professor Daniel

---

[*]James Martin, *An Information Systems Manifesto,* Prentice-Hall, Inc., Englewood Cliffs, N.J., © 1984, p. 101. Reprinted by permission of Prentice-Hall, Inc.

[†]James Martin, *Application Development Without Programmers,* Prentice-Hall, Inc., Englewood Cliffs, N.J., © 1982, p. 298. Reprinted by permission of Prentice-Hall, Inc.

[‡]Daniel Teichroew and Ernest A. Hershey, III, "PSL/PSA: A Computer-Aided Technique for Structured Documentation and Analysis of Information Processing Systems," *IEEE Transactions on Software Engineering,* Vol. SE-3, No. 1, January 1977, pp. 41–48.

Teichroew, has been to produce executable computer code from a problem statement for any information system. Figure 21-10 presents the major components of ISDOS.

FIGURE 21-10. Major components of ISDOS.

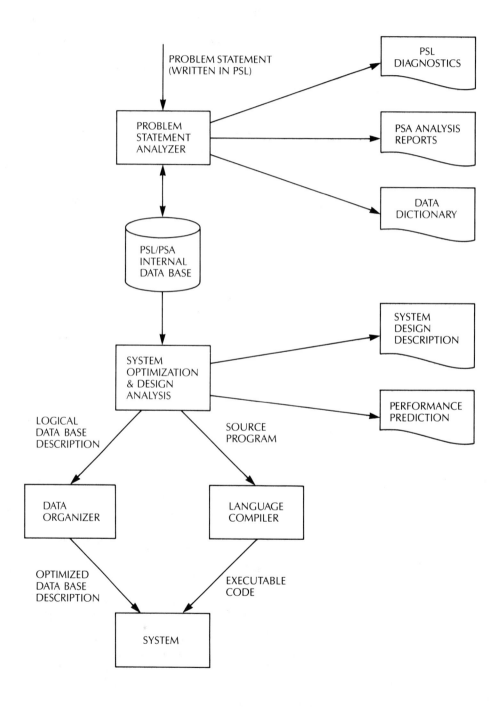

The analyst uses a Problem Statement Language (PSL) to build a data base describing an information system. The objects in the data base model the components that comprise the system, along with their attributes and values. Relationships between objects are also described with PSL. The problem statement in PSL is submitted to the Problem Statement Analyzer (PSA). The PSA checks the syntax and completeness of the PSL problem statement. It can also produce a data dictionary and other analysis reports. The System Optimization and Design Analysis (SODA) portion of ISDOS accesses the PSL/PSA data base to generate source code in some programming language (COBOL or FORTRAN, for example) and to produce a logical data base description. The source code is translated by a compiler into executable code. The logical data base description is translated into an optimized physical data base description. SODA can also produce a description of the system design and an associated prediction of the system's performance.

Another automated system development technique, Higher Order Software (HOS), provides components similar to those of ISDOS.* The important contribution of HOS is its basis in formal mathematics, which permits a rigorous verification of system requirements. This attempt to provide mathematically provable software is typical of newer automated system development tools.

We expect important contributions to automating system development from artificial intelligence research in the areas of expert systems, knowledge engineering, and automatic programming.

An **expert system** can be defined as a program, or system of programs, that performs like a human expert, exhibiting similar knowledge and reasoning skill in some domain. Expert systems exist today that approach the performance of human experts. In the domain of medical diagnosis, the CADUCEUS project at Carnegie-Mellon University maintains a data base that in 1982 contained approximately 100,000 associations between diseases and symptoms in internal medicine. MYCIN, a project at Stanford University, addresses the problem of diagnosing and treating infectious blood diseases.

Knowledge engineering is the term used to describe the branch of artificial intelligence that deals with expert systems. Figure 21-11 gives the generic categories for knowledge engineering applications.

Some aspect of each of these categories is applicable to automating system development.

Artificial intelligence researchers use the term **automatic programming** to refer to systems that assist people with program development. The automation can range from generating machine language instructions from a procedural language to assisting the analyst with the preparation of a requirements specification.†

Program transformation is an inportant aspect of automatic programming. Program transformation means changing a program description, or part of a pro-

---

*M. Hamilton and S. Zeldin, "Higher Order Software—A Methodology for Defining Software," *IEEE Transactions on Software Engineering,* Vol. SE-2, No. 1, 1976, pp. 9–32.

†*The Handbook of Artificial Intelligence: Volume 2,* Avron Barr and Edward A. Feigenbaum, editors, William Kaufmann, Inc., Los Altos, Calif., 1982, Chapter X.

FIGURE 21-11. Categories for knowledge engineering applications.*

CATEGORY	PROBLEM ADDRESSED
Interpretation	Inferring situation descriptions from sensor data
Prediction	Inferring likely consequences of given situations
Diagnosis	Inferring system malfunctions from observables
Design	Configuring objects under constraints
Planning	Designing actions
Monitoring	Comparing observations to plan vulnerabilities
Debugging	Prescribing remedies for malfunctions
Repair	Executing a plan to administer a prescribed remedy
Instruction	Diagnosing, debugging, and repairing student behavior
Control	Interpreting, predicting, repairing, and monitoring system behaviors

gram description, into another form. The reasons for the transformation may be quite diverse. For example, transformations may result from the automated diagnosis of problems with completeness, correctness, or consistency. Alternatively, transformations may be used to convert a specification from a form understandable by humans to a form that is more efficient for machine processing.

THE IMPACT ON USERS

In Chapter 4, we identified four types of users and enumerated the principal concerns of each type of user. Each type of user will be impacted by the trends in information system development just described. Some of these impacts are summarized briefly in this section.

The System Owner

Recall that the system owner is a high-level manager and decision-maker with profit-and-loss responsibility. The system owner is concerned with wise deployment of company resources. We can expect an increasing number of system owners to consider user-driven computing and the information center as cost-effective alternatives to complement, and perhaps eventually replace, prespecified computing. User-driven computing and the information center will require top management's support as well as an understanding of their benefits and limitations. With users taking a more active role in system development, there should be greater compatibility between the objectives and functions of information systems and the goals of the organization. These approaches hold great potential for delivering working systems on

*Frederick Hayes-Roth, Donald A. Waterman, and Douglas R. Lenat, **Building Expert Systems,** © 1983, Addison-Wesley, Reading, Mass., Table 1.1, p. 14. Reprinted with permission.

schedule and within the budget, while satisfying the other types of users. However, since users often do not know in detail what they want, the potential mentioned here can only be realized as their requirements are made explicit. There is no substitute for clearly understood requirements, whether in a prespecified or user-driven computing environment. However, there are alternative methods by which requirements may be developed. The system owner needs to know that these alternatives exist.

The Responsible User

The responsible user is a low- to middle-level manager with direct day-to-day responsiblity for the functioning of a system. Perhaps the greatest impact of the trends in systems development will be felt by these users.

It is a well-understood management principle that control and responsibility should be closely matched in an organizational structure. With more traditional system development strategies, control over system development is administered more centrally. This sometimes leads to systems that are unresponsive to local needs.

The more modern system development strategies encourage more local control. Thus control and responsibility are more closely aligned. The reasons for this are sound. The responsible user knows the appropriate system scope and the required support for the operations that he supervises. With more control and responsibility comes more user involvement in system development—a critical constituent for success. Furthermore, there is more potential for adaptation to local needs. This has a positive impact on reliability, dependability, and productivity. Perhaps most importantly, systems can be made more responsive to the needs of hands-on and beneficial users because of local autonomy. However, this autonomy must be constrained by the need for sharing information within the organization. The shared data base imposes a restriction that the system development tools be integrable throughout the organization. In this regard, the responsible users and the system owners must work together closely to ensure integration.

The Hands-on User

The hands-on user, the person who interacts directly with an automated system, should find automated systems easier to use, more reliable, and more dependable. This is due to two factors—the benefits of nonprocedural languages and the responsiveness achieved through local control over systems. Hands-on users should be able to communicate directly with responsible users to effect system changes. Thus, hands-on users should experience increased satisfaction as a result of personal involvement in system improvement.

The Beneficial User

Beneficial users should experience more responsive systems in terms of quality of service, understandable and usable system inputs and outputs, and security and privacy for confidential personal information. These benefits should accrue as a result of being able to make their wishes known to the responsible and hands-on users—the people who can transform those wishes into system functions.

THE IMPACT ON ANALYSTS

One of the most challenging aspects of the trends in system development is the increasing number of system modeling tools and techniques. A great deal of research is being devoted to automating system development. Aspiring systems analysts must become skilled in new tools and techniques as they become available. Furthermore, new skills are needed not only for the analysis phase of system development but also for design and construction. In addition to acquiring these new technical skills, successful analysts will learn alternative system development strategies. These strategies will range from those appropriate to the larger and longer prespecified computing projects, which may take months or years, to those for user-driven projects, which are smaller and may take only days or weeks. In order to extract more benefit from their data, users will be called upon to share data more. This will increase the pressure on analysts to produce comprehensive and effective logical data base models to ensure that all aspects of data security, privacy, and integrity are considered. In every case, because of a closer working relationship with users at all levels, analysts will find an increasing requirement for effective communication skills—oral, written, and graphic.

The career opportunities for analysts are exciting, with prospects for greater specialization. There will probably be a demand for prespecified computing for some time to come. However, for all but the larger applications, prespecified systems may shift to the tools that information systems professionals will use to develop applications for and with users. The information center offers challenges to analysts who want to consult with end users. There will also be more opportunity for analysts to join user departmental staff as in-house information systems specialists.

THE CHALLENGE OF THE FUTURE

The cost of computers has decreased dramatically in the past two decades. This has made computers widely available to users without any formal training in programming. In addition, the ability to communicate electronically on a worldwide basis has led to the emergence of distributed computing systems that place a wealth of information and computing resources in the hands of an increasing number of users. The challenge to business is to make effective use of both users and computing resources in furthering an organization's objectives. The challenge to systems analysts is to help users identify their requirements in a clear, complete, and coherent statement that may be transformed into an effective, working system. Increasing automation of the development process itself will assist both users and analysts to meet these challenges with greater sophistication and enhanced productivity.

SUMMARY

The data flow techniques for modeling system requirements introduced in this text are a foundation for future careers in systems analysis. Current trends in structured system development provide clues to that future.

Procedural programming languages like COBOL are being complemented by fourth-generation languages for system development. Rapid prototyping is also becoming increasingly important as a means of eliciting initially ill-defined requirements for systems to support users' decisions. The information requirements of these decisions will drive a variety of new user applications, some of them implemented by users themselves. Current research in computing points toward enhanced productivity in the future through the automation of information system development.

Consequently, the role of systems analysts and the skills they require are changing. There are prospects for more specialization and a greater diversity of career opportunities. Systems analysts will do well to learn alternative techniques for system modeling, including those based on data structure and control flow. Communication skills will remain of utmost importance. Many analysts already work in an information center. Some analysts work directly in user departments. In the future, more analysts may carry out system design and construction as well as analysis. The challenges and opportunities are great for both users and analysts in the future development of computer information systems for business.

REVIEW

21-1. Classify the following system modeling techniques as data-flow, data-structure, or control-flow based:

    a. Structured Analysis and Design Technique (SADT)

    b. Jackson System Development (JSD)

    c. Warnier-Orr

    d. Software Requirements Engineering Methodology (SREM)

21-2. Contrast the characteristics of prespecified and user-driven computing.

21-3. Explain the meaning of the following terms:

    a. procedural language

    b. non-procedural language

    c. first-generation language

    d. second-generation language

    e. third-generation language

    f. information center

21-4. Summarize the characteristics of fourth-generation languages.

21-5. Summarize some of the impacts of user-driven computing, fourth-generation languages, and automated system development on the four types of system users.

21-6. Describe some expected changes in the activities of systems analysts. What skills will these changes require analysts to develop and enhance?

GLOSSARY

abstraction. (1) The extraction of essential information relevant to a particular purpose. (2) The essential information thus selected.

access. A flow of information to or from a data store.

actigram. A process model in SADT.

action matrix. The portion of a decision table that shows the actions to be taken for each combination of conditions.

action stub. The portion of a decision table containing the independent actions.

action. A step or instruction in a transform description.

adjusting entry. A bookkeeping term to describe a change to a previous journal entry.

aggregation. A process of assembling a system using a set of elementary components.

algorithm. The step-by-step specification of the operations or actions that accomplish a transformation.

alias. An alternate name for an item, a synonym.

analysis. The process of determining the structure of an object or system, decomposition. (See systems analysis.)

application program. Software that causes general-purpose hardware to carry out transformations required by users.

arc. A component of a network that connects two nodes.

artifact. An object or system devised by human beings.

artificial intelligence. A discipline that deals with computer systems that produce results comparable to that of human intelligence.

attribute.   A named characteristic of an object that may take on a value.

attribute file.   A data store corresponding to an object in a data base description.

authorization.   A means of controlling system inputs by requiring prior approval.

automatic programming.   Computer systems that assist people in computer program development.

automation boundary.   A boundary that partitions an information processing system into an automated and a manual portion.

balanced.   A parent and child diagram are balanced when the data flows entering and leaving the child diagram are equivalent in composition to the data flows entering and leaving the corresponding transform on the parent diagram.

batch number.   A unique identifier of a batch of inputs in a batch system.

batch system.   A system in which inputs are entered as a batch or group.

batch total.   A control total, usually calculated on a data element most characteristic of the batch.

beneficial user.   A person who has no direct contact with an automated information processing system but who receives system output or provides system input.

bit.   A contraction of the term *binary digit*; a unit of information represented by either a zero or a one. [IEEE]*

blitzing.   Applying resources intensively to study complex or important areas or a system in detail.

body (of a loop).   In an iteration control structure, the action(s) to be repeated.

boundary.   (See system boundary.)

boolean operator.   A symbol for one of the logical, or boolean, operations—AND, OR, and NOT.

bottom-up.   Pertaining to an approach that starts with the lowest level components of a hierarchy and proceeds through progressively higher levels to the top level component. Contrast with top-down.

branch.   In a tree structure, a component (arc) connecting two points. Branches are directed away from the root.

case.   A multi-branch selection control structure in which the branches are mutually exclusive.

check digit.   A digit used to check for numeric data entry errors; its value is calculated from the remaining digits in the number.

child diagram.   In a set of data flow diagrams, a diagram at the next lower level corresponding to a specified transform. The child diagram shows the immediate subordinates of that transform.

component.   A part of a system or program.

composition.   (1) The components that constitute a structure. (2) (Of a data flow or data store) The information content and data structure.

conceptual data model.   A comprehensive, logical, essential description of user requirements for a data base.

*Definitions marked [IEEE] are from the *IEEE Standard Glossary of Software Engineering Terminology,* © 1983, The Institute of Electrical and Electronics Engineers, Inc. Reprinted by permission.

condition matrix. The portion of a decision table containing values associated with each condition or decision variable.

condition stub. The portion of a decision table containing the conditions or decision variables.

configuration-dependent (system characteristics). Characteristics of an automated system that are dependent on the choice of hardware and system software.

connection. An association established between functional units for conveying information.

conservation of data flow. The principle that any data flow leaving a transform or system must be derivable as a known transformation of the entering data flows.

constraint. A condition or requirement that must be satisfied (by a proposed problem solution or system design) for the system to be acceptable.

context (of a system). The environment.

context diagram. (1) A diagram of a system, its environment, and their interactions. (2) In a set of data flow diagrams, a diagram showing the system as a single transform, all its inputs and outputs, the origins of the inputs, and the destinations of the outputs.

continuous data element. A data element that takes on a continuous range of values.

control flow. Information that activates or terminates the execution of a transformation or that controls the sequence of execution of the procedure for the transformation.

control structure. A construct that determines the flow of control through a computer program. [IEEE]

control total. A data-entry control, calculated by adding numbers whose sum is meaningful in terms of the application.

correlative file. A data store corresponding to a relationship in a data base description.

cost/benefit analysis. A systematic quantitative and qualitative comparison of the relative advantages and disadvantages of alternatives.

couple. Information flow across a connection.

criterion. (plural, **criteria**) A factor used to evaluate the relative desirability of alternatives.

cybernetic system. A system with feedback control to accommodate shifts in the environment and with objectives that are adapted to changes in the system or the environment.

data. (1) Information. (2) A representation of facts, concepts or instructions in a formalized manner suitable for communication, interpretation, or processing by human or automatic means. [ISO]*

data access diagram. A graphic representation of a data base in terms of objects and relationships.

data base. A systematically partitioned, reusable, integrated collection of data that can be shared by many users.

data base description. A description of the content and structure of a data base.

data base management system. The computer system software that manages a data base, providing for logical and physical data independence, controlling redundancy, and enforcing integrity constraints, privacy, and security.

---

*Definitions marked [ISO] were developed by Technical Committee 97, Subcommittee 1, of The International Organization for Standardization. Reprinted from the ***IEEE Standard Glossary of Software Engineering Terminology*** by permission of the American National Standards Institute.

data dictionary.   The portion of the system dictionary in which the names of data stores, data flows, and data elements are listed and defined.

data element.   A data structure primitive, an item of information that does not require any decomposition in the system of which it is a part.

data flow.   A movement of information within a system or across the system boundary.

data flow diagram.   A graphic representation of an information processing system, showing origins, destinations, data stores, and transformations as nodes and data flows as arcs.

data immediate access diagram.   A data access diagram.

data store.   A time-delayed repository of information.

data structure.   An aggregation of data elements or other data structures using the operations of sequence, selection, and iteration.

data structure diagram.   A data access diagram.

data structure operator.   A symbol that represents a data structuring operation.

data structuring operation.   One of the three operations used to form data structures: sequence, selection, or iteration.

datagram.   A data model in SADT.

data-capture device.   Hardware that transforms automated system inputs from human-readable form to machine-readable form.

data-display device.   Hardware that transforms automated system outputs from machine-readable form to human-readable form.

decision.   A selection control structure.

decision rule.   A combination of conditions and the actions to be taken when that combination of conditions is true.

decision table.   A presentation in tabular form of a set of conditions and their corresponding actions. [ANSI]*

decision tree.   A graphic representation of a selection control structure in which the combinations of conditions are shown as sequences of decisions.

decision variable.   A data element whose values are used to specify conditions for a selection control structure.

decomposition.   (1) The partitioning of a whole into its component parts. (2) The identification of the component parts of a whole.

design.   (1) A process that creates descriptions of a newly devised artifact. (2) The product of the design process—a description that is sufficiently complete and detailed to assure that the artifact can be built.

destination.   A person or organization or system outside the system that receives a system output.

directed arc.   An arc in which the direction of flow is specified.

---

*Definitions marked [ANSI] were extracted from the ANSI Technical Report, American National Dictionary for Information Processing, X3/TR1-77, September 1977. Reprinted from the *IEEE Standard Glossary of Software Engineering Terminology,* by permission of the American National Standards Institute.

discounted cash flow. A sequence of future payments or receipts adjusted for the time value of money.

discrete data element. A data element that takes on a discrete set of values.

distribution control. A procedure to assure that system outputs are delivered to the intended recipient.

domain of change. The parts of the current system that are expected to be different in the new system.

entry (in system dictionary). All the information recorded in the system dictionary about a named system component.

environment (of a system). Whatever lies outside the system. Often, something outside a system that affects or interacts with the system.

essential transformation. A transform that is necessary in an information processing system because it is required either (1) to carry out a necessary transformation or (2) to maintain an essential data store.

expert system. A computer system that exhibits knowledge and reasoning skill similar to that of a human in some domain of expertise.

extended-entry decision table. A decision table in which entries in the condition matrix are permitted to take on more than two values.

external entity. An origin or destination.

feasible. Satisfying all the constraints on a system.

feasibility report. The product of a feasibility study (the activity of identifying user needs), containing both technical and management guidelines for a system development project, including a summary of system alternatives.

feedback. A measurement of system output used to adjust the operation of a system so that the system objectives are met.

first-generation language. A machine language for computer programming.

flow. Movement. In a network, movement occurs on the arcs.

flow of control. The sequence of operations performed in the execution of an algorithm. [IEEE]

fourth-generation language. A nonprocedural programming language.

function. (1) A specific purpose of an entity or its characteristic action. [ANSI] (2) (of a system) A task, activity, or work that a system performs.

functional primitive. A primitive transformation.

general ledger. The primary set of accounts in which business transactions are classified and recorded for reporting on the financial statements. It provides the classified information for the financial statements including: assets, liabilities, owner's equity, revenue and expenses.

hands-on user. A person who interacts directly with the data-capture and data-display devices for an automated information processing system.

hash total. A data-entry control, calculated by adding numbers whose sum is not meaningful in terms of the application. (See control total.)

hierarchy. A system structure in which every system component except the one at the top is subordinate to exactly one immediate superordinate.

Higher Order Software (HOS). An automated technique for information processing system development.

higher-level language. A programming language that usually includes features such as nested expressions, user defined data types, and parameter passing not normally found in lower-level languages, that does not reflect the structure of any one computer or class of computers, and that can be used to write machine-independent source programs. A single, higher-level, language statement may represent multiple machine operations. [IEEE, higher order language]

if-then. A form of the selection control structure in which no action is taken if the condition is false.

if-then-else. A form of the selection control structure with alternative actions.

industrial firm. Manufacturing firm.

information. (1) Data. (2) A pattern of binary digits (bits). (3) An aggregation of data or information that is significant.

information center. A group within an organization that serves users' application development needs directly and promptly.

information processing system. An information system that performs transformations.

information resource management. The management of information as a valuable corporate resource that is as important as people, material, equipment, and money.

information system life cycle. A conceptual framework for understanding and managing the activities involved in information system development and use.

Information Systems Design and Optimization System (ISDOS). An automated sytem for the development of information processing systems.

input. Data flow that enters a transformation.

instance. A specific member of the collection of information stored for an object or relationship; an individual occurrence of an object or relationship.

instruction. (computer) A program statement that causes a computer to perform a particular operation or set of operations.

interface. (1) A shared boundary. [ANSI] (2) A connection or interaction between two components or systems.

iteration. (1) A control structure for transform description: the process of repeatedly executing a given sequence of programming language statements until a given condition is met or while a given condition is true. (2) A single execution of a loop. (3) A data structuring operation: the repetition of a data structure or data element.

Jackson System Development (JSD) A data-structure approach to information system description.

journal. The original book of entry for recording accounting transactions.

JSP. A program design method developed by Michael Jackson.

key. A data element or concatenation of data elements within a data store (or data base object) whose value uniquely identifies a specific record (or instance).

knowledge engineering. A branch of artificial intelligence that deals with expert systems.

leaf. A point in a tree structure with no subordinates, thus with no branches leaving it.

leveled set (of data flow diagrams). A hierarchically organized set of data flow diagrams that shows the decomposition of a system.

limited-entry decision table. A decision table in which the entries in the condition matrix are limited to the values YES and NO.

logical. (1) Pertaining to the content or to essential, abstract, or implementation-independent aspects of a system. Contrast with physical. (2) Boolean.

logical current system description. An abstract, essential, implementation-independent description of an existing information processing system.

logical data base description. (1) A conceptual data model. (2) A description of a data base in terms of objects, relationships, operations on objects and relationships, and preconditions and postconditions on the operations.

logical new system description. An abstract, essential, implementation-independent description of a new, changed, or proposed information processing system.

machine language. A representation of instructions and data that is directly executable by a computer. Contrast with assembly language, higher-level language. [IEEE]

man-machine boundary. Automation boundary.

manufacturing firm. A business that produces goods.

matrix. An array or table: a structure in which the location of a component can be specified by two or more indices.

module. A named, bounded, contiguous set of executable program statements.

net flow. (data store convention) A convention that distinguishes three types of data flow between a transform and a data store: read-only access, write-only access, and access in which the transform requires information from the data store as an input in order to derive an output that is written to the data store.

network. A structure comprising a set of nodes connected by arcs.

node. A point within a network.

nonprocedural language. A language that allows its user to specify the solution to a problem without stating the details of the solution procedure.

notation. A set of symbols used to represent concepts or logical relationships.

object. A component of a logical data base description that represents a real-world entity about which information is stored.

objective (of a system). A human goal or purpose that a system serves.

on-line system. A system in which each data flow entering the automated portion is processed as soon as it is received.

operating system. Software that controls the execution of programs. An operating system may provide services such as resource allocation, scheduling, input/output control, and data management.

origin/destination dictionary. The portion of the system dictionary in which names of the origins and destinations are listed.

origin. A person or organization or system outside the system that provides the system with an input.

output. Data flow that leaves a transformation.

output verification control. A control used to account for printed forms or documents.

packaging. (1) The process of collecting and organizing the products of analysis into the structured specification. (2) The process of partitioning a system description into implementation-related units.

parallel decomposition (of data and transformation). The simultaneous partitioning of both data and transformations on a lower-level data flow diagram.

parent diagram. In a set of data flow diagrams, a diagram at the next higher level that contains the transform corresponding to a specified diagram (the child).

partition. Subdivide.

physical. Pertaining to the form or to non-essential, concrete, or implementation-dependent aspects of a system. Contrast with logical.

physical current system description. A description of an existing information processing system that incorporates concrete, nonessential, or implementation-dependent details.

physical new system description. A description of a new, changed, or proposed information processing system that shows what part of the system will be automated. It may also include the manual portion of the system.

policy. A high-level statement of rules or guidelines for decision making.

post. To record information in a data store.

postcondition. A condition that the results of a transformation must satisfy in order to be valid.

precondition. A condition that the inputs to a transformation must satisfy.

present value. The value today of future payments or receipts, discounted for the time value of money. (See discounted cash flow.)

prespecified computing. A system developed from formal requirements statements.

primitive. An elementary component of a system—one that is not decomposed further.

primitive transform. A transform that is not or cannot be partitioned further.

Problem Statement Analyzer (PSA). A command language in ISDOS for creating, maintaining, and modifying an automated model of an information processing system.

Problem Statement Language (PSL). A language in ISDOS for describing an automated model of an information processing system.

procedural language. A language that is higher than machine or assembly language but that still requires its user to give a detailed, step-by-step specification of an algorithm.

procedure. (1) A set of steps to be followed to accomplish a task each time the task is to be done. (2) The rules for a transformation.

process. (1) A transformation. (2) To transform inputs into outputs.

program flowchart. A graphic representation of an algorithm.

program transformation system. A system for changing a program description from one form to another.

prototyping. An approach in which a computer program simulates a proposed information processing, especially its interaction with the users.

range check. An input control to determine whether a continuous data element has a value within the allowable range.

real-time. Pertaining to the processing of data by a computer in connection with another process outside the computer according to time requirements imposed by the outside process. [ISO]

record. An iterated constituent of a data store.

relational operator. A symbol for one of the relations—less than, equal to, greater than, less than or equal to, or greater than or equal to.

relationship. A component of a logical data base description that represents associations or interactions between objects.

repeat-until. A form of the iteration control structure in which the test follows the body of the loop.

representation. A graphical, mathematical, or other physical presentation of concepts or logical relationships.

request for proposal (RFP). A request for suppliers of goods or services to state (usually in a competitive situation) how they propose to meet a need of a prospective purchaser of the goods or services.

requirement. (1) A condition or capability needed by a user to solve a problem or achieve an objective. (2) A condition or capability that must be met by a system or system component to satisfy a contract, standard, specification, or other formal document. [IEEE]

Requirements Engineering and Validation System (REVS). Automated tools for completeness and consistency checking in SREM.

Requirements Statement Language (RSL). A formal language used in SREM for specifying real-time system requirements.

responsible user. A low- to middle-level manager with direct day-to-day responsibility for the business functions supported by an information processing system.

root. In a tree, the point to which all other components are subordinate.

schema. A version of the conceptual data model coded to be understandable to a data base management system.

scope of automation. The automated portion of an information processing system.

second-generation language. A symbolic assembler language.

selection. (1) A data structuring operation: the choice of exactly one of two or more mutually exclusive data structures or data elements. (2) A control structure for transform description in which the next action to be performed is determined by the outcome of a decision in which a condition is tested for truth or falsity.

sequence. (1) A data structuring operation: concatenation. (2) A control structure for transform description: two or more actions or other control structures performed so that each immediately follows its predecessor.

service firm. A business that sells or distributes products or provides services.

sink. Destination.

Software Requirements Engineering Methodology (SREM). A method for specifying requirements for real-time systems.

source. Origin.

storage schema. A physical model of a data base.

structure chart.   A representation of the overall structure of an automated information processing system.

Structured Analysis and Design Technique (SADT).   A data-flow approach to system modeling.

Structured English.   A subset of English used to specify transform descriptions.

structured specification.   The product of structured analysis; the definitive statement of all the users' requirements for a new or changed information processing system.

subschema.   A subset of a schema.

symbolic assembly language.   A machine-specific language whose instructions are usually in one-to-one correspondence with computer instructions. Constrast with machine language, higher-level language. [IEEE, assembly language]

synthesis.   The creation of a whole from components.

system.   An interrelated set of components that are viewed as a whole.

system boundary.   The limit dividing the interior of the system from what is outside the system.

system design.   (1) The process of defining the structure and components of a system that satisfies the specified users' requirements. (2) The result of the system design process; a description of a system that satisfies the specified requirements and that is adequate for constructing the system.

system dictionary.   The portion of the structured specification in which the names of all system components are listed and defined or specified.

system input.   A data flow that crosses the system boundary to enter a system.

system model.   An abstraction of a real-world system that shows the significant components and their relationships. This abstraction is often used to study and predict the behavior or performance of the system.

system output.   A data flow that crosses the system boundary to leave a system.

system owner.   A high-level manager and decision maker for the business area supported by an information processing system.

System Optimization and Design Analysis (SODA).   A portion of ISDOS that generates program statements from a system description.

systems analysis.   The process of studying user needs to arrive at a definition of system or software requirements. [IEEE, requirements analysis]

third-generation language.   A procedural, or higher-level, language.

topological repartitioning.   A technique for reducing the number of interfaces in a data flow diagram by examining the connectedness of the network.

top-down.   Pertaining to an approach that starts with the highest level component of a hierarchy and proceeds through progressively lower levels. Contrast with bottom-up. [IEEE]

transaction analysis.   A strategy for producing an initial structure chart from a set of data flow diagrams.

transaction count.   The number of transactions in a batch.

transform.   A process that changes incoming data flows into outgoing data flows.

transform analysis.   A strategy for producing an initial structure chart from a set of data flow diagrams.

transform description.   A specification of the policy, procedure, or algorithm for a primitive transform.

transform dictionary.   The portion of the system dictionary in which the transform names are listed and in which the primitive transforms are defined.

transformation.   A transform.

tree structure.   A hierarchical structure, especially one represented in the form of a tree.

trial balance.   A list of all the accounts of a business along with their balances as of a particular date. It is used to summarize the effect of all these transactions for a period of time.

user view.   A task-related perspective for identifying a user's information requirements.

user-driven computing.   A system to support a decision maker, developed without a formal requirements specification.

value check.   An input control to determine whether a discrete data element has a value from the allowable set.

walkthrough.   A group review of a work product for the purpose of judging its completeness, correctness, consistency, and adherence to standards.

Warnier diagram.   A diagram for modeling actions or data in the Warnier-Orr method for information system description.

Warnier-Orr method.   A data-structure based approach to information system description, named for its developers, Jean-Dominique Warnier and Kenneth Orr.

while.   A form of the iteration control structure in which the test precedes the body of the loop.

INDEX

Abstraction 15
Access, storage or retrieval 77
Action 126
Actigrams (in SADT) 378
Aggregation 15, 18
Aggregation of related transforms
   204–207
Algorithm 119, 193
Alias 101, 102, 103
Alexander, Christopher 6
Analysis 19
Analysis process, characteristics 58
Analyst
   functions and responsibilities 52
   skills 53
Application program 33
Application programmer 159
Application software 33
Arc 9
Athey, Thomas 313, 374
Artificial intelligence 389
Attribute 159
Attribute file 161
Audit requirements 304
Authorization 304
Automatic programming 389
Automation alternatives 288–291
Automation boundary 271, 274
Automation of system development
   387–390

Backup 305
Balancing 81, 182
Batch 275
Batch number 304
Batch total 304
Bit 34
Blitzing 326
Boolean operator 128
Bottom-up 18
Branches 9
Brooks, Federick P. 328

Case (control structure) 124, 128
Change 312–317
   analysis of 320
   to structured specification 317–320
Check digits 304
Child diagram 81
Communicability 200–201
Communication, principles 350
Completeness 197–199, 229
Components 7
Composition 18, 96, 193
Conditional 127
Configuration-dependent 275
Connection 357, 361
Conservation of data flow 188–190
Consistency 199
Constraints 29, 334

Context 14
Context diagram 14
Control flows 300
Control structures 121–132
   combinations 130
   expressing 126–129
   indicating scope 130–132
Controls, system 304
Conversion 302–304, 373
Correctness 199–200
Correlative file 163
Cost/benefit analysis 276–277
Couple (structure chart) 361
Criteria 29, 334
Cybernetic system 313

Data 35
Data access diagram (DAD) 158
Data base 155
Data base administration 157
Data base description 61
   example 247–260
   techniques 63
   testing 168–172
Data base management system 157
Data decomposition, guidelines 110
Data dictionary
   data element entry 103–104
   data flow entry 101–102

data store entry 102–103
data structure notation 98
design phase 361
attribute values 161
attributes 161
objects 161
relationships 163
Data-capture device 275
Data-display device 275
Data element 109
continuous 104
discrete valued 104
Data flow
conservation of 188
definition 77
directness 203
internal 186
Data flow diagram (DFD) 61, 193
balanced 81
balancing 240
child diagram 81
completing 180–181
connection 182
continuing 179–180
definition 77
initial 177–179
leveled set 81
parent-child relationship 81
principles for drawing 182–190
vs. program flowchart 86
Data immediate access diagram (See data access diagram)
Data model, conceptual 157
Data store
conventions 84, 185–190
definition 77
essential 229, 232, 233, 235
file 78
local 186
shared 186
Data structure, constituents of 96
Data structure operators 98
Data structuring operations 97–98
Datagrams (in SADT) 378
Decision table 138–143
action matrix 139
action stub 139
condition matrix 139
condition stub 139
construction 141–143
decision rule 139
decision variable 140
extended entry 140
indifference 143
limited-entry 139
Decision tree 144–145
construction 144–145
decision rule 144
path 144
sequence of decisions 144
Decisions 86

Decomposition 15, 18, 19, 240
DeMarco, Tom 49, 57, 193, 235, 312
Derived 31
Design 19
definition 355
description 355
goals 356
methods 358
phase 356–358
process 355
tools 359
Designer 356
Destination 77
Dialogue diagrams 361
Domain of change 272, 283–287

Editing and error checking 299
Entity (See Object)
Entry in the system dictionary 95
Environment 14
Error messages 299
Essential
data dictionary 237
data store 229, 232, 233, 235
transform 229, 232, 234–235
transform description 237
Existence 164
Expert system 389
External entity (See Origin, Destination)

Feasibility report 337–339
Feasible 334
Feedback 37
File (See Data store)
Flow of control 86
Flowchart, program 86
Flows (in network) 9
Function 7
Functional primitive (See Primitive transform)

Hierarchies of systems 10
Hierarchy 8, 81
and decomposition 183
Higher Order Software (HOS) 389

If-then 123, 127
If-then-else 123
Imperative sentence 126
Information
definition 34–35
evaluating 346
reporting 349–352
sources 342
Information center 386
Information content of data flow 78
Information dependencies in DFD 90
Information flows between modules 357

Information gathering 342–347
techniques 63
Information management 347–349
automated 348
manual 347
Information processing system 31
Information resource management 36
Information System Design and Optimization System (ISDOS) 387
Information system life cycle 39
overview 40
phases 42–44
Input 31, 77, 232
Instance of object 160
Instance of relationship 161
Instructions 35
Integrity constraints 157
Interface 14, 77
Interview 342–344
Iteration
control structure 86, 124, 129
data structure 98

Jackson System Design (JSD) 380

Key 161
Kipling, Rudyard 350
Knowledge engineering 389

Language, generations 386
Layout (report, screen, source document) 361
Leaves 9
Logical 34
Logical current system description 60, 234–237, 263–268
Logical data base description 158–161
Logical new system description 60, 272–274, 277–287
Logical independence 157
Loop 124, 129

Man-machine boundary (See Automation boundary)
Matrix 9
Martin, James 385, 387
Model, characteristics 16
Models, performance 16
Module 357, 361
Module dictionary 351
Movement 193

Naming components 191–192
Nested 8
Nested data 99
Net flow 85
Net system input (See system input)
Net system output (See system output)

Network 9
Nodes 9
Notation 18
Numbering conventions 81

Object 158, 159
Objective 7
Observation 345
On-line 275
Operating system software 33
Operations on objects and relationships
    163–166
Origin 77
Origin/Destination dictionary 112
Orr, Kenneth T. 382
Otherwise 129
Output 31, 77, 232–233

Packaging 305–306
Parallel decomposition 263
Parent diagram 81
Partitions 19
Performance targets 305
Physical 34
Physical current system description 60
Physical independence 157
Physical new system description 60,
    231–237, 274, 288–291
Policy 119
Postcondition 164
Postimplementation review 374
Precondition 164
Presentations 351–352
Prespecified computing 385
Primitive function (See primitive
    transform)
Primitive transform 84
Primitive 18
Problem Statement Analyzer (PSA) 389
Problem Statement Language (PSL) 389
Procedure 31, 119
Process 31, 79
Profile 324
Program transformation system 389
Program unit 357
Program Design Language (PDL) 361
Prototyping 327–329
Pseudocode 361

Quality control, checklists 213–215
Questionnaire 344

Real-time 275
Record 161
Redundancy 192–193
Referential integrity 164
Relational operator 128
Relationship 158, 161–162
Repeating group 249–251

Repeat-until 124
Report 349
Representation 15, 17
Request for proposal (RFP) 327
Requirements Engineering and Validation
    System (REVS) 384
Requirements Statement Language
    (RSL) 383
Review 374
Root 9

Scenario 336, 361
Schema 158
Scope of automation 274
Scope of control structure 126
Security 157, 305
Selection
    control structure 122, 127
    data structure 97
Selectivity 15
Sequence
    control structure 86, 122, 126
    data structure 97
Simon, Herbert 354
Sink (See Destination)
Software Requirements Engineering
    Methodology (SREM) 382
Source (See Origin)
Specifications 16
Stimulus 233
Storage 30
Storage schema 158
Structure 7, 9
Structure chart 361
    afferent branch 363
    central transform 363
    control flag 361
    couple 361
    data couple 361
    efferent branch 363
Structured analysis, activities 59
Structured Analysis and Design
    Technique (SADT) 378
Structured English 121–138
    guidelines 136–137
    parts 133–134
Structured specification
    basis for design 365–366
    outline 306–310
Subschema 158
Survey 344
Symbols 77
Synthesis 18
System 7
System acceptance tests 369
System boundary 13–14
System context diagram (See Context
    diagram)
System change 374
System description
    physical and logical 33

completing 181
System design (See Design)
System dictionary 61
    entries 95
System enhancement 375
System input 77
System life cycle, analysis phase 59–61
System model
    generation 18
    limitations 19
    need for 15
System Optimization and Design Analysis
    (SODA) 389
System output 77
System owner 49
System structure 8–9
Systems analysis 19

Target document 298
Timing and response requirements 302
Topological repartitioning 195, 201–203
Top-down 19
Transaction 364
Transaction analysis 362, 364
Transaction center 364
Transform 31, 77, 79
    decomposition of 180–181
    essential 229, 232, 234–235
    primitive 84
Transform analysis 362
Transform center 363
Transform description 118, 193
    goals and objectives 118
    hybrid 147
    essential 237
Transform dictionary 112
Transformation (See transform)
Transforms, independence 201
Tree structure 8

Uniqueness 164
User view 154
User-driven computing 385
User
    beneficial 50, 391
    functions and responsibilities 51
    hands-on 50, 391
    responsible 49, 391
    system owner 49, 390

Walkthrough 63, 207–213
Warnier, Jean-Dominique 381
Warnier diagram 381
Warnier-Orr method 381
Wasserman, Anthony I. 327
Weinberg, Victor 137
While 124, 125, 129

Yourdon, Edward 230, 315